Advance praise for *Scientific Authorship*:

"'What is an author?' is a question that has been central to cultural and literary studies for almost thirty years. This collection of essays opens up genuinely new perspectives on this question and shows us how much science studies has to contribute to fundamental issues in the humanities."

—Arnold I. Davidson, University of Chicago

Scientific Authorship

Credit and Intellectual Property in Science

EDITED BY
Mario Biagioli & Peter Galison

Routledge
NEW YORK AND LONDON

Published in 2003 by
Routledge
29 West 35th Street
New York, NY 10001
www.routledge-ny.com

Published in Great Britain by
Routledge
11 New Fetter Lane
London EC4P 4EE
www.routledge.co.uk

Routledge is an imprint of the Taylor & Francis Group.
Printed in the United States of America on acid-free paper.
Design and Typography: Jack Donner.

"Uncommon Controversies" is abridged and adapted by permission of the publisher and the author from *Who Owns Academic Work?: Battling for Control of Intellectual Property* by Corynne McSherry, pp. 68–100, Cambridge, Mass.: Harvard University Press, Copyright © 2001 by the President and Fellows of Harvard College.

10 9 8 7 6 5 4 3 2 1

Cataloging-in-Publication data is available from the Library of Congress.
ISBN 0-415-94292-6—0-415-94293-4 (pbk.)

Contents

Acknowledgments

This volume has developed from the papers presented at the conference on "What Is a Scientific Author?" held at Harvard in March 1997 with support from the National Science Foundation, the Mellon Foundation, and Harvard University.

We would like to thank Elizabeth Lee, Beatrice Lewin Dumin, Colin Milburn, Alisha Rankin, Elly Truitt, and especially Kristina Stewart for all their help in preparing the manuscript for publication.

Introduction

MARIO BIAGIOLI
AND PETER GALISON

More than thirty years ago Michel Foucault reframed the analysis of authorship by asking questions that were simultaneously theoretical and mundane, questions that cut across disciplines as diverse as bibliography, philosophy, literary studies, history, and library management. What does it mean to classify a certain body of texts as belonging to a certain author? How do categories of *author* and *work* relate to and constitute each other? Is the name of the author like any other personal noun, or does it have other functions? How does the relationship between the author's name and the epistemological status of claims vary across different disciplines? What does that tell us about the discursive regimes in which those disciplines operate?[1]

Studies of authorship have covered much ground and have spread in many different directions since the appearance of Foucault's "What Is an Author?" The function of the author has become a standard research question in literary studies, history of the book, legal studies, art and film history, postcolonial studies, anthropology, and gender studies.[2] In recent years these issues have moved from academic to public discourse, propelled by the development of Internet culture, the aggressive use of intellectual property law by corporations eager to regiment global markets, and growing concerns about the appropriation of public knowledge and cultural heritages resulting from these corporate trends.[3] However, despite the increasing visibility of these issues across disciplines and cultural spheres, scientific authorship has attracted very little attention—at least from the humanities and social sciences.

But if scientific authorship has yet to become an object of scholarly and philosophical discourse, it has been the cause of great professional concern among the scientists, especially in today's vast, multiauthor collaborations. The assignment of credit in large-scale biological, computer science, and physics collaborations is a major issue for young scientists; it hits every aspect of their career trajectories, from thesis writing to hiring and promotion. Physicists have long been debating what the Nobel Prize means in an age when two thousand scientists sign a single paper. The hundreds of authors listed in the byline of a biomedical paper reporting the result of a large clinical trial ask the same questions about the meaning and definition of authorship. And so do the editors of the journals who review those manuscripts, and the academic committees, departmental chairs, and deans who have to make hiring or tenure decisions based on CVs listing those publications. For all these reasons, the authorship debate in science has been entirely practical, and has generated thousands of administrative memos, policy statements, guidelines, articles, editorials, reports, and heated letters to the editor.[4]

These issues, however, have important theoretical implications. Given the combination of literary, philosophical, economic, historical, and scientific issues raised by scientific authorship, it would seem a perfect site for an encounter between many kinds of scholarship. And motivations for this convergence would not seem to be lacking given the remarkable concerns about intellectual property found at all levels of science and technology policy and, increasingly, in everyday university life.[5] While the U.S. government has become concerned with the intellectual property status of the genome, universities engage more intimately with the private sector for funding and collaborations, and rely more frequently and aggressively on intellectual property law to mobilize their "knowledge capital."[6] In a time when whole academic departments sell intellectual rights of first refusal to pharmaceutical companies, any university ignoring such issues would be hiding its collective head in the sand. And the trend is not limited to science and engineering. Some universities, for instance, are expanding their intellectual property claims down to course syllabi in the hope that they may be turned into profitable resources for distance-learning ventures. This means that, for the first time, academics from all ranks and disciplines are witnessing the arrival of intellectual property law on campus and have to confront, usually with unease, the tensions between the traditional ethos of academic authorship and the logic of property and the market. But if some of the problems and tensions that have long characterized scientific authorship are now beginning to be experienced or perceived by a much wider range of academics, this has not translated into cross-disciplinary reflections on the nature and problems of academic and scientific authorship.

For instance, most discussions of scientific authorship have not shared many questions or tools with other kinds of authorship studies. To the extent that history of science and science studies have taken up the question of authorship, it has largely been to reduce the function of authorship to the construction of authority: Whose word counts in presenting scientific results? How does this authority shape the conduct and resolution of scientific disputes? In this sense, science studies has asked questions that are comparable to those posed by literary studies when it analyzes the construction of the authorial voice. For example, how does it come to pass that the voice of a given author becomes authoritative, and how do his or her works enter or fail to enter the canon? Unlike science studies, however, literary studies has not stopped here. That is, it has not limited itself to looking at specific instances of the textual and social construction of specific authors and their auras, but has asked more fundamental questions about the very institution of authorship, how it came about, the aporias that underlie it, the economy it supports, and the legal constructs invoked to justify it. Similarly, legal studies has not simply looked at the contextual and sociopolitical factors behind the closure of certain debates about intellectual property rights, the adjudication of specific patent or copyright infringement cases, or the disputes concerning the extension of intellectual property to new objects such as DNA sequences or computer algorithms. While doing all that, legal studies has also pointed to and analyzed the unresolvable tensions within the logic of intellectual property law and of its constitutive elements, such as the figure of the author.

If we are going to tease apart the construction of scientific authority from the genealogy of the role of the authorial name in science and the economies of credit that hinge on it, we need to cross the boundaries between science, literary, legal, and policy studies. That process has already started in specific, limited areas.[7] Historians of science, literary historians, and historians of the book have begun sharing insights into the co-emergence of scientific and literary authorship in the early modern period. Others have looked at the social and gender politics of the name in science, that is, at the conditions under which one is (or is not) allowed to affix his or her name to a publication, or publishes anonymously, pseudonymously, or under a collective name. History of technology, which has always taken a keen interest in patents and patent disputes as a source of evidence, is now becoming more concerned with intellectual property law as a driving force and tactical framework for technological change. Scientists and science administrators have become keenly concerned (occasionally obsessed) with the order of authorship of scientific publications, the distribution of credit in large-scale collaborations, and the relationship between academic publications, patent applications, and intellectual property law in general. Recent well-publicized findings of scientific fraud have drawn attention to the link between authorship and responsibility, forcing universities, funding agencies, and journals to articulate and enforce that bond. In very different circles, politicians and their legal advisors are struggling with how to conceptualize the possible application of Western notions of intellectual property law to non-Western bodies of knowledge about plants and their pharmaceutical properties, that is, the ways in which indigenous people may or may not be construed as authors of that knowledge.[8]

It is not clear whether all these analyses deal with different expressions of the same concept of author, that is, whether there is a unified notion of authorship to be uncovered under its many disciplinary or historical articulations. Probably not. What is clear, however, is that no matter what *author* means or may end up meaning in the future, the authorship of knowledge about nature (and the ways in which it differs from authorship of fiction, art, music, patents, or trademarks) is central to all these discussions. We think of this volume as a space where these conversations can begin, hopefully to continue in other forums and texts.

The essays in this volume map some aspects of the author-function in science from the emergence of scientific academies in the seventeenth century to today's vast multiauthor collaborations. Our contributors represent several (but certainly not all) of the perspectives to be found in science studies, and bring different questions to the analysis of scientific authorship. Our intent has been to place emphasis on the early historical development of scientific authorship and on those recent author-functions that can be related most directly to that genealogy. The volume is structured around three principal sections: Emergence of Authorship, Limits of Authorship, The Fragmentation of Authorship, and then a final set of commentaries intended to draw together some of the issues across disciplinary and chronological boundaries. A significant number of essays focus on the early modern period, for it was at this point that scientific publications began to assume their modern traits (such as the peer review system and the periodical journal format), thereby differentiating themselves from fictional literary genres. The topics of Emergence of Authorship include analyses of the unusual authorial strategies of Newton.

Through the early modern period, academies, laboratories, and universities drew increasingly firmer boundaries delineating who could—or could not—be counted as a scientist. We follow the demarcation of these new boundaries (as well as some protests against them), in the course of which other critical issues immediately arise: how assumptions about gender roles, for example, framed the authorial possibilities for women philosophers outside of the academies, or how skilled technicians were allowed in and kept out of the privileged circle of authorship. Creating and certifying particular authors as paradigmatic had immense consequences. Not only did the canonization of a particular author (such as James Clerk Maxwell) frame the education of generations of young mathematical physicists, it also had a direct impact on the perceived legitimacy of the laboratories in which that education took place.

After this first, primarily historical, section, the volume moves to more recent or contemporary scenarios in which scientific authorship has encountered the limits of its applicability, has clashed with other notions of intellectual property, or has forced us to rethink its assumptions. Part II, Limits of Authorship, opens with the problems that emerge when conflicts over scientific authorship are adjudicated in nonacademic settings (the court of law) according to intellectual property law, not the conventions of academic authorship. It continues by looking at a similar clash, but in different settings, that is, the problems generated by applying Western categories of intellectual property to how indigenous people develop and use their knowledge of the healing properties of plants. Western notions of authorship have traditionally been cast in the language of rights—property rights. But the complex roots of property rights in social relations become more evident if we take seriously the implications of new genetic research for our notion of selfhood, as we are asked to do by a third contribution to this section. Scientific authorship, then, comes back into the picture from the other side: the issue is no longer who is the author of knowledge about nature, but rather how knowledge about our genetic nature may change our own notion of *person*—the very notion that makes authorship thinkable.

The Fragmentation of Authorship, Part III, focuses on the contemporary contexts of science and their impact on what it means to be an author. That fragmentation occurs even for individual researchers as the mode of scientific writing varies radically from the first tentative scribblings at the laboratory bench all the way to the anonymous textbooks that utterly erase the kind of writing that is exploratory, hesitant, and reversible. This creates a fragmentation of genre. At the same time, scientific authorship at the end of the twentieth century and at the beginning of the twenty-first has often become multiauthorship—a fragmentation of scale undreamed of even a generation earlier. To cope with this second kind of fragmentation, collaborations reach for increasingly more elaborate systems to integrate their subgroups and participants into a whole. How do we distinguish who or what is an author in such collaborations? Defining the author is an ever more difficult, tricky business as increasingly specialized and interdisciplinary work casts authorship in a different light within the diverse species of Big Science. Academic laboratories, nuclear weapons laboratories, and industrial sites all carry dramatically different, if not contradictory, values of openness, secrecy, publication, and credit. Accordingly, each develops its own, often divergent, standards of authorship. By focusing on the relationship among authorship, collaboration, and division of labor, it is even possible to explore what happens when scientific collaboration goes sour and allegations of plagiarism fly, or to explore with more analytic care the varied levels of skill and authority embedded in the production of different kinds of scientific texts, from instrument reports to review articles or cutting-edge research papers.

Finally, in Part IV, we have included two commentaries on the volume's contributions from the perspectives of literary and legal studies. These reflections add further cross-disciplinary dimensions to the collection while suggesting some starting points for the discussions we hope this volume will elicit among its various audiences.

"Who owns science?" is a question we are bound to confront more frequently in the future. From designer drugs to the decay of the Higgs boson, from the rain forest to delocalized "mobile agents" of recent computer science, authorship and ownership rights will determine the way the next generation of scientific research is conducted and distributed. We hope that this volume will be seen as an invitation for further work. The varieties of scientific authorship and the sites where it interacts or clashes with other regimes of intellectual property are as numerous as the possibilities for cross-disciplinary analyses. No list could be exhaustive, but surely the longer-term investigation into scientific authorship would include science studies, legal studies, and literary studies, as well as history, anthropology, philosophy, and science policy. Together, perhaps, we can begin to unravel what it is going to mean, in this century, for people to author nature.

Notes

1. Michel Foucault, "What Is an Author?" in *Language, Counter-Memory, Practice: Selected Essays and Interviews,* edited by Donald Bouchard (Ithaca, N.Y.: Cornell University Press, 1977), 113–38.
2. The literature on the topic is vast. These are just some pointers: Martha Woodmansee, "The Genius and the Copyright: Economic and Legal Conditions of the Emergence of the 'Author,'" *Eighteenth-Century Studies* 17 (1984): 425–48; Peter Jaszi, "Toward a Theory of Copyright: The Metamorphoses of 'Authorship,'" *Duke Law Journal* (April 1991): 455–502; David Saunders, *Authorship and Copyright* (London: Routledge, 1992); Mark Rose, *Authors and Owners* (Cambridge, Mass.: Harvard University Press, 1993); Martha Woodmansee and Peter Jaszi, eds., *The Construction of Authorship* (Durham, N.C.: Duke University Press, 1994); James Boyle, *Shamans, Software, and Spleens: Law and the Construction of Information Society* (Cambridge, Mass.: Harvard University Press, 1995); Jane Gaines, *Contested Culture: The Image, the Voice, and the Law* (Chapel

Hill: University of North Carolina Press, 1991); Jeremy Braddock and Stephen Hock, eds., *Directed by Allen Smithee* (Minneapolis: University of Minnesota Press, 2001); Bernard Edelman, *Ownership of the Image: Elements for a Marxist Theory of Law* (London: Routledge & Kegan Paul, 1979); Joan DeJean, "Lafayette's Ellipses: The Privileges of Anonymity," *PMLA* 99 (1984): 884–902; Carla Hesse, *Publishing and Cultural Politics in Revolutionary Paris, 1789–1810* (Berkeley: University of California Press, 1991); Catherine Gallagher, *Nobody's Story: The Vanishing Acts of Women Writers in the Marketplace, 1670–1820* (Berkeley: University of California Press, 1994); Roger Chartier, *The Order of Books* (Stanford, Calif.: Stanford University Press, 1994); Peggy Kamuf, *Signature Pieces: On the Institution of Authorship* (Ithaca, N.Y.: Cornell University Press, 1988); Kevin Dunn, *Pretexts of Authority: The Rhetoric of Authorship in the Renaissance Preface* (Stanford, Calif.: Stanford University Press, 1994); Jeffrey Masten, *Textual Intercourse: Collaboration, Authorship, and Sexualities in Renaissance Drama* (Cambridge: Cambridge University Press, 1997); Svetlana Alpers, *Rembrandt's Enterprise: The Studio and the Market* (Chicago: University of Chicago Press, 1988); Caroline Jones, *Machine in the Studio* (Chicago: University of Chicago Press, 1996); Sally Price, *Primitive Art in Civilized Places* (Chicago: University of Chicago Press, 1989).

3. Larry Lessig, *Code and Other Laws of Cyberspace* (New York: Basic Books, 1999); Larry Lessig, *The Future of Ideas* (New York: Random House, 2001); Jessica Litman, *Digital Copyright* (New York: Prometheus Books, 2001); Siva Vaidhyanathan, *Copyrights and Copywrongs* (New York: New York University Press, 2001); Rosemary Coombe, *The Cultural Life of Intellectual Properties* (Durham, N.C.: Duke University Press, 1998); Ronald Bettig, *Copyrighting Culture* (Boulder, Colo.: Westview, 1996); Bruce Ziff and Pratima Rao, eds., *Borrowed Power: Essays on Cultural Appropriation* (New Brunswick, N.J.: Rutgers University Press, 1997); Michael Ryan, *Knowledge Diplomacy: Global Competition and the Politics of Intellectual Property* (Washington, D.C.: Brookings Institution, 1998); Lise Buranen and Alice Roy, eds., *Perspectives on Plagiarism and Intellectual Property in a Postmodern World* (Albany: State University of New York Press, 1999); and The Duke Conference on the Public Domain, 9–11 November 2001, <www.law.duke.edu/pd>.
4. A bibliography on scientific authorship in biomedicine can be found at <http://www.councilscienceeditors.org/services>.
5. On the turn toward an entrepeneurial ethos in the university, and its implications for science policy, see Sheila Slaughter and Larry Leslie, *Academic Capitalism: Politics, Policies, and the Entrepeneurial University* (Baltimore: Johns Hopkins University, 1997); Philip Mirovski and Ester-Mirjam Sent, eds., *Science Bought and Sold* (Chicago: University of Chicago Press, 2002); David Guston, *Between Politics and Science: Assuring the Integrity and Productivity of Research* (Cambridge: Cambridge University Press, 2000).
6. On issues of intellectual property on campus, see Corynne McSherry,

Who Owns Academic Work? (Cambridge, Mass.: Harvard University Press, 2001); and the "Symposium on Intellectual Property," *The Journal of College and University Law* 27, no. 1. (Summer 2000). For earlier contexts of intellectual property and academic science see Dorothy Nelkin, *Science as Intellectual Property* (New York: MacMillan, 1984). For an assessment of contemporary scenarios, see John Golden, "Biotechnology, Technology Policy, and Patentability: Natural Products and Invention in the American System," *Emory Law Journal*, 50, no. 1: 101–92; and Rochelle Dreyfuss, Diane Zimmerman, Harry First, eds., *Expanding the Boundaries of Intellectual Property: Innovation Policy for the Knowledge Society* (Oxford: Oxford University Press, 2001).

7. This trend is more visible in the early modern field. See, for instance, Steven Shapin, "The Invisible Technician," *American Scientist* 77 (1989): 554–63; Adrian Johns, *The Nature of the Book: Print and Knowledge in the Making* (Chicago: University of Chicago Press, 1998); Pamela Long, *Openness, Secrecy, Authorship: Technical Arts and the Culture of Knowledge from Antiquity to the Renaissance* (Baltimore: Johns Hopkins University Press, 2001); Marina Frasca-Spada and Nick Jardine, eds., *Books and the Sciences in History* (Cambridge: Cambridge University Press, 2000); Domenico Bertoloni Meli, "Authorship and Teamwork around the Cimento Academy," *Early Science and Medicine* 6 (2001): 65–95; Paula Findlen, "Translating Science," *Configurations* 6 (1996): 235–62; Christian Licoppe, *La formation de la pratique scientifique: le discours de l'experimente en France et en Angleterre (1630–1820)* (Paris: Editions La Decouverte, 1996); Owen Gingerich and Robert Westman, "The Wittich Connection: Conflict and Priority in Late Sixteenth-Century Cosmology," Part 7, *Transactions of the American Philosophical Society* 78 (1988).
8. Stephen Brush and Doreen Stabinsky, eds., *Valuing Local Knowledges: Indigenous People and Intellectual Property Rights* (Washington, D.C.: Island Press, 1996); Vandana Shiva, *Biopiracy: The Plunder of Nature and Knowledge* (Boston: South End Press, 1997).

PART I
Emergence of Authorship

1.

Foucault's Chiasmus

Authorship between Science and Literature in the Seventeenth and Eighteenth Centuries

ROGER CHARTIER

What is a scientific author? Such a topic obliges us to look back at the question that Michel Foucault posed in 1969 in his famous lecture «Qu'est-ce qu'un auteur?» and to the distinction he proposed between the "sociohistorical analysis of the author as an individual" and the more fundamental problem of the construction of an "author-function," that is to say, "the manner in which a text apparently points to this figure [the author] who is outside and precedes it."[1] Although it was not his prime objective, Foucault sketched in his lecture a history of the conditions of the emergence and the variations of the author-function. He outlined two series of conditions that refer to two different chronological stages. The first context, and the only one that many commentators of Foucault's text have stressed, is given by the "moment when a system of ownership and strict copyright rules were established," that is to say, for him, toward the end of the eighteenth and beginning of the nineteenth century, when the "social order of propriety which governs our culture" was codified. A strong tie is thus established between the juridical construction of the authorship and the legal definition of the bourgeois conceptions of the individual and private property. But, according to Foucault, the status of texts as property is historically second to "what one could call its legal appropriations" («*ce que l'on pourrait appeler l'appropriation pénale*»). Therefore, the author-function is first rooted in the effects of the censorship of churches and states: "Speeches and books were assigned to real authors, other than mythical or important religious figures, only when the author became subject to punishment and to the extent that his discourse was considered transgressive." Foucault did not propose any chronology for the "penal appropriation" that linked the author-function to the exercise of power by an authority endowed with the right to censor, judge, and punish. But it is clear that, thus defined, the author-function predates the early modern period.

In order to illustrate the point that the author-function is not universal and present in all discourse, Foucault evokes the radical reversal, or chiasmus, that, according to him, occurred in the seventeenth or eighteenth century. During these two centuries, rules for the identification of texts belonging to scientific and literary discourse were exchanged. For him, before that watershed moment, only the scientific statements owed their authority to the name of their author: "There was a time when these texts which we call 'literary' (stories, folk tales, epics and tragedies) were accepted, circulated, and valorized without any question about the identity of their author. Their anonymity was ignored because their real or supposed age was a sufficient guarantee of their authenticity. Texts, however, that we call 'scientific' (dealing with cosmology and the heavens, medicine and illness, the natural sciences or geography) were only considered truthful in the Middle Ages if the name of the author was indicated. Statements on the order of 'Hippocrates said . . . ' or 'Pliny said . . . ' were not merely formulas for an argument based on authority; they marked a proven discourse."

In the early modern period, according to Foucault's loose chronology, this distribution of the author-function is turned upside down. On the one hand, "scientific texts were accepted on their own merits and positioned within an anonymous and coherent conceptual system of established truths and methods of verification. Authentication no longer required reference to the individual who had produced them; the role of the author disappeared as an index of truthfulness and, where it remained as an inventor's name, it was merely to denote a specific theorem, a [substance], a group of elements or pathological syndrome." From that moment on, a rule of anonymity commanded the production and accreditation of scientific statements.

In the opposite direction, from the same moment on, "literary discourse was acceptable only if it carried an author's name; every text of poetry or fiction was obliged to state its author and the date, place, and circumstances of its writing. . . . If by accident or design a text was presented anonymously, every effort was made to locate its author. Literary anonymity was of interest only as a puzzle to be solved as, in our day, literary works are totally dominated by the sovereignty of the author."

Such a chiasmus (highly debatable as we shall see later) enlightens, at first, the very function attributed to the author-function: to guarantee the unity of a body of texts by ascribing it to one sole source of expression and to neutralize the eventual unevenness or contradictions between different works of the same author. And even if we cannot accept the chronology it proposes for the principle of identification of the scientific texts, Foucault's statement leads to the recognition of the existence of the author-function for certain classes of texts as far back as the Middle Ages. A hasty reading must not reduce Foucault's thoughts to oversimplified formulas: in no way does he postulate an exclusive and determinant connection between the system of property that characterizes the modern societies and the construction of the author as the fundamental principle for identifying certain classes of discourse. By moving the figure of the author back in time and by linking it with mechanisms for controlling the circulation of texts or for lending them authority, Foucault invites us to a retrospective investigation focusing on the articulation between the role of (or absence of) the function of the author within different discursive practices with the history of the conditions for production, dissemination, and appropriation of these discursive formations.

For discussing the validity of the reversal between scientific and literary discourse that Foucault situated in the seventeenth and eighteenth centuries, it is necessary to observe that he was very cautious when he alluded to the distinction between these two classes of discourse. He used such expressions as "these texts we now call 'literary'" (*«ces textes qu'aujourd'hui nous appellerions 'littéraires'»*) or the "texts we call 'scientific'" (*«les textes que nous dirions maintenant 'scientifiques'»*). These expressions are clear symptoms of the tensions existing between the inertia of vocabulary, which implicitly supposes the universality of the categories that allow us to distinguish between different genres of discourse and the historical variations of such distinctions. Behind the lazy convenience of vocabulary, what we need to recognize are singular demarcations, specific distributions, and particular systems of exclusion. From this Foucaultian perspective, the distinction between science and literature cannot be taken as universal or stable.

In his lecture, Foucault did not give any importance in the trajectory of the author-function to the difference we can and must establish for each historical moment between the ancient and canonical authors (such as Hippocrates and Pliny) and the contemporary writers. When this difference is taken into account, it invites us to substitute another hypothesis for the chiasmus described by Foucault, which considers the construction of the author-function as the process by which the texts in the vernacular (whatever they may be) were assigned to a principle of designation and attribution that had long been characteristic only of works that were referred to as ancient *auctoritates*, religious or not, continually cited and tirelessly commented upon. Such a revision of Foucault's perspective will be placed at the core of my essay, which will put more emphasis on this difference between traditional authorities and modern authors and less on the distinction between scientific and literary discourse.

Foucault proposed in his lecture and some months later in *L'ordre du discours* (translated into English as *The Discourse on Language*) the project of an "historical analysis of discourse" which required a radical shift in the questionnaire about texts: "We should ask: Under what conditions and through what forms can an entity like the subject appear in the order of discourse? What position does it occupy? What function does it exhibit? And what rules does it follow in each type of discourse? In short, the subject (and its substitutes) must be stripped of its creative role and analyzed as a complex and variable function of discourse."[2]

In this essay I would like to present some scattered reflections that are focused on the different forms of classification and publication of some fundamental textual genres in the early modern period and the late Middle Ages.

The Invention of Copyright

First of all, it is necessary to recall that the construction of the author as proprietor, far from arising from a particular application of the bourgeois definition of property right, was deeply rooted in the defense of the booksellers' privileges. According to Mark Rose : "It might be said that the London booksellers invented the modern proprietary author, constructing him as a weapon in their struggle with the booksellers of the provinces."[3] In 1710 the statute passed by the Parliament overturned the old publishing system which, since 1557, had granted to the London booksellers and printers of the Stationers' Company first a monopoly on obtaining copyrights (before 1701 the common term was *rights in copies*) and, secondly, the perpetuity of their ownership of the titles they had "entered" in the Register of the Company. The 1710 *Act for the Encouragement of Learning by Vesting the Copies of Printed Books in the Authors, or Purchasers, of Such Copies during the Times therein Mentioned* authorized authors to demand copyrights for themselves and limited the duration of the copyright to fourteen years (plus a fourteen-year renewal if the author were still alive).

The only way that the booksellers of the Stationers' Company could reassert their traditional ownership was to plead for the recognition of the author's perpetual right over his own work, hence the equally perpetual right of anyone who had acquired that work. Thus, they had to invent the author as proprietor of his works and the author-function as a fundamental characteristic of the works they published. The lawyers defending the London booksellers against their provincial Irish or Scottish colleagues, who tried to take advantage of the statute in order to reprint titles for which the Londoners claimed to hold a perpetual right, developed a dual line of argumentation.

The first line was founded on a theory of property derived from Locke that sustained that a man, as the proprietor of his own person, is also the owner of all the products of his labor. Since literary compositions are a product of labor, their authors have therefore a natural right of property in their works. The second argument was based on an aesthetic category: originality made literary compositions incompatible with mechanical inventions were subject to patents. (The 1710 Act granted a monopoly on their exclusive exploitation only for a limited period of time—the fourteen years precisely mentioned by the Statute.) Consequently, literary compositions were not identified with any of their material forms; according to Blackstone, their identity was given by the irreducible singularity of their "style, sentiment, and language" present in every duplicate of the work. The inalienable right of the author was thus transformed into an essential characteristic of the discourse itself, whatever the vehicle of its transmission might be: a manuscript, a printed book, or a performance.

Such a reappraisal of the invention of the copyright suggests a challenge of Foucault's assertions concerning the relationship between the author-function and the modern system of property. First, the construction of authorship on the basis of the right of property of the author to his or her work cannot be dated to the late eighteenth or early nineteenth centuries. Rather, it emerged from or series of cases opened in England in the 1720s after the vote of the Statute of Anne. Second, in England and later in France, the invention of the author as proprietor is directly linked with the claim for the perpetuation of an old system of privileges, guaranteed either by a guild or by the king. It did not derive from a new bourgeois conception of the property and the market free from the regulations of the trades or the state. Third, the distinction between the essential identity of the work and the diversity of the "mere accidents" or "vehicles" that convey it plays a fundamental role in the process through which the author is constructed as a principle of unity and a singular source of expression.

Was the logic of the author as proprietor unknown before 1710? Let us take two examples. When he published the 1616 folio of his *Workes,* Ben Jonson broke with the traditional practice that gave the ownership of the play to the theatrical company as if the very "author" was the director of the company and not the playwright.[4] In Golden Age Castile, the *autor de comedias* was not the playwright (called *poeta* or *ingenio*) but the man who received the license authorizing the company, who bought plays from the writers, who rented the *corrales* where they would be performed and who, as God in Calderón's *El gran teatro del mundo* (*The Great Theater of the World*), was responsible for the distribution of the characters, the scenery and the costumes, and the production itself. By selling his masques and plays to publishers directly, Ben Jonson exploited the resources of the printed book in order to affirm his proprietary relationship to his own works.[5] In the contract of the "Induction" of *Bartholomew's Fair*, the author usurped the company's rights by signing an agreement (fictional of course) with the spectators directly: "The author promiseth to present them, by us [that is to say, the actors] with a new sufficient play called *Bathol'meuw Fair*, merry, and as full of noise and sport, made to delight all, and to offend none." The theatrical performance was no more thought of as a contribution to a collaborative production of the play but was considered a mere vehicle (by us) for transmitting the author's creative work.[6] Such a construction of authorship directly linked with the marketplace was not at all exclusive of the patronage system. But the word *Workes* used in the title page of the 1616 folio, which was used for the classics of the antiquity and inspired by the 1611 folio of Spenser's *Works of England Arch-Poet*, expressed a strong desire for capturing the canonical *auctoritas* of the ancient poets and higher genres and for shaping a distinctive authorial persona thanks to the printed book.

The reality of the logic of the author as proprietor before 1710 is also demonstrated by Milton's contract with the printer Samuel Simmons for the publication of *Paradise Lost*.[7] This contract, signed in 1667, was often quoted as evidence of the rapacity of publishers taking advantage of vulnerable authors, since Milton received only five pounds upfront and an additional five pounds for each of the editions of his work. But this document could be understood in another manner as shown by Peter Lindenbaum. Different clauses of the contract anticipated some features that will later characterize copyright: for example, the stipulation that placed an upward limit of fifteen hundred copies on each printing, the payment to Milton of a sum of money for every edition at a time in which the publisher's habit was to buy the manuscript outright and to remunerate the author with copies of the book and not money, or the clause that specified that Milton could demand a formal document giving an accounting of sales of his work at reasonable intervals. Moreover, sometimes an author or his heirs kept some residual property interests in a work after it was entered by a bookseller or a printer in the Register of the Stationers' Company. This was the case with the contract between Milton's widow and Joseph Watts signed in 1695 that concerned some works already entered by Watts in the Register Book of the Company.[8] In spite of their claim to a perpetual right on the copies they registered, the Stationers sometimes seemed to recognize that something of the author's initial property right in his work did not completely disappear with the sale of his manuscript. In sum, the process that links the construction of the author-function with the recognition of the author's copyright did not begin with the Statute of 1710.

Propriety and Property

Nevertheless, the juridical and judiciary debates emerged after the passing of the Statute of 1710 in England defined a new position for the author. This new definition was opposed, term for term, either to the aristocratic figure of the "gentleman-writer" or the ethos and practices of the contemporary Republic of Letters.[9] The gentlemen-amateurs and the members of the community of scholars shared some common values. They were disdainful of the bookselling trade, which, according to them, corrupted at the same time the integrity of the texts, distorted by hands of "rude mechanicals" (as says Puck in *A Midsummer Night's Dream*); the literary code of ethics by introducing into the commerce of letters cupidity and piracy; and the evidence of meaning by allowing an uncontrolled circulation and possible misunderstanding of the works.[10] Gentlemen-writers and erudite scholars preferred the circulation of manuscripts for their works because they were addressed to the chosen public of the peers and they embodied the ethos of personal obligations and communal politeness that characterized aristocratic civility as well as the ethics of reciprocity of the *Res Publica Literatorum*. The concealment of the proper name behind anonymity or the collective work of the learned community was another characteristic of such ethics. This practice did not mean the absence of the author, since the Republic of Letters celebrated its more illustrious members through the writing of their biographies and *eloges*, or the publication of their correspondence. But it means that this specific modality of the author-function was built on a value system largely unconcerned with the monetary rewards promised by the author's propriety.

In a similar manner, it is only after the passing of the Statute of 1710 and the cases it triggered that the author-function was attributed to forms of writing that were excluded from the logic of literary property before the eighteenth century. This was the case with the correspondence. In 1741, the decision *Pope v. Curll* established that the writer of a letter kept an intangible property in his copyright. Consequently, a letter could not be printed without the consent of its author.[11] This judgment is important for two reasons. First, it stressed the essentially immaterial nature of the object of copyright since the propriety of the receiver of the physical letter did not allow him or her a claim to the right of publishing its text. Secondly, it mingled two different definitions of the concept of intellectual property: the right of the author to control the publication of his own texts in order to preserve his privacy, honor, and reputation, and the property right in the sense of an economic interest in an alienable commodity. In spite of the instability of the ancient vocabulary, Mark Rose has proposed to call "propriety" the first right and "property" the second. Traditionally, they were separated. When Lope de Vega protested against the pirated editions of his *comedias*, it was not for economic reasons but because these editions published without his consent and control damaged his honor (his *honra* or *reputación*) in two ways: by presenting to the reader corrupted texts of his own plays, and by attributing to him works (supposedly bad) that he had never written. A century later, Alexander Pope's suit against Curll dissolved matters of propriety into matters of property and entangled the code of honor proper to the gentleman-writer with the economic aspirations of the author as proprietor.

Political Authority and Scientific Authorship in the Sixteenth and Seventeenth Centuries

Such a reappraisal of the history of literary authorship through the seventeenth and eighteenth centuries allows us to reframe our initial question: What is a scientific author in the same period? Is it less certain that, as Foucault thought, a rule of anonymity has commanded the scientific statements since this historical moment? For a notable length of time, the validation of an experiment or the accreditation of a discovery presupposed the guarantee provided by a proper name—but the proper names of those who, by their position in society, had the power to proclaim the truth.[12] The fact that technicians disappeared behind aristocratic authority did not imply the anonymity of discourse whose possible acceptance as true was not exclusively dependent on its compatibility with an already constituted body of knowledge. During the seventeenth and the eighteenth centuries, a number of scientific texts displayed a characteristic that Foucault reserved (wrongly) for medieval works alone: they were "only considered truthful [. . .] if the name of the author was indicated"—an author, however, who was long understood as someone whose social position could lend "authority" to knowledge in a time in which the distribution of credibility reproduced the hierarchy of society.

The validation of experiments and the authentication of experimental narratives by an aristocratic or princely testimony was the rule in all early modern Europe.[13] The rhetoric of dedications clearly expressed this transfer of the author-function from the actual writer or scholar to the sovereign or the minister to whom the work is addressed. The dedicatee is celebrated as the primordial inspiration and the first author of the work presented to him, as if its actual author was offering him a work that was in fact his own. The dedication of the *Sidereus Nuncius* by Galileo to Cosimo de Medici, which made the prince the discoverer and owner of the natural reality (the Medicean Stars) dedicated to him, is a perfect example of this rhetorical device.[14] Possessing not only what he gave but also what he received, the prince was thus praised as a true scholar.

The model of aristocratic validation of the experiments framed the definition of scientific authorship and the posture of all scholars, whether gentlemen or not. For them, as Steven Shapin wrote about Boyle, "a disengaged and nonproprietary presentation of authorial self" was necessary for distancing themselves from the mercenary practices of the book trade and for securing the truth of their knowledge claims.[15] The trope of reluctant authorship and resistance to printed publication was very common in early modern culture, but it acquired a particular meaning with scientific texts: it assured their credibility since it proved that there was no economic interest attached to the published knowledge claims: "A gentleman's word might be relied upon partly because what he said was without consideration of remuneration."[16] This does not mean, however, that anonymity was the rule for the construction of scientific knowledge, but that the conception of authorship in science remained more strongly linked to propriety than to property in a time in which the two notions began to merge in literary compositions.

Censorship, Printing, and Authorship

Foucault attached the emergence of the author-function not only to the invention of copyright in the eighteenth century but also to the judicial responsibility of the writer as it was defined in an earlier period by the censorship of churches and states: "Speeches and books were assigned to real authors . . . when the author became subject to punishment and to the extent that his discourse was considered transgressive." For examining such a statement that concerns both scientific and literary texts, I shall take the example of the Spanish Inquisition. In the 1612 *Index librorum prohibitorum et expurgatorum* published by the General Inquisitor Bernardo de Rojas y Sandoval, the category of the author and the presence of the proper names are fundamental principles for designating the prohibited works.[17] All three classes of forbidden books distinguished by the *Index* referred to the author-function. The first condemned all the works (*todas sus obras*) of the authors considered to be heretics—that is to say, not only the works they had already written but also the works they were to write and publish in the future (*no solo las que hasta aora han escrito i divulgado, mas tambien las que adelante escrivieren i publicaren*). The author was thus clearly constructed as a unique source of thought and expression that was equally manifested in all his texts—even in the texts that were not yet composed. The second class, which prohibited titles but not authors, was still linked to the author-function since the identification of these titles was possible through the mentioning of the names of their authors. This is the reason why booksellers were obliged to send an inventory of the books they stocked to the inquisitors within a period of sixty days after the publication of the *Index*, listing them in alphabetical order by authors' names. Finally, the third class of the prohibited works also expressed a clear recognition of the author's responsibility since, according to the tenth rule of the previous Spanish *Index* published in 1584 by the General Inquisitor Quiroga; it censored (allowing for exceptions) all the books printed after this date that did not include author's and printer's names. The author-function was thus constituted in the late sixteenth and early seventeenth centuries as an essential weapon in the battle of the Catholic Church against the diffusion of texts suspected of heresy and heterodoxy. From this point of view, there was no difference between literary and scientific texts, between ancient and contemporary.

Since the mid-fifteenth century, however, different devices reinforced the first emergence of the author-function in the case of the literary writers. Analyzing the example of the Parisian *rhétoriqueurs* (Jean Molinet, André de La Vigne, Pierre Gringore, etc.), Cynthia J. Brown identified between 1450 and 1530 the "self-promotional strategies . . . which underscore the author's development from a conventionally medieval secondary stance to a growing authoritative presence."[18] On the one hand, vernacular writers tried to control the distribution of their works by asking for *privilèges* for their publication and by initiating lawsuits against the printers who had published their texts without their consent[19]—the first instance was in 1504. On the other hand, the author's identity was more clearly advertised by title pages and colophons, since the authorial naming developed from hidden or metaphoric signatures into nonfictional and personal ones. Finally, authorial identity was also promoted by the shift that replaced the dedication scene in the frontispiece in which the author, on his knees before a prince seated on his throne, offered to him a richly bound book with a portrait of himself, sometimes represented as composing his work.[20] In Paris between 1450 and 1530, "the author's consciousness of a need to adopt a protective posture vis-à-vis the book producers and his audience and the public's awareness and recognition of increasing authorial concern for literary property and propriety led to a greater focus on the writer's individuality."[21] These different authorial shifts can be considered a decisive step, previous to the seventeenth or eighteenth century mentioned by Foucault, for the definition of literary discourse as "acceptable only if it carried an author's name."

Book, Author, and Work in the Fourteenth Century

Must we link this first emergence of the author-function to the invention of print? For Cynthia J. Brown (and many other historians) there is no question. The transformations in the presentation and self-fashioning of authorship was a result of the impact of print, the development of a buying public, and "commodification" of the book. In my opinion, it is possible to challenge such a perspective by arguing that the relationship of patronage did not disappear with the print culture—far from it—and that the affirmation of the author's identity and the authorial function predated the invention of the printed book.

The fourteenth century, a disenchanted and melancholic age haunted by the impermanence of things and the fragility of words, was indeed a time of inventions fundamental to written culture.[22] The first was in language: the new meanings attributed to three words. First, *author* (in French *acteur,* transformed later into *autheur*) endowed the *actores*, that is to say, the contemporary writers long thought of as mere compilers and commentators (according to the etymology of the word deriving from *agere*, "to do something") with the authority traditionally reserved to the ancient *auctores* (a word coming from *augere,* which meant "to give existence, to create something"). Two centuries later, Hobbes played with these two different etymological meanings of *actors* and *authors* in the sixteenth chapter of *Leviathan*: "Of persons artificial, some have their words and actions owned by those whom they represent. And then the person is the actor, and he that owns his words and actions is the AUTHOR."[23] Second, in the fourteenth century, *writer* (*écrivain*) began to designate the person who composes a work as well as one who copies a book. Third, *invention* came to mean an original creation rather than solely the discovery of what God had produced. In the manuscript book, miniatures showed the portrait of the author pictured in the process of writing (in both senses of the word). A traditional depiction of *auctoritates* was thus shifted to contemporary writers who expressed themselves in vernacular rather than in Latin, and who composed poems, romances, and histories rather than theological, juridical, or encyclopedic works.

Another transformation accompanied the linguistic ones: the invention of *literature* as the very matter of the poetic gesture. In a time that considered poetic material to be drying up and inspiration exhausted, writing, literature, and the book became the subject matter of works aimed at entertaining or teaching. Literature was born of that return of writing onto itself, of that secondary effort to constitute a repertory of genres and a canon for "moderns." This *mise en abyme* of the creative act, reflected in literary creation itself, took several forms: narratives about the mythical inventors of writing (Orpheus, Thôt, Cadmus, or Carmentis), funerary eulogies by a writer to the memory of another poet whom he recognized as his master, or literary cemeteries gathering respected and canonized authors. In the *Livre du coeur d'amours épris*, René d'Anjou mentioned six tombs set apart where rest the bodies of Ovid, Guillaume de Machaut, Jean de Meung, Petrarch, Boccacio, and Alain Chartier—four of six authors from the fourteenth century.[24]

A last but fundamental invention of the underestimated fourteenth century is the alliance set up among the book as an object, the work, and the name of its author. The traditional and durable form of the manuscript book was that of a collection of texts of varying genres, dates, and authorship. From the eighth century on, the model of the miscellaneous book had become a dominant form for the manuscripts (except for the canonical *auctoritates*).[25] Moreover, according to Francisco Rico, commenting upon the *Libro del Caballero Zifar*, this particular form of book constituted a textual paradigm used for the composition of literary works that juxtaposed disparate and unconnected genres and fragments.[26] For a long time, the unity of a book had nothing to do with an absent author-function. Such a unity depended upon the will of a reader who desired to join diverse works in a single book or upon the activity of a scribe who decided to copy and associate a series of heterogeneous texts. This conception of the book was still present in the sixteenth and seventeenth centuries in the framework of the scribal culture that survived the invention of print. The handwritten collections of poems that made up works composed by different authors (without necessarily naming them) and that added to them texts copied or composed by the owner of the manuscript are a good example of the lasting importance of the miscellaneous books.[27]

Nonetheless, it was during the fourteenth century that, at least for certain works, this traditional and dominant definition of the codex was replaced by a new conception of the book, offering the works of only one author or even just one of his works. This was a somewhat paradoxical development in an age in which writing was often thought of as compilation or reuse and in which its most frequent image was that of the gleaner gathering strands of grain left by the great harvesters of the past. The unity established between the material integrity of the book and the singularity of the works that came from the same pen clearly shows that some authors of the time (who wrote in the vernacular), enjoyed the same "codicological dignity" as the older authorities. It is, thus, before the age of the printed book (which, incidentally prolonged the tradition of composite collections in a number of genres) that the connection between a material unit and a textual one ascribed to an author became true in certain works in the vulgar tongue.[28]

Foucault's Chiasmus Revisited

The genealogy of literary authorship was longer than Foucault thought. A fundamental element in its construction was the existence of the *libro unitario* (according to Armando Petrucci's expression), that is to say, the book considered as an entity embodying a work or as a series of works written by the same author. From such a perspective, the technique of reproducing the text (it can be copied by hand or printed) is not decisive in itself. What is important is the material and intellectual relationship established between an object, a work (or a series of works), and a proper name.

Conversely, the genealogy of scientific authorship is much more complex than a simple shift from *auctoritates* to anonymity. First, in the Middle Ages and Renaissance a large part of the discourse we can label as "scientific," since it procured knowledge on the natural phenomena, was not referred back to canonical *auctoritates*, but was a collective and anonymous knowledge. Such was the case with the books of secrets; the *Kunstbüchlein*, or craft manuals;[29] and the *libri di bottega*, which were handwritten technical handbooks used in the workshops.[30] This was also the case with commonplace books that neutralized the individuality of the proper names of their compilers or the authors they quoted in favor of an anonymous body of universally accepted knowledge.[31]

Secondly, and conversely, it is clear that the scientific revolution of the seventeenth century—whatever its definition may be—was not synonymous with the expulsion of proper names from knowledge claims. The authentication of experiments or discoveries required the guarantee given by an authority progressively displaced from princely or aristocratic power to scientific authorship.[32]

Finally, against a too simplistic opposition between different genres of discourse or practices, it can be argued that it is the patents granted for the invention of machines, processes or devices from the fourteenth century on, that constituted a definitional matrix of intellectual property. The fact that in the eighteenth century the lawyers pleading for the Stationers established a strong distinction between copyright and patent as an argument for refusing their assimilation by the Statute of Anne does not invalidate the idea that literary authorship was thought of and constructed according to the right of property recognized for the inventors of new techniques.[33]

The terms of the provocative chiasmus proposed by Foucault are no longer acceptable. We have to discuss them and revise them. But as Foucault stated in "What Is an Author?" the "return" to any work that has instituted a new discursive practice necessarily means a transformation of its theoretical construction. According to him, the "initiators of discursive practices cleared a space for the introduction of elements other than their own, which, nevertheless, remain within the field of discourse they initiated." Foucault was such an initiator of discursive practices. It is the categories he proposed and the questions he raised in 1969 that have rendered possible different reappraisals that today challenge his own.

The "elements" I propose to introduce into his own canvas came mainly from a sociology of texts defined as the study of "the text as a recorded form."[34] Such a perspective allows us to understand that the author-function is not only a discursive function, but also a function of the materiality of the text. D. F. McKenzie wrote: "New readers make new books . . . and their new meanings are a function of their new forms."[35] Paraphrasing such a remark, we could say that new books make new authors, that is to say, that the construction of authorship is a function of a new form of the book of which the unity is both physical and authorial. It is this new form that established in Western culture, much before Gutenberg, the necessary, unstable, and conflictual relationship between the writer as an individual and the author as fiction.

Notes

1. Michel Foucault, "Qu'est-ce qu'un auteur?" in Foucault's *Dits et écrits 1954–1988*, Tome 1, *1954–1969*, eds. Daniel Defert and François Ewald, (Paris: Gallimard, 1994), 789–821; English translation: "What Is an Author?" in Foucault, *Language, Counter-Memory, Practice: Selected Essays and Interviews*, ed. Donald F. Bouchard (Ithaca and London: Cornell University Press, 1977), 113–38. I have dedicated a previous essay to Foucault's lecture in Roger Chartier, "Figures of the Author," in *The Order of Books: Readers, Authors, and Libraries in Europe between the Fourteenth and Eighteenth Centuries* (Stanford, Calif.: Stanford University Press, 1994), 25–59.
2. Michel Foucault, *L'ordre du discours* (Paris, Gallimard, 1979); English translation: "The Discourse on Language," in Foucault's *The Archeology of Knowledge and the Discourse on Language* (New York: Pantheon Books, 1972), 215–37; quotation from "What Is an Author?" 137–38.
3. Mark Rose, "The Author as Proprietor: Donaldson v. Becket and the Geneaology of Modern Authorship," *Representations* 23 (1988): 41 (quotation), 51–85; and *Authors and Owners: The Invention of Copyright* (Cambridge and London: Harvard University Press, 1993).
4. Joseph Loewenstein, "The Script in the Marketplace," *Representations* 12 (fall 1985): 101–14; and Mark Bland, "William Stansby and the Production of *The Workes* of Benjamin Jonson, 1615–16," *The Library* 20, sixth series, no. 1 (March 1998): 1–31.
5. Louis A. Montrose, "Spenser's Domestic Domain: Poetry, Property, and the Early Modern Subject," in *Subject and Object in Renaissance Culture*, eds. Margreta de Grazia, Maureen Quilligan, and Peter Stallybrass (Cambridge: Cambridge University Press, 1996), 83–130.
6. Jeffrey Masten, "Beaumont and/or Fletcher: Collaboration and the Interpretation of Renaissance Drama," in *The Construction of Authorship. Textual Appropriation in Law and Literature*, eds. Martha Woodmansee and Peter Jaszi (Durham and London: Duke University Press, 1994), 360–81, and *Textual Intercourse: Collaboration, Authorship, and Sexualities in Renaissance Drama*, (Cambridge: Cambridge University Press, 1997).
7. Peter Lindenbaum, "Milton's Contract," in *The Construction of Authorship. Textual Appropriation in Law and Literature*, 175–90; and "Rematerializing Milton," *Publishing History* 41 (1997): 5–22.
8. Peter Lindenbaum, "Authors and Publishers in the Late Seventeenth Century: New Evidence on Their Relations," *The Library* 17, sixth series, no. 3 (September 1995): 250–69.
9. Alvin Kernan, *Printing Technology, Letters, and Samuel Johnson* (Princeton, N.J.: Princeton University Press, 1987); Ann Goldgar, *Impolite Learning: Conduct and Community in the Republic of Letters, 1680–1750* (New Haven and London: Yale University Press, 1995).
10. Adrian Johns, "History, Science, and the History of the Book: The Making of Natural Philosophy in Early Modern England," *Publishing*

History 30 (1991): 5–30; and *The Nature of the Book: Print and Knowledge in the Making* (Chicago: Chicago University Press, 1998).

11. Mark Rose, "The Author in Court: Pope v. Curll (1741)," in *The Construction of Authorship*, 211–29.
12. Steven Shapin, "The House of Experiment in Seventeenth-Century England," *Isis* 79 (1988): 373–404.
13. Paula Findlen, "Controlling the Experiment: Rhetoric, Court Patronage, and the Experimental Method of Francesco Redi," *History of Science* 31, part 1, no. 91 (March 1993): 35–64, and *Possessing Nature: Museums, Collecting, and Scientific Culture in Early Modern Italy* (Berkeley and London: University of California Press, 1994).
14. Mario Biagioli, "Galileo the Emblem Maker," *Isis* 81 (1990): 230–58; and *Galileo Courtier: The Practice of Science in the Culture of Absolutism* (Chicago: University of Chicago Press, 1993).
15. Shapin, *A Social History of Truth: Civility and Science in Seventeenth-Century England* (Chicago and London: Chicago University Press, 1994).
16. Shapin, "The House of Experiment in Seventeenth-Century England," 395.
17. This document is quoted by José Pardo Tomás, *Ciencia y censura. La Inquisición Española y los libros científicos en los siglos XVI y XVII* (Madrid: Consejo Superior de Investigaciones Científicas, 1991), 373–74.
18. Cynthia J. Brown, *Poets, Patrons, and Printers: Crisis of Authority in Late Medieval France* (Ithaca and London: Cornell University Press, 1995), 7.
19. Elizabeth Armstrong, *Before Copyright: The French Book-Privilege System, 1498–1526* (Cambridge: Cambridge University Press, 1990).
20. Roger Chartier, "Princely Patronage and the Economy of Dedication," in *Forms and Meanings: Texts, Performances, and Audiences from Codex to Computer* (Philadelphia: University of Pennsylvania Press, 1995), 25–42.
21. Brown, *Poets, Patrons, and Printers*, 6.
22. Jacqueline Cerquiglini, *La Couleur de la mélancolie. La fréquentation des livres au XIVe siècle 1300–1415* (Paris: Hatier, 1993). English translation: *The Color of Melancholy: The Uses of Books in the Fourteenth Century* (Baltimore and London: Johns Hopkins University Press, 1997).
23. Thomas Hobbes, *Leviathan* (London: Penguin Books, 1985), 217–22, 218 (quotation).
24. Cf. Jacqueline Cerquiglini's *Colour of Melancholy*, 35–36 (on the literary eulogies), 110–22 (on the inventors of writing), and 132–44 (on cemetery of love).
25. Armando Petrucci, "Del libro unitario al libro miscellaneo," in *Società e imperio tardoantico*, vol. 4, *Tradizioni dei classici, trasformazione della cultura*, ed. Andrea Giardina (Bari, 1986), 173–87. English translation: "From the Unitary Book to the Miscellany," in Armando Petrucci's *Writers and Readers in Medieval Italy: Studies in the History of Written Culture*, ed. Charles M. Radding (New Haven and London: Yale University Press, 1995), 1–18.
26. Francisco Rico, *Entre el códice y el libro (Notas sobre los paradigmas misceláneos y la literatura del siglo XIV)*, 1997 (unpublished).

27. Harold Love, *Scribal Publication in Seventeenth-Century England* (Oxford: Clarendon Press, 1993); Max W. Thomas, "Reading and Writing in the Renaissance Commonplace Book: A Question of Authorship?" in *The Construction of Authorship*, 401–15.
28. See, as an example, Gemma Guerrini, "Il sistema di communicazione di un 'corpus' di manoscritti quattrocenteschi: I 'Trionfi' di Petrarca," *Scrittura e Civiltà* 10 (1986): 122–97.
29. William Eamon, *Science and the Secrets of Nature: Books of Secrets in Medieval and Early Modern Culture* (Princeton, N.J.: Princeton University Press, 1994).
30. Carlo Macagni, "Leggere, scrivere e disegnare. La 'scienza volgare,'" *Scrittura e Civiltà* 15 (1991): 267–88.
31. Ann Blair, "Humanist Methods in Natural Philosophy: The Common Place Book," *Journal of the History of Ideas* 53 (1992): 541–51; and *The Theater of Nature: Jean Bodin and Renaissance Science* (Princeton, N.J.: Princeton University Press, 1997), 49–81, 180–224; Ann Moss, *Printed Commonplace-Books and the Structuring of Renaissance Thought* (Oxford: Clarendon Press, 1996).
32. Christian Licoppe, *La Formation de la pratique scientifique. Le discours de l'expérience en France et en Angleterre (1630–1820)* (Paris: Editions La Découverte, 1995).
33. Pamela Long, "Invention, Authorship, 'Intellectual Property,' and the Origins of Patents: Notes toward a Conceptual History," *Technology and Culture* 32, no. 4 (1991): 313–35.
34. D. F. McKenzie, *Bibliography and the Sociology of Texts*, The Panizzi Lectures 1985 (London: The British Library, 1986), 4.
35. Ibid., 20.

2.

Butter for Parsnips

Authorship, Audience, and the Incomprehensibility of the *Principia*

ROB ILIFFE

In this paper I examine how Newton developed a specifically mathematical style in natural philosophy that owed a great deal to the narrative form and mode of proof of the mixed mathematical sciences. The implied audience of such difficult texts stood in stark contrast to that emplotted within experimental reports of the late seventeenth century. Whereas the source of belief in the former is supposed to be primarily visual and based on an unconstrained act of will, the force of the demonstration of a mathematical text exerts something like an irresistible compulsion on those that are held to comprehend it.

Mathematical texts are intrinsically difficult and their very impenetrability raises significant issues about the kinds of people who have the right and skill to judge their quality, and by extension, the competence of their authors. An analysis of the immediate reception of the *Principia* shows how readers were implicitly and explicitly classified according to whether they claimed or were held to understand the *Principia*, as well as by the *extent* to which they claimed or were held to understand it. I am not concerned with whether readers *really* understood the *Principia*, but rather with the way the term *understanding* functioned in establishing the reputation of the book, its author, and his acolytes. This is analyzed in terms of a mutual relationship between master and disciples as well as by the real training undertaken by potential disciples to grasp the text to varying degrees. Such processes constituted a sort of concentric ring of competence extending inward from those who were held to understand nothing, to the center point inhabited—perhaps—by the author himself.

Perhaps paradoxically, a number of great works and theories are as notorious for their obscurity as they are for their aesthetic simplicity, and it is a feature of epoch-making discoveries that they are held to be incomprehensible to any but a select few. The reception of the general theory of relativity provides a well-known example.[1] The magisterial *Principia Mathematica* of Isaac Newton was surrounded by similar tales when it was first released to the public in 1687, and its impenetrability has become legendary. In its own day, it was couched in a language that few were able to begin to read; it was hard going for even the most adept of contemporary practitioners, and the number of people who could offer expositions of its most abstruse sections was miniscule. Anecdotally, even its author did not fully grasp its meaning. On the other hand, in an implicit homage to the difficulty of the book and to the talent and skill of the gifted reader, a number of stories circulated in the eighteenth century about recently deceased individuals who were—allegedly—said by Newton to have "understood" the *Principia* as well as anyone.[2]

The category of understanding is integrally related to the way in which mathematical texts are made and approached by readers. For example, René Descartes told Marin Mersenne in May 1637 regarding the status of his own *Géométrie* that he expected that Pierre de Fermat, "if he is an honest and open man, will be one of those to make the most of the work and that he will be one of those most capable of understanding it." For, he continued, "I will tell you honestly that I feel there will be very few people who will understand it." To make his *Meditations* the source of a quasi-devotional odyssey, Descartes decided to cast it in a particular style which followed the order used in geometry while avoiding what he called the geometrical "method," so as to force readers to meditate seriously and slowly. Nevertheless, he decided at the behest of Mersenne to recast the major arguments in a geometrical format at the end of his replies to the Second Objection, so that the work could be read in a way more redolent of a mathematical work.[3]

Despite the *Principia*'s unique status within its genre, mathematical writings *in general* are paradigmatic of esoteric texts and their readership is elite, small, and well-defined, since only those with a specific specialist training can comprehend the work to any great degree. Beginning with definitions or axioms that are taken for granted, a mathematical text is succinct, rigorous, and highly prescriptive with regard to what it expects of its audience. It implicitly specifies what the reader needs to know to read it, in what order, how much effort is required, and indeed how quickly one is expected to go through it. Only when all these factors have been satisfied can a reader be said to "understand" a text.[4]

However, in opposition to many of the features of a mathematical or geometrical presentation, the dominant discursive form promoted by the evangelists for the early Royal Society was explicitly based on the generally empiricist works exemplified by the writings of Robert Boyle. Readers of Boyle's printed works were participants in an extended forum who might see engravings of real instruments rather than the idealized, schematic representations favored by mathematicians such as Blaise Pascal. For many experimental philosophers, the overconfident epistemological claims made by mathematicians were fundamentally inadequate for representing a natural world that was not everywhere written in the language of mathematics. The rational consumers of empiricist literature were invited to see for themselves, either by means of engravings or drawings of real entities in the text, or by means of an honest, naked form of writing, and willingly give their assent as a result of this. In stark contrast, the obscure and abstract mathematical format *compelled* assent only from readers who understood the demonstrations. Prominent proselytizers for the Royal Society endorsed features of the new philosophy that gave primacy to the simplicity of the language used, as well as to the intelligibility of the conceptions involved. The implied empiricist audience was widely extended; that of mathematics greatly restricted.[5]

Early modern mathematical texts contained dedicatory epistles and prefaces that were more accessible than the mathematical content, which formed the bulk of the work. Authors such as Galileo and Descartes were steeped in the different rhetorical practices appropriate for different philosophical or mathematical audiences, and when it suited them, their virtuosity and versatility allowed them to publish in various forms, and in Latin or vernacular.[6] Copernicus explicitly addressed his *De Revolutionibus Orbium Cœlestium* to at least two different sorts of audiences among the ecclesiastical milieu of which he was a member. As Robert Westman has pointed out, Copernicus's address to Pope Paul III in the preface showed "a rigorous knowledge of common epistolographical and rhetorical resources, such as understatement of his own achievements and exaggerated modesty: all characteristic strategies in *captatio benevolentiae* designed to capture an audience's attention and good will." The mathematically unlearned might be convinced by the philosophical arguments in Book One that derided the monstrous state of astronomy and that offered plausible physical accounts consistent with a heliocentric cosmology. However, the remainder of the book was written in mathematics and was intended only for mathematicians. It was thus mathematically adept divines alone who could accept the proofs offered therein, admit the changing relationship between the disciplines of astronomy and physics that resulted therefrom, and who would as a consequence authorize the change in biblical exegesis that had to result from this transformation.[7]

In so far as they are seen to be disinterested seekers of truth whose goal is to donate their findings to a wider community of scholars, such self-effacing strategies have always been vital for scientific authors.[8] In the early modern period, a number of resources, such as the genteel distaste for print, were available for this.[9] In addition, the tradition of the *prisca sapientia* meant strictly that individuals were merely rediscoverers of lost knowledge. Newton himself argued that he was merely recovering what God had given to mankind at the beginning and in that sense, he was not its author. In the seventeenth century, appeal to the *prisca* tradition was one of the only ways in which a writer in natural philosophy could author something that could not possibly be his, namely, divinely created nature, but whose credibility did not rest on a named ancient authority. Nevertheless, the significance of the author in giving authority to texts in natural philosophy did not decline in the seventeenth and eighteenth centuries, and authorial codes in locations such as the Royal Society transplanted credibility from the ancient authorities—most especially Aristotle, Plato, and Galen—to modern individual authors of specific reports. As new *auctores* of knowledge, such authors were entitled to receive credit from the relevant community and to be cited in future works, although the authority of their work was also said to rest on its relation to the external world. In a significant sense, the knowledge that they produced belonged to all mankind.[10]

Ambitious among the Crowd

Newton was well acquainted with classical rhetorical techniques and in his early dealings with the wider public and print culture he adopted conventional authorial strategies such as pleading for anonymity and slighting the significance of his own work. In natural philosophy, however, his penchant for the mathematical style of presentation set him deeply at odds with the conventional contemporary codes of scientific writing that stressed openness and probabilism (the stance that statements about nature could at best be probable rather than absolutely certain) as proper authorial and epistemological stances. From his very first dealings with a wider readership outside Trinity College, he claimed to set little value on his own writings and, indeed, he was loath to put his name to some of his mathematical work.[11] In February 1670 he told the London mathematical intelligencer John Collins that publication in the *Philosophical Transactions* "would perhaps increase my acquaintance, ye thing wch I chiefly study to decline . . . I see not what there is desirable in publick esteeme, were I able to acquire & maintaine it." Although he found it difficult in practice to withdraw from the scientific community in the mid-1670s, he refused to compromise his attitude to authorship.[12]

Newton was elected to the Lucasian chair on October 29, 1669, and chose to make geometrical optics the topic of his Lucasian lectures when he began giving them in January 1670. His confidence in the mathematizability of nature represented a continuation of the sentiments that Isaac Barrow, the first incumbent in the Lucasian chair, had expressed in the lectures (and which Newton almost certainly attended). A version of Newton's lectures was in an advanced state of preparation for publication by October 1671, while at the same time he embarked on substantial revisions of his work on infinite series and fluxions. The expanded fluxional treatise, "De Methodis serierum et fluxionum," was a major exposition that put him at the forefront of European mathematicians, and it formed the basis of a great deal of his more extensively circulated work, including the two epistolae sent to Henry Oldenburg (the secretary of the Royal Society for Leibniz) in 1676. By Christmas 1671, Newton had already begun a revision of his optical lectures for publication and, in its finished state, the second version (the "Optica") was almost half as long again as the first. With precise descriptions of relevant physical magnitudes such as angles and lengths, the lectures were addressed to readers with instructions on how to perform experiments successfully, and the format of the work, especially in the revised "Optica," explicitly imitated the conventional presentation for mixed mathematical texts.[13]

Whatever plans for publication he may have had, they were all shelved in the months following the appearance of his reflecting telescope at the Royal Society at the end of 1671. Newton boasted to Oldenburg that the natural philosophy that underlay the design of the telescope was "in my Judgment the oddest if not the most considerable detection wch hath hitherto beene made in the operations of nature." The paper he sent to Oldenburg as a result of this was published (with some changes) in the *Philosophical Transactions* and marked Newton's first exposure before a broad audience; more significantly, it constituted his first foray into print. Newton began in the historical narrative style favored by Boyle and told how he had allowed sunlight to pass through a prism, following the "celebrated Phænomena of colours" mentioned in Descartes's *Meteorology*, onto a wall twenty-two feet away. He was surprised to see an oblong figure instead of the circular shape predicted under the older modification theory, and after measuring the angles involved, resolved the cause of the phenomenon by means of a "crucial experiment," which showed "that *Light* consists of *Rays differently refrangible*." At this point, he went on, he had realized that the development of refracting telescopes was limited by chromatic aberration, and he was led to consider reflections as a basis for the improvement of telescopes.[14]

At this stage in the narrative, Newton changed his style of exposition and argued that the science of colors could become mathematical, with "as much certainty in it as in any other part of Opticks." His doctrine was not a mere hypothesis, "but evinced by ye mediation of experiments concluding directly & wthout suspicion of doubt." To pursue the "historical narration" any further, he continued, "would make the discourse too tedious & confused, & therefore I shall rather lay down the *Doctrine* first, and then for its examination, give you an instance or two of the *Experiments*, as a specimen of the rest." He added a number of propositions to illustrate the "doctrine," and concluded with an experiment for the delectation of the members of the Royal Society. Before he removed the unacceptably dogmatic and offensive portion of the text from "hypothesis" to "confused" for its appearance in the *Philosophical Transactions* (under his editorship), Oldenburg told him that the fellows had applauded his "Ingeniosity, as well as [his] high degree of francknesse in this matter and in the communication thereof." Newton reciprocated with the compliment that he took it to be a great privilege "that instead of exposing discourses to a prejudic't and censorious multitude (by wch many truths have been bafled and lost) I may wth freedom apply myself to so judicious & impartiall an assembly."[15]

A Prejudic't and Censorious Multitude

Given what he knew of the audience at the society, as well as what he took to be the readership of and stylistic conventions associated with the *Philosophical Transactions*, Newton immediately recognized that there was a problem with the way in which he had presented his work. When Oldenburg told him that the society had voted for his paper to be printed in the *Transactions*, Newton acquiesced but added that he

> should have thought it too straight & narrow for publick view. I designed it onely to those that know how to improve upon hints of things, & therefore to shun tediousnesse omitted many such remarques & experiments as might be collected by considering the assigned laws of refractions; some of wch I believe wth the generality of men would yet bee almost as taking as any of those I described.

He offered to send some further experiments to "second" the letter, but in the following months was increasingly put out by comments made by Robert Hooke and Ignace Pardies. From then on in his correspondence he chose to stress the mathematical certainty of his accounts of light and color, as opposed to the conjectural nature of the physical causes of these phenomena. Against Pardies, Newton stressed that the properties of light he had discovered were true, and if he did not know them to be such, would "prefer to reject [them] as vain and empty speculation, than acknowledge them as my hypothesis."[16]

Hooke agreed with Newton's novel description of the phenomena of light and colors, "as having by many hundereds of tryalls found them so," but disagreed that Newton's crucial experiment showed that the colored rays produced after refraction by the prism actually constituted the original ray of light *before* it met the prism. He explained: "How certaine soever I think myself of my hypothesis, wch I did not take up without first trying some hundereds of expts; yet I should be very glad to meet wth one Experimentum crucis from Mr Newton, that should Divorce me from it." What Newton took to be rigidly demonstrated by means of his experiments, namely, the heterogeneity of white light, Hooke took to be a mere hypothesis, not at all implied by the phenomena. Having proferred some important criticisms of his own, Hooke remarked: "Nor would I be understood to have said all this against his theory as it is an hypothesis, for I doe most readily agree wth him in every part thereof, and esteem it very subtill and ingenious, and capable of salving all the phænomena of colours." However, he went on, "I cannot think it to be the only hypothesis; not soe certain as mathematicall Demonstrations."[17]

Newton's response constituted a masterful demolition of Hooke's critique. In an early draft, he professed that he had long been satisfied with his own theory, "& it was not for my own sake that I propounded it to others, & therefore I cannot esteem interest or concernement to defend it." In the letter that was actually sent, his tone was more strident, and indeed he later asked Oldenburg to "mollify any expressions that may have a shew of harshnesse" before it appeared in the *Philosophical Transactions*. Hooke, he stated, "knows well y^t^ it is not for one man to prescribe Rules to ye studies of another, especially not without understanding the grounds on wch he proceeds" and he should have obliged Newton with a private letter after which "I would have acquainted him wth my successes in the tryalls I have made of that kind." Moreover, while Hooke harped on about the need for hypotheses, Hooke's own explanations were "not onely *insufficient*, but in some respects *unintelligible*." Stung by the accusation that his explanation might be inscrutable, Hooke responded in kind, telling a senior member of the Royal Society that he did not doubt that Newton "doth perfectly understand by what he alledges, but to me they seem altogether as difficult to understand."[18]

Newton had decided that the mathematical format and his treatment of a ray of light as an abstract mathematical entity defined merely by its degree of refrangibility were the only ways by which he could publish without causing unnecessary disputes. Yet, ironically, this very form of brevity was causing problems for readers and gave rise to the contention he professed to despise, not least because the sketchy description of his experimental procedures made replication difficult. Newton realized this and in a letter to Oldenburg of July 8, 1672, he repeated his comments regarding the deleterious effects of his paper's obscurity and brevity. Against his wishes, the effect of his work being disputed, he protested, had been to cause him to lose the ability to work on subjects of his own choosing and in his own time, which he had enjoyed in the privacy of his college. Although he went into print at this very point, having performed the minor task of "describ[ing] schemes" for his edition of Bernard Varenius's *Geographia Universalis*, he told Collins that he had found "by that little use I have made of the Presse, that I shall not enjoy my former serene liberty till I have done with it" and that this had dissuaded him from publishing his optical lectures. The nature and extent of criticisms from people like Pardies, Hooke, and Christiaan Huygens continued to take their toll and by March of the following year, he signified to Oldenburg that he wished to withdraw from being a fellow of the Royal Society.[19]

Newton's approach clashed with accepted practice in another, related way. As Peter Dear has shown, for a number of reasons the laying of stress on one experiment as a basis for making statements about the natural world clashed with the conventional way in which experimental evidence had come to be corroborated in Europe throughout the seventeenth century. Most scholars, whether or not they believed that probabilist or higher-level degrees of certainty were appropriate for their presentations, came to agree that to be made universal and evident, "contrived experiences" had to be repeated many times, in many places, and in front of competent or high-status witnesses. Newton's almost complete unwillingness to agree that witnesses were relevant to the validation of mathematically certain knowledge set him apart even from Galileo and the remaining scholars who still worked in Aristotelian paradigms. This is shown by his exchanges with a number of Liège Jesuits in the continuing debates over the truth and certainty of his optical work. Newton treated the inability of this group to replicate his experiments with disdain, and suggested that they were calling his word into question when they claimed that they had done many more experiments than he. One of the Jesuits, Anthony Lucas, argued that only by trying related experiments in a number of circumstances could the true cause of phenomena be uncovered:

> This I conceive, was the reason why severall worthy members of ye Royall Society have bottomed new Theorys upon a *Number* of experiments, particularly the ingenious Mr Boyle strongly asserting the weight of the Atmosphere by a vast number of new experiments, each whereof, is deservedly conceived to add new strength to this Theory.

On the other hand, Newton insisted that his crucial experiment was by itself sufficient to prove his case and when another Jesuit, Anthony Lucas, proposed further experiments to decide the issue, he told Oldenburg in August 1676 that Lucas should "instead of a multitude of things try only the *Experimentum Crucis*. For it is not number of Exp^ts^, but weight to be regarded; & where one will do, what need of many?" Newton pointed out to Oldenburg that in general the dispute "was not about any ratiocination, but my veracity in relating an experiment." His credibility and competence had become key issues in the dispute but a single, schematically represented experiment, unwitnessed by any named individuals, could not command assent in the extended empiricist polity of the Royal Society.[20]

Despite the criticisms made by the Liège Jesuits, Newton's style probably had more in common with theirs than it did with the approach favored by Boyle, and in fact he was later highly critical of the latter's tendency to be too open and too desirous of fame. Despite Newton's continuous attempts to restrict the arena to conversation among "friends," violent disagreements were less liable to take place if one abided by the rules that required authors to publish descriptions of easily replicable experiments in print form. By the end of 1676 he had had his fill of print culture and had long since tired of his critics' attitude to his optical theories, although he was still prepared to engage in a correspondence about the best sort of apple trees for making cider. Being dragged into public disputes denied him the liberty of responding to private letters in his own time; if he got free of this "business," he complained apocalyptically to Oldenburg, "I will resolutely bid adew to it eternally, excepting what I do for my privat satisfaction or leave to come out after me. For I see a man must either resolve to put out nothing new or to become a slave to defend it." If there was to be any audience for his work other than himself, then it was not one that existed in his own lifetime.[21]

Philosophical Litigation

Newton's problems with the reception of his theory of light and colors set the pattern for his authorial relationship with the audience in the philosophical Republic of Letters. In the manner of a genteel humanist, he professed a dislike for the barren format of the dispute, but when he had to, he relished the chance to display his prowess as a philosophical litigator and crushed his opposition with aplomb. From an ideal collegiate space, in which an author was obliged to present his productions to a coterie of discerning and magnanimous friends, the philosophical arena could collapse into a court of law, with an ethos and etiquette that revolved around debate and dispute. Newton largely succeeded in avoiding philosophical intercourse in the last few years of the 1670s, preferring to engage in private alchemical and theological pursuits. Indeed, when Hooke approached him to make a contribution to the ailing society in November 1679 on the topic of celestial dynamics, Newton presented himself as a country recluse who had been away from philosophical conversation for too long to be considered capable of saying anything important. After Hooke had unwisely made public a letter that Newton had explicitly asked to remain private, he retreated yet again to the safety of Trinity College, where he continued his studies in alchemy and theology.[22]

This hermitage changed when Edmund Halley visited him in the summer of 1684 and coerced the Lucasian professor to produce a demonstration of the link between the ellipticity of planetary orbits and a general inverse square distance law. Newton sent him a nine-page treatise, "De motu corporum in gyrum," in November and developed it over the following year and a half, incorporating physical and mathematical insights such as the calculus and the laws of motion, and the concepts of inertial and gravitational mass, centripetal force, and universal gravitation. Most importantly, and with unrivaled mathematical skill, he produced a demonstration that an inverse square law lay behind the *elliptical* orbits of planets and their satellites. By autumn 1685 he had completed a work consisting of two books (the *De Motu Corporum, Liber Primus* and *De Motu Corporum, Liber Secundus*). Subsequently, he expanded the *Liber Primus* into drafts of what were to become Books One and Two of the *Principia*, while the initial *Liber Secundus* was transformed into the published Book Three. The rejected *Liber Secundus* had lacked a "good method" to serve as the basis of calculations and contained only a rudimentary treatment of comets; it also lacked the notion of universal gravitation and did not take the heliocentric system for granted. In its original state, it began with an account of the view of the ancients regarding the nature of the heavens and the paths of comets, in which he argued that they had correctly understood that comets were celestial phenomena.[23]

Nevertheless, at this very moment, the issue of understanding became central to his concerns and Hooke again forced him to consider the wisdom of publishing the *Principia*. Hooke seems to have functioned at every turn for Newton as an authorial Antichrist and on this occasion, despite his very reasonable claims for some form of recognition from Newton, the latter discharged his bile over Hooke's tattered credibility with barely controlled passion. The issue of Newton's understanding of his own text became the focus of proprietorial concerns when he was informed by Halley that Hooke was claiming precedence for the proof that elliptical orbits were a consequence of the inverse square relation. Hooke, Halley told Newton in May, "sais you had the notion from him, though he owns the Demonstration of the Curves generated therby to be wholly your own." Halley suggested that Hooke only seemed to want a mention in the preface, and rather hopefully claimed that he was sure Newton would act with the greatest candor, "who of all men has the least need to borrow reputation." Halley, who still believed that the *Principia* was composed of two books, also stressed the importance of its accessibility to philosophers and on June 7 asked Newton: "I hope you will please to bestow the second part, or what remains of this, upon us, as soon as you shall have finished it; for the application of this Mathematical part, to the System of the world; is what will render it acceptable to all Naturalists, as well as Mathematiciens; and much advance the sale of ye book."[24]

Believing himself to be once more pressed into a public dispute, Newton moved as before to establish his priority and gave Halley his own history of the epistolary exchanges with Hooke in 1679–80. In May, June, and July 1686 he told Halley that Hooke's claims were tantamount to saying that Newton did not understand the implications of his own doctrines; he therefore took it upon himself to show that he *had* understood his own work for well over a decade and that this understanding constituted the only relevant form of intellectual property. In June 1686, just before he threatened to suppress Book Three, he told Halley

> That wn Hugenius put out his *Horol. Oscil.* a copy being presented to me; in my letter of thanks to him I gave those rules in ye end thereof a particular commendation for their usefulness in Philosophy, & added out of my aforesaid paper an instance of their usefulness in comparing ye forces of ye Moon from ye earth & earth from ye Sun in determining a Probleme about ye Moons phase & putting a limit to ye Sun's parallax. Which shews that I had then my eye upon comparing ye forces of ye Planets arising from their circular motion *& then understood it.*

Newton then referred to the "Hypothesis"—an exposition of his views on the causes of light and colors that had been sent to the Royal Society in 1675—and told Halley that in that production "I hinted a cause of gravity toward ye earth Sun & Planets wth ye dependence of ye celestial motions thereon." In this system, gravity had to decrease from the surface of the Earth according to the inverse square law, although Newton was forced to admit that "for brevities sake" the mathematical relation had not explicitly been expressed in the paper. "I hope," he went on, "I shall not be urged to declare in print that I understood not ye obvious mathematical conditions of my own

Hypothesis." Hooke's behavior had changed the rules of the game, and Newton told Halley that he did not want to engage in a public disputation. Philosophy, he wrote, "is such an impertinently litigious Lady that a man had as good be engaged in Law suits as have to do with her. I found it so formerly & now I no sooner come near her again but she gives me warning." He also mentioned that he had thought of changing the title of his book back to *De Motu Corporum Libri Duo* but had decided to retain the heading under which it eventually appeared: "Twill help ye sale of ye book wch I ought not to diminish now tis yours."[25]

The pose of avoiding disputes could not be kept up for long and the issue of understanding gnawed away at Newton's pride. In a postscript to the same letter, he raged that Hooke did *not* understand the issue, but instead "make[s] a great stir pretending I had all from him & desiring they [i.e., the Royal Society] would see that he had justice done him." As he warmed to the judicial style in which he performed so well, he related to Halley just how Hooke had stolen Borelli's "hypothesis" from its author. The profligate sower of "hints," Newton sarcastically noted, was to take all the credit while "I must now acknowledge in print I had all from him & so did nothing my self but drudge in calculating demonstrating & writing upon ye inventions of this great man." He even wove a story about how Hooke might have stolen the idea from perusing his original letter to Oldenburg for Huygens, but noted that Hooke had *mis*understood what he had tried to thieve. Nor was the issue quite finished and, in a further letter of July 27, he told Halley that he had by chance alighted upon his original letter to Huygens in the hand of his roommate John Wickins, "& so it is authentick." Although he did not express the inverse square relationship in the letter, if it were compared to passages in the "'Hypothesis you will see y^t^ I then understood it." All Newton owed to Hooke was "ye diversion he gave me from my other studies to think on these things, & for his dogmaticalnes in writing as if he had found ye motion in ye Ellipsis, wch inclined me to try it after I saw by what method it was to be done." *Understanding* in this case meant precisely the capacity to give geometrical demonstrations of heavenly and terrestrial phenomena, and it was this facility that, as Newton related at length, his opponent sorely lacked.[26]

Hooke's intervention almost certainly shaped the way in which Newton presented the text and he also suppressed a daring "Conclusio" that gave rein to his alchemical views as well as to his more mainstream conception of the role of active principles in the world. In the new and final version of Book Three, entitled "The System of the World," he reiterated that the preceding books had laid down principles which were mathematical and not philosophical and referred to his suppression of the *Liber Secundus*, a book that, he said, had been composed "in a popular method":

> that it might be read by many; but afterwards, considering that such as had not sufficiently entered into the principles could not easily discern the strength of the consequences, nor lay aside the prejudices to which they had been many years accustomed, therefore, to prevent the disputes which might be raised upon such accounts, I chose to reduce the substance of this Book into the form of Propositions (in the mathematical way), which should be read by those only who had first made themselves masters of the principles established in the preceding Books.

That said, for an intermediate group of readers Newton specified a different route into the book and a means of getting through it in a reasonable amount of time. For this sort of reader it was not, he wrote, worth going back over *every* proposition in the first two books, "for they abound with such as *might cost too much time*, even to readers of good mathematical learning." It is enough, he went on, "if one carefully reads the definitions, the laws of Motion, and the first three sections of the first Book. He may then pass on to this [i.e., the Third] Book, and consult such of the remaining Propositions of the first two Books, as the references in this, and his occasions, shall require." Despite the fact that the new Book Three now contained the doctrine of universal gravitation and had a much more sophisticated theory of comets, the last-minute decision to dress it up in a "mathematical way," with propositions, scholia, theorems, and lemmas, hardly expressed the major changes that had taken place from the *Liber Secundus*.[27]

By couching Book Three in a mathematical form, Newton ran the risk of being seen to treat mathematical and not physically real objects, even though the function of the last book was supposedly to deal with the real system of the world. At the last moment, to section II of Book One, he added the phenomenalist rider that he would consider centripetal forces as "attractions though perhaps in a physical strictness they may more properly be called impulses." The following propositions, he stated, "are to be considered as purely mathematical; and therefore, laying aside all physical considerations, I make use of a familiar way of speaking, to make myself the more easily understood by a mathematical reader." However, in Book Three the mathematical format created no small problem, since the initial point of a work on the system of the world was to show that the laws that governed it were so close to those demonstrated in the mathematical constructs of his first two books that the latter could be held to apply similarly in the physical world. Whereas one might have been sure earlier that the gravitation of Book Three was the physical correlate of the notion of attraction in Books One and Two, now the situation was more complicated. Although experiments and well-attested physical and astronomical observations might provide the building blocks for speaking publicly about the laws of nature, Newton later adopted a *physicalist* phenomenalism that left obscure his views on the basic ontological structure underlying conceptual entities such as gravitation and force.[28]

The Incomprehensible as Ineffable

Nearly all readers—including some of the most able mathematicians—found the going unbearably tough, and the question of the abstruse nature of the *Principia* was central in early accounts of its power and authority. Correspondents with Newton and reports of the contents of the book invariably remarked on the total otherness of the text, and its author never discouraged such claims. Even before its publication Halley referred to his "divine Treatise," having earlier told him: "You will do your self the honour of perfecting scientifically what all past ages have but blindly groped after." In the summer, Halley promised James II (to whom he presented the book) that he would explain anything that was too obscure or difficult, while in a review of its contents for the *Philosophical Transactions* he concluded that "it may justly be said, that so many and so Valuable Philosophical Truths, as are herein discovered and put past Dispute, were never yet owing to the Capacity and Industry of any one man."[29]

Although we do not have evidence of James II's reaction to the book, the confidence of his advisor on political arithmetic, Sir William Petty, was shattered almost as soon as he ventured beyond its initial definitions. In a letter to Robert Southwell of July 9, 1687, he noted that "Mr Newton's excellent book is come out," and three days later Southwell asked him whether he was satisfied with Newton's account of the tides. On the 23rd, a defeated Petty told Southwell: "I would give £500 to have been the author of it, and £200 that Charles [his son] understood it. My bad eyes disable mee to make the most of it, for diagrams cannot bee read by others." His friend reassured him that "whereas you would give £500 to have beene Author of Mr Newton's booke, I would give a 1000 to be the Author of the least of yours" and when Petty sent him his "Bible for Ireland," *The Weight of the Crown of England*, in early August, Southwell was again enthusiastic:

> The things are mighty, and call unto my mind that when Paul reasoned of Righteousnesse, Temperance, and Judgement to come, Felix trembled. You know Columbus made the first offer to us of his Goulden World, and was rejected; That the Sybill's Bookes, tho never so true, were undervalued; and Mr Newton's demonstrations will hardly be understood. The market rule goes fair in everything else, *Tantum valet quantum vendi potest*.[30]

In response, Petty compared the lack of recognition from the king for his own magnum opus to the reception of the *Principia*: "I desired the King to pick out of the whole one Article which he wished to be true and another which he thinketh to be false, and Comand me within 24 houres, and within one sheet of paper, to show him my further thoughts Concerning them." This, he continued, was all "very well taken, but without getting better Buter to my Parsnips or hobnayles for my Shoes; and poor Isaac Newton will certainly meet with the same fate in the world, for I have not met with one Man that puts an extraordinary value upon his Book."[31]

The retired Cambridge mathematician Gilbert Clerke was quick to give Newton his opinion of the *Principia*'s merits, although he too could penetrate only a short distance into the book. At the end of September 1687 he wrote to Newton concerning some problems in terminology, though he confessed that he did not yet "understand so much as your first three sections, for wch you doe not require y^{t} a man should be *mathematice doctus*; but if I must not tell you, till I understand those sections & your third booke very well; perhaps I must never tell you." Newton replied that he was not surprised "that in reading a hard Book you meet wth some scruples & hope y^{t} ye removal of those you propound may help you to understand it more easily." Clerke was rightly indignant at the notation used in Proposition 11 in which addition signs apparently possessed the force of multiplication (i.e., meaning "conjoined with"), and he light-heartedly berated Newton for this and other inaccuracies: "Your booke is hard enough, make it as easie as you can: so there should have been two prickd lines from ye center to ye tangents in prop. 5. p.44 & you should have had marginal references to Eucl. & Apoll." In short, Newton should have been "prodigal of *per* this & *per* that, of your *nempe's* & *quoniam's* & enlargd your scholium's." In early November he lamented that he had been out of the loop for too long: "I have long lived in an obscure village, in worldly business & field-recreations & have not been acquainted wth ye brave notions of Galileus, Hugenius &c. & so despaire of understanding your booke well." He was also losing his own memory, a faculty that was of prime importance for the mathematician if he were not to be forced, tediously and laboriously, to check every proof in Euclid or Apollonius. In a postscript he remarked that he felt he needed to understand Newton's eleventh section to grasp the whole and "lookd over some of ye foregoing sections & thought I could understand y^{m}, but they would take up too much time & would be easily lost: as I experienced in two or three props."[32]

The category of understanding conditioned the response of every putative reader of the *Principia*, a fact borne out by contemporary evidence as well as by Newton's octogenarian reflections. Locally, Newton sent his amanuensis (Humphrey Newton—whom he later blamed for causing transcription errors because he "understood not what he copied") to give copies to some twenty or so acquaintances and heads of Cambridge colleges, "some of w^{ch} (particularly Dr. Babington of Trinity) said that they might study seven years, before they understood anything of it." According to Jean-Théosophile Desaguliers, Newton told him that John Locke got the gist of the work from Huygens and was reassured that the mathematics and mechanics were sound. As

Bernard Cohen points out, after a statement advising good mathematicians not to read every proposition of the first two books before turning to the third, Newton added a sentence in his interleaved version of the second edition of the *Principia* to the effect that "those who are not mathematically learned can read the Propositions also, and can consult mathematicians concerning the truth of the Demonstrations." However, few did have sufficient mathematics to comprehend the more abstruse truths of the book, and an anonymous student attained immortality by remarking (in a statement related by Newton to Martin Folkes) as the Lucasian professor passed by: "There goes a man that hath writt a book that neither he nor any body else understands."[33]

Apostolic Success

If Newton's apparently superhuman labor had led to the book's creation—and it was his own "industry and patient thought" that he himself repeatedly stressed—then potential disciples had to work just as hard to master its contents. Accordingly, much of the evidence from the period points to the sheer toil required to excavate its treasures and young acolytes were not lacking to undergo this supreme test. Newton's credibility, and the authority of the book, was coextensive with his capacity to generate disciples. From the moment he submitted the first version of *De Motu Corporum* to the Royal Society at the end of 1684, his reputation—previously restricted to some *cognoscenti* among the mathematical community but now fostered by Halley and others—grew exponentially. With willing disciples, the incomprehensibility of the *Principia* became a source of power, the reverse of the fate that had befallen his apparently obscure optical work a decade earlier.

The most adept Scottish mathematicians quickly took up the challenge though a willing spirit was not enough. John Craig told Colin Campbell at the end of 1687 that he was finding it almost impossible to send him a satisfactory account of the *Principia*: "Tho I have not yet perused on [sic] quarter of it, and have an unsatiable desire to know the wholl: yet knowing how extreamly acceptable it will be unto you I have sent the book it selfe." In a postscript he noted: "Be not hastie in reading over Mr Newtons book for I shall lett you have it as long as you please," while David Gregory also counseled Campbell that

"Newton will take you up the first month you have him." Just over a year later, Craig asked the befuddled Campbell to return the book as securely as he could, saying that he was

> sorry to understand that you have been at so much pains with Mr Newtons book, if I had not been preingag'd, I would have prevented all that trouble in letting you keep it; but now I must give you notice that it is my desire that you would send it to me with the first convenient occasion for I have given it to a freind who is impatient to have it, but be carefull to send it with a save [sic] hand.

The more adept Abraham de Moivre opened the book and "deceived by its apparent simplicity persuaded himself that he was going to understand it without difficulty. But he was surprised to find it beyond the range of his knowledge and to see himself obliged to admit that what he had taken for mathematics was merely the beginning of a long and difficult course that he had yet to undertake." De Moivre displayed his faith by tearing out pages of the "'divine Treatise' in order to carry them in his pocket and to study them during his free time."[34]

Tyros, like de Moivre, took on the Herculean task of grasping the contents of the book, not least because they were aware of the intellectual kudos that would ensue. From Newton's point of view, the creation of disciples who could mediate between the meaning of the text and wider publics solved the problem of how this incomprehensible text and its inaccessible author could become credible and authoritative. Thus did ambitious men flock to the high priest and his secular bible. In the rush to become one of what John Flamsteed decried as Newton's "darlings," the classicist and Trinity fellow Richard Bentley approached Craig for advice on approaching the text. Craig disarmingly suggested that he embark upon an inordinately long list of works to read before he could come to terms with the book's contents, in particular the "Method of Tangents," among many of which he recommended Leibniz's paper in the *Acta Eruditorum* as the best hors d'oeuvre. For the method of indivisibles he suggested Cavalieri and Barrow's *Geometrical Lectures*, and for Newton's mechanics he named works by Barrow, Galileo, Torricelli, and even Hobbes, telling William Wotton: "Nothing less than a thorough knowledge of all that is yet known in most curious parts of the Mathematicks can make him capable to read Mr. newton's book." To come to terms with the totality of Newton's output, Craig continued, would require mastery of works by Descartes, James Gregory, Fabri, and Tacquet for optics, Archimedes, Borelli, and Wallis for hydrostatics, and "because much of Master Newton's book refers to the Quadratures of Figures, He must read what has been written on this Subject, by *Dr Wallis & Mr David Gregory*."[35]

Put out by the same list but desperately anxious to join the ranks of Newton's inner circle, Bentley approached his maestro colleague directly and was given a marginally less demanding set of readings. Clearly Newton did not expect Bentley to get beyond the basic doctrines of the *Principia*, and he advised him not to read over all the commentaries he was suggesting, but "only ye solutions of such problems as you will here & there meet with ... [all] These [works] are sufficient for understanding my book: but if you can procure Hugenius's *Horologium Oscillatorium*, the perusal of that will make you much more ready." That this was actually by no means sufficient to comprehend the *Principia* at any deep level was indicated by the next paragraph, in which Bentley was informed that

> At ye first perusal of my Book it's enough if you understand ye Propositions wth some of ye Demonstrations wch are easier then the rest. For when you understand ye easier they will afterwards give you light into ye harder. When you have read ye first 60 pages, pass on to ye 3^d Book & when you see the design of that you may turn back to such Propositions as you shall have a desire to know, or peruse the whole in order if you think fit.

The similarity between this and the *caveat lector* in *The System of the World* suggests that Bentley came close to the ideal type of one audience for Newton's book. In mathematical culture, the difference between the dilettante and the initiate was immense, and there is no indication that Newton seriously believed that Bentley could become one of the latter. Yet Bentley soon became a Newtonian apostle and as one of the executors of Boyle's will, Newton helped him gain the honor of giving the first Boyle lectures in 1692, in the process allowing Bentley unique insights into his underlying metaphysics.[36]

Another central figure in disseminating the Newtonian philosophy was David Gregory, who composed an extended series of notes on the *Principia* over a six-and-a-half-year period from his first acquaintance with the book in September 1687. He also did not hesitate to point out its unprecedented, unanticipated, and virtually incomprehensible achievements to the author and soon after its publication, he told Newton that he had been the first to create a mathematical underpinning to natural philosophy that was adequate to the real world. As if to a divinity, he gave thanks to Newton for

> having been at the pains to teach the world that which I never expected any man should have knowne. for such is the mighty improvement made by you in the Geometry, and so unexpectedlie successful the application thereof to the physicks that you justlie deserve the admiration of the best Geometers and Naturalists, in this and all succeeding ages.

Such an accomplishment did not, Gregory concluded, mean that Newton should stop his endeavours, and he urged that although the *Principia* was "of so transcendent fineness and use that few will understand it, yet this will not I hope hinder you from discovering more hereafter to those few who cannot but be infinitly thankful to you on that account."[37]

Gregory was shortly rewarded with a powerful reference from Newton, who told the election committee for the Savilian chair at Oxford that Gregory was highly skilled in both the new and the old analysis and geometry, and that "He has been conversant in the best writers about Astronomy & understands that science very well." In the coming years, as is evident from the remarkable candor shown by Newton in allowing him access to some of his most unorthodox beliefs in theology, Gregory was probably Newton's most preferred disciple before Samuel Clarke. In the mid-90s he noted the extraordinary changes that Newton was prepared to make, especially to Book Three, and like many others he let it be known that he was the favored son to undertake the second edition (although this ultimately came to nothing).[38]

Although many acolytes ran him close, Fatio de Duillier contorted himself more than any other to genuflect at the feet of the Almighty

Author. Fatio immediately recognized the possibility of developing a patron-client relationship with Newton when he first arrived in London in 1687 and offered himself as a political, theological, and philosophical ally.[39] To the great mathematicians of continental Europe, Fatio proferred himself as the gatekeeper to the *Principia*'s treasures, coextensive with his access to the thoughts of Newton himself. Fatio did not hesitate to offer an account of gravity to his non-British colleagues, and claimed that it came with the Newtonian *imprimatur*, although Gregory recorded that Newton and Halley laughed at the effort.[40] At the same time, he set himself up as the official compiler of its errata in preparation for a second edition, of which he would be the editor. Accordingly, he told Huygens that he had found Newton ready to correct his book on so many occasions on issues that he had mentioned to him, that he could not sufficiently admire Newton's skill, especially in the places Huygens had attacked. In November 1691, Fatio claimed that nobody had so deeply comprehended as much of the book as himself, "thanks to the pains I have taken and the time I have taken up to surmount its obscurity." In any case, he could "easily make a trip to Cambridge" and get from Newton an explanation of anything Huygens had been unable to understand.[41]

By the end of 1691, Fatio apparently believed that Newton would allow him to include his theory of gravity in a new edition of the *Principia* alongside a list of errata, a view corroborated both by a note made by David Gregory at the end of December and by a reference to Fatio's theory in one of Newton's own drafts. In the same series of remarks in which he (later) added evidence of Newton's and Halley's disdain for Fatio's theory, Gregory noted (presumably taken from Fatio's own lips) that Fatio "designs a new edition of Mr Newtons book in folio wherin among a great many notes and elucidations, in the preface he will explain gravity acting as Mr Newton shews it doth." Fatio had also apparently claimed that he had satisfied Newton, Huygens and Halley about his theory and boasted that Newton knew "the inverted problem of the tangents [presumably integration] better than Libnitz." As for Fatio himself, he had greatly overestimated his own ability to come to terms with the work and he confessed to Huygens that he had encountered insurmountable obstacles in his voyage through the text. He had thus far only been able to get through sections 1 to 5 and section 9 of Book One, and the final part of Book Three that dealt with comets, but was finding further progress difficult due to distractions and the time and effort required to master the text. Because of this he proposed that Huygens undertake the sections that he himself had not been able to work through thus far, after which would not be too difficult to come to the end of the work: "Nous pourrions nous rendre conte l'un à l'autre des difficultez que nous aurions rencontrées et nous faciliter reciproquement l'etude d'un livre qui est assurement fort excellent mais en mesme temps fort obscure."[42]

From Divinity to Genius

At the same time, Fatio drew attention to the remarkable research on which Newton was now engaged in order to produce a series of scholia (to Propositions 1 to 9 of Book Three) that described the ways in which the ancients had made mystical references to the true theory of the world. Drawing on the commonplace of the *prisca sapientia*, which had galvanized his efforts to produce a much more ambitious *Principia* in the mid-1680s, the proposed new edition would offer a complex display of hermeneutic analysis. Newton could decipher what the ancients "really" meant when they spoke in code, because he had "rediscovered" this mystery, while the ancients' work also served the further function of allowing him to state publicly—in the guise of the beliefs of others—his own opinions on issues that otherwise would lead to unnecessary disputes. If this *prisca sapientia* was what the ancients had believed, then it followed that the *Principia* was also true, and one needed the great modern work to understand the classical remains. While the *Principia* had been turned into a handbook for classicists, the ancient poets, priests, and prophets had provided a model for how elite knowledge could be passed on to a wider, ignorant audience.[43]

His treatment of the *Principia*, and his management of the dissemination of its truths among differently literate mathematically audiences, paralleled what he had discovered was practiced among the learned ancients. The work gained in stature among a wider but still restricted audience as Newton consolidated links with those prepared to engage with its more abstruse doctrines and techniques. Soon, the Newtonian doctrine was taught to genteel audiences in the streets and coffeehouses of London as well as in the Scottish and English universities. Newton's personal visibility increased too. He entered public service and moved from his position as member of Parliament for Cambridge University in the Convention Parliament of 1689 to that of Warden and then Master of the Mint. Despite this visibility and his phenomenal success in attracting public honors in the first decade of the next century, he continued to cultivate authorial distance. His private life qua scholar attracted the interest of other scholars, and continental European intellectuals supposedly wondered whether Newton shared any traits with his fellow humans. National treasures, Newton and his work remained distant. He wrote for an élite readership approaching unity and he told William Derham that "mainly to avoid being baited by little Smatterers in Mathematicks," he had "designedly made his *Principia* abstruse; but yet so as to be understood by able Mathematicians, who he imagined, by comprehending his demonstrations, would concurr with him in his Theory."[44]

Newton's stipulations within the text, and his subsequent negotiations with readers over how it was to be read, created a four-dimensional object into which readers could dip or enter to various extents or in certain places. The *Principia* had its own unique spatiotemporality; different readers were to navigate and indeed actually did journey through its terrain at various speeds, recognizing connections and jumping the demonstrative links in the text at different rates. The *Principia* expressed the superhuman nature of Newton's expertise and, in order to achieve this, other mathematicians had to suffer the same tribulations of industry, patient thought, and continuous thinking that its restless creator had undergone in its production. Given its unique status, understanding it required knowledge of *all* of mathematics, as befitted a work that embodied the fullness of the learning that mankind had been given at the beginning of time. As for Newton, so for God: on the front page of a manuscript exposition of Apollonius's *Conics*, Isaac Barrow wrote: "How great a Geometrician art thou, O Lord! Thou art acquainted with them all [i.e., mathematical theorems] at one View, without any Chain of Consequences, without any Fatigue of Demonstrations;" for Halley, this facility had brought Newton as close to God as anyone could come.[45]

The *Principia* came to be treated as a canonical and multiple-layered holy text, not least because its (modern) author had uncovered the way in which God had inscribed laws in the book of nature. Disciples jostled to gain his favor and to publish *Principia*-related material in the *Philosophical Transactions* or to edit a second edition, and they elevated the status both of themselves and of their patron as they disseminated his thoughts to others who lacked their access. As befitted such a text, it spawned both lists of errata, correcting problems of various degrees of seriousness, and also commentaries that popularized or explicated sections of the text. To the eighteenth century, acolytes bequeathed the credo that henceforth all philosophical and scientific productions would be said to be "implied" by the *Principia*, or "hinted" by the Master as he dropped his inscrutable utterances to disciples. While he lived, the meaning of passages in this philosophical bible could only be resolved by going to Newton or to followers who had the requisite authority to act as both his and its interpreters. Britain now had a cultural and intellectual hero—but as yet there was no word or concept to describe what he had become. For many of the contemporaries (who cared), the *Principia* was literally a divine treatise. In time its author would be termed a "genius" and indeed would become the exemplar of first a rational and then a temporarily mad scientific genius. For now—and just like with the Italian artist-engineers of the Renaissance—only comparison with divinity could describe the man and his achievement.

Notes

1. J. Bernstein, *Einstein* (Glasgow: Fontana/Collins, 1980), 97, and 192 n.2; J. Crelinsten, "Einstein, Relativity and the Press: the Myth of Incomprehensibility," *The Physics Teacher* (February 1980): 115–22, 187–93; A. Pais, *Einstein Lived Here* (New York, 1994), 149; and A. J. Friedman and C. C. Donley, *Einstein as Myth and Muse* (Cambridge: Cambridge University Press, 1985), 13, 16–20. For comments on earlier versions of this paper I'd like to thank Moti Feingold, Andrew Warwick, and Mario Biagioli.
2. For the status of mathematical texts, see R. S. Westman, "The Astronomer's Role in the Sixteenth Century: A Preliminary Study," *History of Science* 18 (1980): 105–47; N. Jardine, *The Birth of the History and Philosophy of Science: Kepler's* Defence of Tycho versus Ursus, *with Essays on Its Provenance and Significance* (Cambridge: Cambridge University Press, 1984), chap. 7; M. Biagioli, *Galileo Courtier: The Practice of Science in the Age of Absolutism* (Chicago: University of Chicago Press, 1993); and P. Dear, *Discipline and Experience: The Mathematical Way in the Scientific*

Revolution (Chicago: University of Chicago Press, 1995), 34–57, 162–78, 180–207.

3. See, in particular, M. Mahoney, *The Mathematical Career of Pierre de Fermat, 1601–1665*, 2d ed. (Princeton: Princeton University Press, 1994), 6–8, 23–25; Biagioli, *Galileo Courtier*, 159–209, esp. 179–80, 205–9; B. Rubidge, "Descartes' *Meditations* and Devotional Meditations," *Journal of the History of Ideas* 28 (1990): 27–49; P. Dear, "Mersenne's Suggestion: Cartesian Meditations and the Mathematical Model of Knowledge in the Seventeenth Century," in *Descartes and His Contemporaries: Meditations, Objections and Replies*, eds. R. Ariew and M. Grene (Chicago: University of Chicago Press, 1995), 44–58, esp. 44–46; Dear, *Discipline and Experience*, esp. 75; and, in particular, J. Rée, *Philosophical Tales: An Essay on Philosophy and Literature* (London and New York: Methuen, 1987), 9–27, esp. 24–27. Approximately 500 of the 600 pages of the *Meditations* were made up of *Objections* and *Replies*.
4. Mahoney, *Fermat*, 174. For the relationship between textual format and specified audience behavior with regard to difficult philosophical and mathematical texts, see Biagioli, *Galileo Courtier*, 114–15. For the "ideal" audiences of texts see in general W. J. Ong, "The Writer's Audience Is Always a Fiction," *PMLA* 90 (1975): 9–21; and S. R. Suleiman and I. Crosman, eds., *The Reader in the Text: Essays on Audience and Interpretation* (Princeton: Princeton University Press, 1980). For an analysis of the techniques and social structures that allow individuals to acquire complex languages such as those based on mathematics, see A. C. Warwick, *Masters of Theory* (Chicago, 2003); more specifically, see Warwick's essay in this volume, which deals with similar themes to my own paper.
5. For the narrative style promoted by Boyle and propagandists for the Royal Society such as Thomas Sprat and Henry Oldenburg, see P. Dear, "'Totius in Verba': Rhetoric and Authority in the Early Royal Society," *Isis* 76 (1985): 145–61; S. Schaffer and S. Shapin, *Leviathan and the Air-Pump: Hobbes, Boyle, and the Experimental Life* (Princeton: Princeton University Press, 1985), 22–79; S. Shapin, "Robert Boyle and Mathematics: Reality, Representation, and Experimental Practice," *Science in Context* 2 (1988): 23–58; S. Shapin, *A Social History of Truth: Civility and Science in Seventeenth-Century England* (Chicago: University of Chicago Press, 1994), 118, 310–53; and see also R. -M. Sargent, *The Diffident Naturalist: Robert Boyle and the Philosophy of Experiment* (Chicago: University of Chicago Press, 1995).
6. Dear, "Mersenne's Suggestion," 44–47, and Rée, *Philosophical Tales*, 9–27. For Galileo, see M. Finocchiaro, *Galileo and the Art of Reasoning: Rhetorical Foundations of Logic and Scientific Method* (London: D. Reidel, 1980); R. S. Westman, "The Reception of Galileo's *Dialogue*," in *Novità Celesti e Crisi del Sapere*, ed., P. Galluzzi (Florence: 1984), 331–35; and esp. Biagioli, *Galileo Courtier*, 114–17, 130, 138–39, 178–86, 205–9, 224–26, 304–6.
7. R. S. Westman, "Proof, Poetics and Patronage: Copernicus's Preface to *De Revolutionibus*," in *Reappraisals of the Scientific Revolution*, eds. D. C. Lindberg and Westman (Cambridge: Cambridge University Press, 1990),

167–88. See also P. L. Rose, *The Italian Renaissance of Mathematics* (Geneva: Droz, 1975); J. -D. Moss, *Novelties in the Heavens: Rhetoric, Science in the Copernican Controversy* (Chicago: University of Chicago Press, 1993); and Laird, "Archimedes among the Humanists," *Isis* 82 (1991): 628–38.

8. For authorial techniques in this period, see inter alia J. R. Henderson, "Erasmus on the Art of Letter-Writing," in *Renaissance Eloquence: Studies in the Theory and Practice of Renaissance Rhetoric*, ed. J. J. Murphy (Berkeley and Los Angeles: University of California Press, 1983), 331–55; D. Quint, *Origin and Originality in Renaissance Literature* (New Haven: Yale University Press, 1983); A. J. Minnis, *The Medieval Theory of Authorship* (London: Scolar Press, 1984); and J. Miller, *Poetic License: Authority and Authorship in Medieval and Renaissance Contexts* (Oxford: Oxford University Press, 1986), esp. 4–31.
9. For the importance—or otherwise—of appearing in print, see J. W. Saunders, "The Stigma of Print: A Note on the Social Bases of Tudor Poetry," *Essays in Criticism* 1 (1951): 139–64; E. Eisenstein, *The Printing Press as an Agent of Change: Communication and Cultural Transformations in Early Modern Europe,* 2 vols. (Cambridge: Cambridge University Press, 1979); A. Grafton, "The Importance of Being Printed," *Journal of Interdisciplinary History* 11 (1980): 265–86; and A. D. S. Johns, *The Nature of the Book: Print and Knowledge in the Making* (Chicago: University of Chicago Press, 1998).
10. See M. Foucault, "What Is an Author" in *The Foucault Reader*, ed. P. Rabinow (Harmondsworth, 1986), 101–20, 113; and for criticism, see R. Chartier, *The Order of Books: Readers, Authors, and Libraries in Europe between the Fourteenth and Eighteenth Centuries*, trans. L. Cochrane (Oxford: Blackwell, 1994), 58–59. More generally, see R. Iliffe, "Author-Mongering: The 'Editor' between Producer and Consumer," in *The Consumption of Culture,* vol. 3, *Word, Image, Object*, eds. A. Bermingham and J. Brewer (London: Routledge, 1996), 166–92; M. Woodmansee and P. Jaszi, eds., *The Construction of Authorship: Textual Appropriation in Law and Literature* (Durham and London, 1994). For gift-giving in early modern natural philosophy, see Biagioli, *Galileo Courtier*, 16–63, esp. 52–54, 151–53; and, in general, see W. O. Hagstrom, "Gift-Giving as an Organizing Principle in Science," in *Science in Context: Readings in the Sociology of Science*, eds. B. Barnes and D. Edge (Milton Keynes: The Open University Press), 21–34.
11. See H. W. Turnbull et al., eds., *The Correspondence of Isaac Newton*, 7 vols. (Cambridge: Cambridge University Press, 1959–77) (hereafter cited as *Newton Correspondence*); D. T. Whiteside, ed., *The Mathematical Papers of Isaac Newton*, 8 vols. (Cambridge: Cambridge University Press, 1967–81); I. B. Cohen, *Introduction to Newton's "Principia"* (Cambridge: Cambridge University Press, 1971); A. E. Shapiro, ed., *The Optical Papers of Isaac Newton,* vol. 1. *The Optical Lectures 1670–1672* (Cambridge: Cambridge University Press, 1974); I. B. Cohen, *The Newtonian Revolution, with Illustrations of the Transformation of Scientific Ideas* (Cambridge: Cambridge University Press, 1980); R. S. Westfall, *Never at Rest: A Biography of Isaac Newton* (Cambridge: Cambridge University

Press, 1984); J. E. McGuire and M. Tamny, *'Certain Philosophical Questions': Newton's Trinity Notebook* (Cambridge: Cambridge University Press, 1983); A. E. Shapiro, *Fits, Passions, and Paroxysms: Physics, Method, and Chemistry, and Newton's Theories of Colored Bodies and Fits of Easy Reflection* (Cambridge: Cambridge University Press, 1993).

12. Newton to Collins, 18 February 1669/70; *Newton Correspondence*, 1: 27; and S. Shapin, *A Social History of Truth: Civility and Science in Seventeenth-Century England*, (Chicago: University of Chicago Press, 1994), 150–51, 176–79, 223. For anonymity, see Chartier, *Order of the Book*, 38–39, 58–59; M. Laugaa, *La Pensée du Pseudonyme* (Paris: Presses Universitaires de France, 1986); Mahoney, *Fermat*, 23–24; and Biagioli, *Galileo, Courtier*, 63–65. For the codes governing the private circulation of scripts and manuscript "publication" in the period between 1400 and 1750, see A. F. Marotti, *Manuscript, Print, and the English Renaissance Lyric* (Ithaca, N.Y.: Cornell University Press, 1995); H. R. Woudhuysen, *Sir Philip Sidney and the Circulation of Manuscripts, 1558–1640* (Oxford: Clarendon Press, 1996) and H. Love, *The Culture and Commerce of Texts: Scribal Publication in Seventeenth Century England* (Amherst: University of Massachusetts Press, 1998), esp. 9–10, 15–17, 27–28, 35–56, 177–92, 290–98.
13. Shapiro, *Optical Papers*, 1: 46–279 (*Lectiones Optica*) and 280–603 (*Optica*); Whiteside, *Mathematical Papers*, 3: 5–6, 28–31, 32–353 (*De Methodis Serierum*); Barrow, *The Usefulness of Mathematical Learning Explained and Demonstrated: Being the Mathematical Lectures Read in the Publick Schools of the University of Cambridge* (1734; reprint, London: Frank Cass & Co., 1970), 27. The *Lectiones Opticae* and *Optica* are now CUL Add. Ms. 4002 and Ms. Dd.9.67, respectively (deposited in the university library as his 1674 professorial lectures as per the Lucasian statutes). For relations between Barrow and Newton in the 1660s and early 1670s, see M. Feingold, "Newton, Leibniz, and Barrow Too: An Attempt at a Reinterpretation," *Isis* 84 (1993): 310–38.
14. Newton to Oldenburg, 18 January and 6 February 1671/2; *Newton Correspondence*, 1: 82–83, 92–107, esp. 92, 95. See also J. A. Lohne, "Experimentum Crucis," *Notes and Records of the Royal Society of London* 23 (1968): 169–99; Z. Bechler, "Newton's 1672 Optical Controversies: A Study in the Grammar of Scientific Dissent," in *The Interaction between Science and Philosophy*, ed. Y. Elkana (Atlantic Highlands, N.J.: Humanities Press, 1974), 115–42; R. Laymon, "Newton's *Experimentum Crucis* and the Logic of Idealization and Theory Refutation," *Studies in the History and Philosophy of Science* 9 (1978): 51–77; A. E. Shapiro, "The Evolving Theory of Newton's Theory of White Light and Colour," *Isis* 71 (1980): 211–35; S. Schaffer, "'Glass Works': Newton's Prisms and the Uses of Experiment," in *The Uses of Experiment: Studies in the Natural Sciences*, eds. D. Gooding, T. Pinch, and Schaffer (Cambridge: Cambridge University Press, 1989), 67–104; and A. E. Shapiro, "The Gradual Acceptance of Newton's Theory of Light and Color, 1672–1727," *Perspectives in Science* 4 (1996): 59–139.
15. Newton to Oldenburg, 6 February; Oldenburg to Newton, 8 February; and Newton to Oldenburg, 10 February 1671/2; *Newton Correspondence* 1: 96, 100–2, 107–9. Oldenburg also removed the text referring to the

science of colors becoming mathematical and having as much certainty as any other part of optics.

16. Newton to Oldenburg, 10 February, 1671/2, Newton to Pardies, 13 April 1671/2 and 10 June 1672; *Newton Correspondence*, 1: 108–9, 140–44, and 163–71. Before Newton, the classic statement of this phenomenalist or nominalist attitude to entities such as gravity, marking a significant distinction between Kepler on the one hand, and Newton and Galileo on the other, is to be found in the latter's *Two New Sciences*. See Galileo, *Two New Sciences, Including Centers of Gravity and Force of Percussion*, ed. and trans. S. Drake (Madison: University of Wisconsin Press, 1974), xxvi–xxx, 153–61, esp. P. Machamer, "Galileo and the causes," in eds. R. E. Butts and J. C. Pitt, *New Perspectives on Galileo*, (Boston: D. Reidel, 1978), 161–80; Jardine, "Birth of the History and Philosophy of Science," passim., and for the Newtonian style, see Cohen, *Newtonian Revolution*, xii-xiii, 15–16, 27–37, 65–76, 99–117, 129–40.
17. Hooke to Oldenburg, 15 February, 1671/2; ibid., 1: 110–14. For the notion of demonstration, see N. Jardine, "Epistemology of the Sciences," in *The Cambridge History of Renaissance Philosophy*, ed. C. B. Schmitt (Cambridge: Cambridge University Press, 1988), 685–711; D. W. Hanson, "The Meaning of 'Demonstration' in Hobbes's Science," *History of Political Thought* 11 (1990), 587–626; and P. Mancosu, *Philosophy of Mathematics and Mathematical Practice in the Seventeenth Century* (Oxford: Oxford University Press, 1996), 8–33.
18. Newton to Oldenburg, 11 June 1672, *Newton Correspondence*, 1: 171–93, esp. 191 n.20; Hooke to Lord Brouncker (?), c. June 1672; ibid., 200–201.
19. Newton to Oldenburg, 8 July 1672, Newton to Collins, 25 May 1672, Newton to Oldenburg, 21 September 1672 and 8 March 1672/3, Oldenburg to Newton, 13 March 1672/3; ibid., 1: 212–13, 161–62, 237–38, 262 and 263. For a robust statement at the end of 1674 to the effect that he had "long since determined to concern my self no further about y[e] promotion of Philosophy," see ibid., 1: 328. For a good account of the restricted role of mathematics in the early Royal Society, see Moti Feingold, "Mathematicians and naturalists: Sir Isaac Newton and the Royal Society," in *Isaac Newton's Natural Philosophy*, eds. J. Z. Buchwald and I. B. Cohen (Cambridge, Mass., 2000), 77–102.
20. Lucas to Oldenburg, 13 October 1676; Newton to Oldenburg, 13 November 1675 and 18 August 1676; ibid., 1: 356–58, 2: 79, 104–108; Dear, *Discipline and Experience*, 13–92. For the inability of his opponents to replicate Newton's experiments to his own satisfaction, see Schaffer, "Glass Works."
21. Newton to Fatio de Duillier, 10 October 1689; Newton to Oldenburg, 14, 18, and 28 November, 1676; *Newton Correspondence*, 2: 181–82, 183, 184, 3: 45.
22. Newton to Hooke, 28 November 1679; *Newton Correspondence*, 2: 300–302.
23. F. Cajori, ed., *Sir Isaac Newton, Principia*, vol. 2, *The System of the World* (London: University of California Press, 1962); 549–50. See also Cohen, "Introduction," 48–53, 56–79, 109–115; and Westfall, *Never at Rest*, 402–68, esp. 433–49, 459–60. The *Liber Secundus* of autumn 1685, in Humphrey

Newton's hand but with additions and corrections by Newton (now CUL Add. Ms. 3990) was published in 1728.

24. Halley to Newton, 22 May 1686 and 7 June 1686; *Newton Correspondence*, 2: 431 and 434. For contemporary codes governing intellectual property, see R. Iliffe, "'In the Warehouse': Privacy, Property and Priority in the Early Royal Society," *History of Science* (1992): 29–68; and M. Rose, *Authors and Owners: The Invention of Copyright* (Cambridge, Mass.: Harvard University Press, 1993).
25. Newton to Halley, 27 May and 20 June 1686; ibid., 2: 433–34 and 436–37 (my italics). A more detailed version of the excuse for not specifying the inverse square relation can be found in the postscript to this letter, where Newton, in fact, refers to an added alchemical account of the cosmos; ibid., 2: 440. Oldenburg omitted the very passage to which Newton referred when he passed on Newton's letter to Huygens in 1673, even though (or precisely because) it was not particularly germane to the point of the letter, which concerned optical matters; cf. Newton to Oldenburg, 23 June 1673, 1: 290–97, esp. 295 n. 3.
26. Newton to Halley, 20 June and 27 July 1686, 2: 438–39 and 446–47. Newton consequently began to downgrade any credit he had already given Hooke; see Cohen, "Introduction," 114–15, 135; and Westfall, *Never at Rest*, 449.
27. See Cajori, *Principia*, 2: 397; and Westfall, *Never at Rest*, 459–60, for a different view. For the "Conclusio" and the preface, which was a shorter version of this (and which, likewise, did not find its way into the final version), see Hall and Hall, "Unpublished Scientific Papers," 321–47, 302–8.
28. Cajori, *Principia*, 1: 164 (and cf. 192); and Westfall, *Never at Rest*, 465 n. 159. For a detailed account of the changes to Book Three as well as of the structure of the whole work, see Cohen, "Transformation," xii, 15–16, 29, 65–67, 71–72, 74–75, 106–10, 110–11, 131–32. For a good recent account of the difficulties posed by the mathematicism of Book Three to continental European readers, see Y. Gingras, "What Did Mathematics Do to Physics?" *History of Science* 39 (2001): 383–416.
29. Halley to Newton, 24 February, 1686/7 and 5 April 1687, and to James II, July 1687; ibid., 2: 470, 473–74, 483.
30. The Marquis of Lansdowne, *The Petty-Southwell Correspondence* (1928; reprint, London, 1967), 277–81.
31. Ibid., 283–88.
32. Clerke to Newton, 26 September 1687; Newton to Clerke, mid- to late September 1687; Clerke to Newton, 7 November 1687, ibid., 2: 485, 487, 492–94. Clerke did force Newton to change elements of his terminology, such as *dimidiata* to mean $A^{1/2} : 1$ when *dimidiate* had been used by Oughtred and others to mean division by 2. Newton changed this to *subduplicata* in his interleaved and annotated versions of his own copies of his first edition and all subsequent editions; see Cohen, "Introduction," 158–60.
33. CUL Add. Ms. 3968, f. 106; *Philosophical Transactions* 16 (1686–87):

291–97; Keynes Ms., 135; Westfall, *Never at Rest*, 468; Cohen, "Introduction," 148. The problem posed by the *Principia* for Locke's empiricist philosophy was encapsulated by the latter's comment to William Molyneux that "Every one could not have demonstrated what Mr. Newton's book hath shewn to be demonstrated"; see Locke, *An Essay Concerning Human Understanding*, ed. P. Nidditch (Oxford: Oxford University Press, 1991), 643–34; Locke to Molyneux, 20 September 1692 in *The Correspondence of John Locke*, 8 vols., ed. E. S. de Beer (Oxford: Clarendon Press, 1978–), 4: 524; J. Marshall, *John Locke: Resistance, Religion, and Responsibility* (Cambridge: Cambridge University Press, 1994), 384–413, esp. 387.

34. Craig to Campbell, 29 December 1687 and 15 January 1688/9; Gregory to Campbell, 16 December 1687; *Newton Correspondence*, 2: 501, 501 nn. 2–3; "Eloge de Mr. Moivre," *Histoire de l'Académie Royale des Sciences*, 1754, 261–62, cited in Westfall, *Never at Rest*, 471. Campbell claimed to be a novice in the reduction of equations into series in a letter to Gregory in 1685, but he was no mean mathematician. He apparently corresponded with Newton, who is recorded to have said of him: "I see that were he amongst us, he would make children of us all," though this is suspiciously similar to remarks Newton was also said to have made about Roger Cotes (the eventual editor of the second edition in 1713) and indeed a number of others; cf. *Newton Correspondence*, 2: 416 and 417 n. 1.
35. Whiston, *Memoirs*, 315–16; John Craig to William Wotton, 24 June 1691; Trinity College Library Ms. R.4.7, no. 9, cited in L. Stewart, *The Rise of Public Science: Rhetoric, Technology, and Natural Philosophy in Newtonian Britain, 1660–1750* (Cambridge, 1992), 102; and *Newton Correspondence*, 3: 150–51. For Craig's copy of the *Principia*, see Cohen, "Introduction," 203–5.
36. See Bentley's annotations to the previous letter, mentioning Descartes, de la Hire, Mercator, Huygens and Bartholinus, and Newton's directions to Bentley; *Newton Correspondence*, 3: 152, 155–56.
37. Gregory to Newton, 2 September 1687, ibid. 2: 484. Bearing in mind the quote attributed to Babington above, Gregory's reading notes indicate that he began to devour the *Principia* in 1687 and finished the book in 1694; see W. P. D. Wightman, "David Gregory's Commentary on Newton's *Principia*," *Nature* 179 (1959): 393–94; and C. M. Eagles, "The Mathematical Work of David Gregory, 1659–1708," (Ph.D. diss., Edinburgh University, 1978), 28–34. The original of Gregory's "Notae in Newtoni Principia Mathematica Philosophiae Naturalis" is currently in the library of the Royal Society.
38. Newton to Charlett, 27 July 1691; Newton to Flamsteed, 10 August 1691, Gregory to Newton, 27 August and 10 October 1691; *Newton Correspondence*, 3: 154–55, 164, 165–66, 169–70.
39. Fatio to Huygens, 24 June 1687, and Leibniz to Huygens, October 1690, in *Oeuvres Complètes de Christiaan Huygens*, 22 vols. (La Haye: Martinus Nijhoff, 1888–1951), 9: 167, 523–26. See I. B. Cohen, "Introduction," 145–61; and E. A. Fellmann, "The *Principia* and Continental

Mathematicians," *Notes and Records of the Royal Society of London* 42 (1988): 13–34.

40. Newton to Huygens, August 1689; Huygens to Fatio, 28 January 1689/90; Fatio to Huygens, 24 February 1689/90; *Newton Correspondence*, 3: 26–27, 67–68, 68–70. For Fatio's theory of gravity, see ibid., 3: 69–70; B. Gagnebin, "Mémoire de Nicolas Fatio de Duillier. De la cause de la pesanteur. Presenté à la Royal Society le 26 Fevrier 1690," *Notes and Records of the Royal Society of London* 6 (1949): 105–60; and H. Zehe, "Die Gravitationstheorie des Nicolas Fatio de Duillier," *Archive for History of Exact Sciences* 28 (1983): 1–23.
41. Fatio to Huygens, 24 February 1689/90, 24 February 1689/90, and 8 September 1691; *Newton Correspondence*, 2: 69, 168, 169 n. 2. Fatio continued to report the details of Newton's emotional response to criticisms from Huygens and others, although he lost contact with Newton for a number of months after 1690; cf. ibid., 70 n. 2, 169 n. 1. Newton had two sets of his own errata; see CUL Add. Ms. 3965, ff. $180^{r\,v}$, 635; Cohen, "Introduction," 186–87.
42. *Newton Correspondence*, 3: 191; Cohen, "Introduction," 178–80, 184; Fatio to Huygens, 29 April 1692, in *Oeuvres*, 22: 158–59. For a good account of the later battle over metaphysics and intellectual property between Newton and Leibniz, see A. R. Hall, *Philosophers at War* (Cambridge: Cambridge University Press, 1980).
43. R. S. Gregory, Ms. 247, fol. 13^{r}; CUL Add. Ms. 3965, fols. 271^{r}–272^{r}; Adv.b.39.1, fols 2^{r}–4^{r}. Compare with the much later Gregory Memorandum of December 1705, in which Newton thought of explaining his views on the cause of gravity by citing the belief of the ancients (i.e., that "they reckoned God the Cause of it, nothing els"); cf. W. Hiscock, ed., *David Gregory, Isaac Newton, and His Circle: Extracts from Gregory's Memoranda, 1677–1708* (Oxford, 1937), 29–30.
44. Westfall, *Never at Rest*, 423–24, 473, 521; Keynes, Ms. 133 (10); "An Account of the Book Entitled *Commercium Epistolicum*," *Philosophical Transactions* 29 (1714–16): 206. See also R. Iliffe, "'Is He Like Other Men?' Newton, the *Principia*, and the Author as Idol," in *Culture and Society in the Stuart Restoration: Literature, Drama, History*, ed. G. Maclean (Cambridge: Cambridge University Press, 1995), 159–78.
45. Feingold, *Before Newton*, 54. The relationship between labor and tedium is often expressed in Newton's correspondence and mathematical writings; see, for example, Newton to Flamsteed, 17 November 1694, in *Newton Correspondence*, 4: 47.

3.

The Ambivalence of Authorship in Early Modern Natural Philosophy

ADRIAN JOHNS

The advent of the modern authorial persona occurred in the eighteenth century, or so historians of the subject tend to believe. The received account traces its emergence in this period to shifts in policing, property, bibliographical classification, and, finally, to understandings of the creative process. It seems reasonable to look to the same period for important developments in scientific authorship. This chapter, therefore, asks how our account of the origins of literary authorship can be used to approach authorial claims in the sciences.

The major focus of the chapter is not on authorship itself, however, but on *transgression*. It seeks to identify and classify various practices of the eighteenth-century publishing world that could be called "anti-authorial," in that they compromised, directly or indirectly, claims to authorship. It argues that these practices were persistent, identifiable, and consequential. They came to impinge not just on the production and circulation of scientific materials, but on their credibility and even, in some circumstances, their content. Further, the chapter maintains that they necessitated the establishment of collective attitudes to creativity and public utterance that became central components of scientific communication.

Many historians have tracked the processes through which responsible, proprietorial, and creative conventions of authorship came to converge after 1700. An important resource for their efforts has been the archive of records left by the major trials of the eighteenth century.

Historians argue, almost certainly rightly, that the changing allegations of transgression to be found in such records can be used to identify by contrast what the culture of the time took authorial to be orthodoxy—or at least what it wished that orthodoxy to be. The chapter builds on this interpretation by showing the ways in which scientific workers in particular sought to mold to their own ends both anti-authorial practices and nascent responses to those practices. But the approach here is historical rather than philosophical or legal, and its focus is on social practice, not juridical doctrine. The chapter argues that would-be scientific authors had to confront a culture of print that they believed to be riven by anti-authorial conduct. These practices and representations took effect in the constitution of authorship itself.

A Fabulous Way of Philosophizing

What problems faced a would-be scientific author at the onset of the modern age? Consider as an example Thomas Burnet, Master of the Charterhouse School in the late seventeenth century. Burnet was England's best-known proponent of theories purporting to elucidate scriptural testimony about Creation by means of the latest natural philosophy. His *Sacred Theory of the Earth* provided just such an account, explicating Noah's flood by means of Cartesian philosophy. It was a controversial work, attracting numerous rebuttals from churchmen and others on the grounds that its explanation in natural-philosophical terms implicitly eroded the miraculous character of the flood as a direct manifestation of divine will. Burnet, therefore, resisted translating it into English.[1] But in his *Archaeologiae Philosophicae*, published in Latin in 1692, he persevered with the same enterprise. The new book attempted to reconcile remaining elements of his "sacred theory" with the Book of Genesis.

At an important stage of his text, Burnet ventured to reconstruct verbatim the conversation that had taken place between Eve and the Serpent in the Garden of Eden. The exchange was a civil one, recited in prose as though it were a dialogue in a rather earnest play. But his distinctly earthbound dramatization was not just an unfortunate lapse into amateur dramatics. Rather, it was a deliberate example of a "fabulous Way of Philosophizing," which Burnet believed he had inherited from the ancients. He affirmed that the Greeks had adopted just such fictional methods to convey profound truths. They had represented the gods themselves as teaching humans in much the same way that adults did their children. No strict criterion of truth had characterized such efforts. It had been "consistent with their Religion" to instruct the common people by means of "Fables and honest Frauds; like to what we call officious Lying." Such manoeuvres had been deemed necessary whenever the wise condescended to "philosophise with the Multitude." This style of pedagogy had given rise to the mythologies of Greece itself, Rome, and other civilizations. Latterly, however, the technique had fallen into disuse, obscured by "a severer Kind of Learning." It was therefore a good idea to make explicit again the conditions of credibility that had given rise to the ancients' myths, so that one might "dispel the Darknes from their Writings, and clear them from false Aspersions." Burnet's reconstruction of the conversation between Eve and the serpent served that end. As well as a retelling of a passage from Genesis, it was a modern re-creation of this ancient communicative method. The dialogue did, Burnet insisted, contain the "Sum and Substance" of the Mosaic story. But Burnet's text rehearsed it "freely," using "other Words," so that, freed from literalism, "we may more freely Judge of the Thing itself." He had wanted his own multitude of readers to approach this account of the Fall "as if it were written by a modern Author."[2]

What did Burnet mean by a "modern Author"? The character of authorship in the late seventeenth century was in flux. In particular, authorial identity had long been subject to conventions within the printing and bookselling community for recognizing "propriety" in the book trade. By this term, the trade community meant both customary property and civil conduct. The Restoration regime's legal machinery for regulating the press, based primarily in prepublication licensing, also relied on the customs of the book trade for its own efficacy, not least because printers and booksellers were expected to execute its provisions themselves. But by the mid-1690s all this had come under serious threat. The Press Act, which had underwritten order in the trade (on and off) since 1662, was allowed to lapse, and prepublication licensing disappeared. With it went legal recognition of the stationers' regime of propriety that had reigned for almost 150 years. The "modern author" was suddenly up for grabs. The consequence was a realm of authorship that was replete with problems of credibility and burgeoning with new rhetorical genres addressing those problems—a rather similar realm, in fact, to that in Burnet's version of antiquity.

It was not that an existing and stable regime immediately disappeared with the lapse of the Press Act. No simple change in the law could abrogate what was a respected cluster of practical conventions, and those conventions, in fact, proved resilient. But their legal recognition was now cast into doubt. As a result, booksellers, printers, and would-be authors alike would be forced to take their debates out of the privacy of stationers' premises and into the legal system, in a bid to define anew the formal rules for authorship and its protection. Successive test cases sought to provide an archive of precedents in terms of which transgression and propriety might jointly be defined. At the same time, a vast expansion occurred in the production of cheap printed materials—newsbooks, diurnals, weeklies, mercuries, and miscellanies—for a readership rapidly extending beyond the limits of urban society. Legitimate printing houses spread into provincial centers, creating for the first time a prosperous national industry. In the fierce world of competition now being created, reprinting, translating, epitomizing, and abridging were honed into instruments of commercial warfare.

Yet on the other hand, the cultural practices of the printing house itself remained remarkably stable. Between the sixteenth century and the advent of the steam press in the nineteenth, printers and booksellers shared what Roger Chartier has termed a common "typographical old regime."[3] A central aspect of that regime was that it accorded printers, and, more importantly, booksellers, substantial powers over the making and maintenance of identities in print. The constancy of *their* culture was what underpinned the consistency of representations in trials. Burnet's modern author—and, in the event, Burnet himself *as* author—stood at the intersection of these trends, and would be defined by them.

Burnet's book initially attracted little attention. A limited circulation and relatively learned readership meant that at first there was no extraordinary concern about its content. But the infamous atheist Charles Blount then seized upon it. What he found was a text that, in his view, acknowledged the emptiness of its own piety, implied the centrality of lying to modern priestcraft, and placed canonical works of scripture alongside pagan myths in the same remarkably sceptical light. In one of his last acts before committing suicide, Blount extracted the conversation with the serpent and issued an English translation, surrounding it with his own radical interpretation.[4] The new version transformed readings of Burnet's original by making it the occasion for explicit pantheism.

Blount's version became widely known, and even more widely denounced. Blount himself dismissed his critics as "Whole-sale Merchants of Credulity," bent upon the corrupt preservation of priestcraft. Burnet, however, taken aback at the ridicule and outrage, resolved to suppress the unauthorized translation. He met with the bookseller of *Archaeologiae Philosophicae*, who by the custom of the trade owned total rights to the work, and bought the copy back from him. Never again, he resolved, would his fabulous philosophizing see print in any language.

Despite the registration of literary rights to the stationer, then, Burnet as a gentleman did have some power to restrain his own writings and frame his own perceived identity. It sufficed to deal with the unfortunate Blount. But it ceased, of course, with his own death in 1715. His estate included not only the *Archaeologiae Philosophicae*, but a number of risky manuscripts that had not yet been published. It was not long before these manuscripts began to filter into print. A treatise *Of the State of the Dead* came first. Translated and issued by the heir to the pirate king himself, Edmund Curll, it appeared complete with an elaborate apparatus identifying its many "heresies." Whatever the state of the dead in general, the state of Burnet in particular was destined to be unenviable. "The Ashes of the venerable Author were poorly and meanly insulted," remarked John Dennis. Such "low and vile Buffoonry" befitted "neither the Gravity of the Christian, nor the Justness and Spirit of a polite Writer, nor the Honour and Humanity of a Gentleman."[5] And Burnet's published work likewise became vulnerable to such enterprising forms of appropriation. The fact of its prior publication in no sense immunized it. Nor did its registration at Stationers' Hall—something that now carried no legal force, whatever residual respect it might command from honorable printers. Sensing an opportunity to capitalize on its remembered notoriety, an alliance (or "conger") of booksellers led by one William Chetwood moved to reissue Blount's translation of the Eve-serpent exchange.

This time the executor—Thomas's brother, George Burnet—decided to act. He requested an injunction against Chetwood's publication. The case seemed to him watertight. Just a few years before, the first so-called Copyright Act had been passed, after much pleading from booksellers. Under its terms, Burnet believed, the text of the *Archaeologiae Philosophicae* ought to be protected until at least 1731. Chetwood was driven to coin an interesting argument in response. He urged that the translation, being Blount's, was a different work altogether from Burnet's. Such a translation, he declared, "may be called a different book, and the translator may be said to be the author." The Act should be taken to encourage, not proscribe, its publication.

Lord Chancellor Macclesfield decided against Chetwood. In effect, authorial propriety won out over the stationer's claimed prerogative. But that should by no means be taken to imply that Macclesfield, repudiated the Stationer's argument about authorship. For Macclesfield, this was at most a secondary consideration. He scarcely even addressed Burnet's authorial concerns, let alone resolved them. In fact, he reflected openly that Chetwood might well be right. "A translation might not be the same with the reprinting of the original," he conceded, "on account that the translator has bestowed his care and pains upon it." But he had nonetheless resolved to uphold the injunction on the ground that this particular book was improper to be read in English. He knew it to be, having applied himself to it at length "in his study." Macclesfield thus considered himself well qualified to conclude that the text "contained strange notions, intended by the author to be concealed from the vulgar in the Latin language, in which language it could not do much hurt, the learned being better able to judge of it." He thus sustained the injunction against Chetwood on grounds of juridical responsibility, not authorial property or propriety. Macclesfield took it for granted that the Court of Chancery held "a superintendency over all books," and hence that he should "in a summary way" prevent the publication of works impugning religion or morals. His decision vis à vis authorship was entirely subordinate to this concern, to him far more important, for the sociology and politics of reading.[6]

The Burnet case seemed as unusual to contemporaries as it does to us. One strange factor was the "pirate" bookseller's insistence on authorial creativity being based in labor. This consideration, clearly Lockean in provenance, was a novelty, and the only element in the case of which Macclesfield explicitly approved. Later in the century, the same kind of argument would prove pivotal in the establishment of legal definitions of authorship, which would prevent exactly the kind of action Chetwood was seeking to defend. But not only was Macclesfield's decision itself innocent of any concern to protect authorial propriety; it also had no such consequence in practice. Printer Edmund Curll, the worst pirate of the age, proceeded to issue a series of unauthorized reprints of Burnet's text. Like Blount's, these editions placed the original work in a new and distinctly unflattering context, this time that of Grub Street. George Burnet had no success whatsoever in quashing them.

But in the meantime, Burnet had brought the first action ever fought under anything like a modern copyright regime. It also seems to have been the first in which a case against a press-pirate was brought, not by the defrauded bookseller who had published the work in the first place, but by (representatives of) the writer himself. This was the author's first day in court.

Categories of Anti-Authorship in the Old Regime

The period following George Burnet's action is the best studied in the history of authorship. Between the first Copyright Act of 1710 and the establishment of the steam press in the early nineteenth century, the author is commonly said to have been invented. One could be an author in 1800—in a sense impossible only a century earlier. Burnet's assertion of authorial identity (if that is what it was) has thus been taken as marking the beginning of a transitional period of central importance to the making of modern authorship—and, therefore, of modernity itself.

That consensus rests largely on records preserved from legal cases. That is, it relies on allegations of transgression brought before a state institution for resolution and recompense. The fact that such debates reached this arena is the only reason why they are now recalled by historians of authorship. This raises three issues. The first is a question of representativeness. Who knows how many similar charges languished unsettled or unrecorded because they never made it into the legal system, or else did make it there but were not written down because they furnished no new precedents. It is hard to ascertain whether the legal record is representative of the quotidian realm of negotiations in more intimate spheres of authorial creativity and reproduction, not least because the purpose of legal recording was precisely to note departures from the norm.[7] It may be that the chronology we have reconstructed is an artifact of the legal system as much as of the culture that it served. The second issue concerns the identity of the cases themselves. Their common characteristic was that they were not about authorship per se, but about *piracy*. That is, they addressed what one may term "anti-authorship." They can only be taken as evidence of authorial identities by invoking a logical process of inversion. That process is not contentious, but its use should nevertheless be noted. And, thirdly, arguments over authorship often took second place to those over reading. In Burnet's case, the distribution of a text in a language accessible to the vulgar took priority over the theoretical question of authorial right. Reception was of greater concern than creativity. The legal consequences of a right were produced, but not because the right itself was recognized or even much discussed.

In part, all this simply reminds us that we should not assume the legal realm to be prior to the cultural. Concepts articulated in the courts are as likely to have been adopted from lay usages as vice-versa. Here, perceptions and practices in the book trade formed both the transgressions identified by legal institutions and the propriety to which they were opposed. So we need to look for such perceptions and practices. With this in mind, the following paragraphs present a brief taxonomy of infringements. What emerges is that natural philosophers and natural philosophies were of central importance in establishing that taxonomy.[8]

The most blatant transgression was outright reprinting, or *comprinting*. This was the unauthorized republication of a work either published already by another or else registered to another at Stationers' Hall. Comprinting could be anything from a sordid, backstreet activity to a sophisticated international enterprise, as in the case of reprinters of Diderot and d'Alembert's *Encyclopédie*. Such a project, Darnton shows, could be both immensely profitable and, with judicious changes to the text, ideologically desirable. As that implies, deciding whether a given book were in fact a literal reprint of another was often problematic. An early version of Mandeville's *Fable of the Bees* was just one work that, after its initial publication, "being soon after Pirated, [was] cry'd about the Streets in a Half-Penny Sheet." The result was a popular misconception (according to Mandeville himself) that "the Scope of it was a Satyr upon Virtue and Morality." Daniel Defoe claimed that comprinting robbed authors and readers alike of "the Prize of Learning," which was as much a reward of honor as of cash. He would have been depressed to discover the radicals described in Adrian Desmond's *Politics of Evolution* still engaged in it in the London of the 1820s through 1840s.[9]

An easier tactic involved producing *supernumerary copies*. If a printer were engaged to print one thousand copies of a work such as (to choose a particularly controversial example) Francis Willughby's *Historia Piscium*, then by printing another two hundred without authorization he could expect to garner extra profit. This was suspected to be a common practice, and the Royal Society certainly thought it to have occurred in the case of Willughby's disastrous "Fish Book." Astronomer Royal John Flamsteed likewise accused Isaac Newton of collaborating with bookseller Awnsham Churchill to subject his *Historia Coelestis* to the indignity. The Royal Society forbade its "printers" (in fact, booksellers) from indulging in supernumerary printing, in a bid to limit such problems. But with the market for learned works insecure, and the investment needed to produce them substantial, it remained a real concern.[10]

Daniel Defoe characterized *epitomizing* as "the first Sort of Press-Piracy." He warned that substantial volumes of any kind—"Philosophy, History, or any Subject"—were prone to gutting. John Dunton agreed, referring to "*Original* and *Abridgement*" as the "Man and Wife" of his trade.[11] A cheap abridgment might easily drive a specialized work from the market, and, as John Wallis put it, "endanger the loss of the [mathematical] author himself." That was one reason why the Royal Society stipulated that its printers must not reprint any work "in epitome." Moreover, unlike a straight reprint, an epitome could convey a "false idea" of the original. Difficult questions of identity and difference between texts, therefore, had to be confronted in this context. How could one decide whether a given book were properly considered an epitome of, say, the Bible, material from which was used virtually everywhere? The book trade took pains to develop conventional mechanisms for determining such issues, and those mechanisms were prized as constituting the trade as a civil community.[12]

Translations—from Latin to English, from English to Latin, and between European languages—were an industry in their own right. Relevant here is the observation that regimes of literary propriety stopped short at national boundaries. Supranational authorities such as the Holy Roman Emperor could issue privileges, but in practice those privileges were violated and, as the exporter of Newton's *Principia* learned from his Dutch contacts, "pirates" could often obtain them more easily than authorized publishers. As with epitomes, the implications extended to content as well as commerce. Translators, not sharing the priorities of the Royal Society itself, often re-ordered or reconstructed society texts in unpredictable ways. They thus threatened the essential purpose of works like the *Philosophical Transactions*.[13]

Imitation too might be accounted a violation of propriety. In this case, a transgression could be alleged even if *none* of the original text were reproduced. Newspapers, for example, frequently found themselves competing against rivals that appropriated their own titles and stated authorship. By the early eighteenth century, imitation of this kind had become a sophisticated game, in which William King and Martinus Scriblerus spoofed the virtuosi in part by regurgitating what were portions of their actual texts. Indeed, following Shadwell, this became the dominant form of satire directed against the experimental philosophy. It became difficult to tell who was imitating whom. There were even printed reports retailing proceedings at Royal Society meetings that had never occurred, prejudicing the reputation of "personages of the first Rank & Character."[14] And the most brazen of all Stationers' interventions took the form of an imitation, when the Athenian Society portrayed itself as a real academy dedicated to answering readers' queries and boasting its own journal and apologetic *History*. In fact, it was nothing more than a stationer's academy, existing only in print. In effect, the Athenians reversed the tactic of the virtuosi, and had *real* witnesses reporting to *virtual* scientists.[15]

A branch of imitation that Jonathan Swift particularly resented (but also employed) was *attribution*. He complained that "there is no book, however so vile, which may not be fastened on me." Physicians John

Freind and Hermann Boerhaave were among many other prominent victims, both being ascribed published works that were not, in fact, of their authorship.[16] No citizen of the Republic of Letters could be guaranteed immunity. "Sometimes I was Mr. John Gay, at other Burnet or Addison," recalled one of Edmund Curll's hacks; "I abridged histories and travels, translated from the French what they never wrote, and was expert in finding new titles for old books."

In natural philosophy, allegations of transgression were almost always of epistemic consequence. That consequence could be positive, as discussed in a moment. More usually they were intensely negative; and they could be serious enough to threaten the accused's very place in the philosophical community. It is therefore worth extending our taxonomy into the intellectual realm. As such, complaints took one form above all others: plagiarism, or, as it was often called, *usurpation*. Accusations of this offense were rife in the mid-seventeenth- and eighteenth-century natural sciences. Many philosophers remarked openly upon the impossibility of becoming an author without "passing for a Plagiary." Indeed, it is difficult to think of any prominent investigator of the natural world who was not accused of violating propriety in this way. As participants in an enterprise strongly assertive, in the face of widespread incredulity, of its polite and useful character, accusations that they had breached these cardinal virtues were indeed damaging. Overtly, at least, experimental philosophers set great store on honesty and openness of utterance; nothing should be concluded by cabal. This made them peculiarly vulnerable to the practices of anti-authorship. Robert Boyle, for example, decried the keeping of recipes as valued secrets, not only because it was a form of ostentation, which allowed "counterfeit" recipes to circulate, but because it permitted "fals usurpations" to cheat true discoverers. Faced with a real culture of print that was riven by transgressive practices, natural philosophers put all possible strategies of accreditation to use to guarantee their own authorship and the status of their claims.[17]

Piracy, Authorship, and Gentility

This taxonomy of transgressions acquired a collective, generic title: *piracy*. Legally, the word referred solely to the unauthorized reprinting of a work already owned by another. By the eighteenth century, however, it also had a broader, lay sense, which encompassed all these offenses. It is possible to suggest why in terms of the contemporary meanings of the term *pirate*.

We now assume that a pirate is a sea going vagabond, and by and large this was so in the seventeenth century too. But a complicating nuance emerged from challenges to monarchical politics. In these challenges, the question was raised of the degree of autonomy legitimately enjoyed by a commercial organization like the Stationers' Company. For Hobbes, the answer was "none": corporations must be entirely subjugate to the crown. For republicans, however, no such simplicity was possible. They insisted that corporations *could* claim a certain independence. But at the same time they maintained that this was only legitimate to the extent that they both recognized their place in a virtuous social hierarchy and instilled in their members a customary civility harmonizing with the interests of the commonwealth. A corporation that achieved these feats successfully could be both self-sustaining and integrated into a perpetual commonwealth. Any corporation that did not was a community, it was said, of pirates. So for a republican like John Streater, a pirate community was, *by definition*, a merchant community lacking propriety and a place in the scale of political being.

That such concepts were not unique to the mid-seventeenth century may be implied by references in Adam Smith's lectures, a century later, to the possibility that piracy might be accorded high honor. A pirate was comparable to a merchant, Smith told his students. In the ancient prudence, where repute was correlated with military prowess, a pirate had often been reckoned more honorable than a merchant. Smith added that in his own day, too, people in a foreign country who owed no allegiance to that country or any other were properly treated as pirates. The association of piracy with a systematic indifference to civility apparently held.[18] It is, I think, interesting to speculate that something of this association was intended in the original coining of the term to indicate authorial transgression. If so, then the relation between authorship, its violation, and the making of modern politics takes on a singularly interesting character.[19]

So does the relation between piracy, authorship, and gentility. Piratical practices made authorship intensely problematic. How could a gentleman, and in particular a natural philosopher, be an author? Conventional advice was mixed. Much of it, however, held that gentlemen should abjure the vocation altogether, as something "meerly Mechanick." It was unclear whether something performed by the rote application of rules—and subject to the whims of pirates—could be consistent with freedom of action, and freedom of action was central to the identity of the gentleman. To become an author one had to subject oneself to tradesmen, at least in conventional terms. Master-printers and booksellers had their own views of proper conduct, focused on their own community. A gentleman often had to become the guest or even lodger of such a person. Every aspect of printing thus militated against the supremacy of one's "pleasure." The gentleman who entered the world of the Stationers was reducing himself to just one participant in a collective of craft operatives. Becoming an author meant losing one's self.[20]

This much could be said, perhaps, of most craftsmen with whom gentlemen had to interact. It could also be said of most gentlemen venturing into print, whether or not they were natural philosophers. But most crafts did not then create objects claiming to instantiate forever the knowledge, wit, character, and virtue of the gentleman concerned. And most gestures of authorship did not extend to matters of knowledge. Gentlemen philosophers were thus unusually "punished," to use Robert Burton's word, by the experience of authorship. The permanence and exposure accorded a printed book by its being produced in around a thousand copies, qualified and circumscribed though that permanence was, meant that the sign of constraint was visible, and that it endured. This was only reinforced by the convention that authors had no right to their work once published, and that it might then be subjected to translation, epitomizing, abridgment, and so on. Whether any of the resulting publications were accounted "authorized" might be determined by the stationers' own court, to which authors had no right of access. Gentlemen thus repudiated authorship not out of simple snobbery, nor from affected repugnance at "the stigma of print," but because the character and civility of the stationer impinged on gentility itself. And this was very much how gentlemen writers themselves spoke of their experiences. "It is now-a-days the hard fate of such as pretend to be Authors, that they are not permitted to be masters of their own works," complained one; "for, if such papers (however imperfect) as may be called a *copy* of them, either by a servant or any other means, come to the hands of a Bookseller, he never considers whether it be for the person's reputation to come into the world." The good name of a gentlemen could never rest safe "in the power of a Bookseller."[21]

How to respond? One way might be to abandon print for some purposes, and vest faith in manuscript circulation. Isaac Newton, Samuel Hartlib, and Henry Oldenburg evidently did this in their different ways, as indeed did that paragon of openness, Robert Boyle, when dealing with alchemical subjects. But manuscripts were in practice scarcely less vulnerable than printed books, as Burnet's case showed. Besides, it was difficult to believe that manuscripts could reach and unite a widely dispersed readership.[22]

In certain circumstances, however, complaints of anti-authorship might actually result in *enhanced* credibility. This was especially likely to prove the case if the "unauthorized" work contained scandalous or secret information. An aura of penetrated secrecy could easily lend authority. Some writers took advantage of this tendency to spread unorthodox views, or to achieve influence without appearing to desire it. Isaac Newton thus repeatedly considered promoting his heterodoxy by means of the "unauthorized" publication of texts questioning conventional viewpoints. In alchemy, too, an apparently unauthorized revelation might be more likely to find acceptance than one of self-proclaimed authorship, because of alchemists' views on the proper character of the practitioner as the silent beneficiary of grace. The strategy was every bit as useful for mathematicians. In each case, the very doubt generated by charges of unauthorized publication created a window of opportunity for increasing the renown and influence of one's ideas.[23]

Two more general practices emerged in response to the received character of authorship. The first was the assumption by academies, like the Royal Society, of a guarantor's role with respect to printed communication and the consistency of authorial recognition. Such places exerted civil disciplining such that violent challenges to authorship were defused. They created "registration" systems, ostensibly to preserve authorial status from challenges altogether, but really to set the rules by which future challenges would be conducted. And in the case of the Royal Society, they adopted licensing practices, appropriating the state's regulatory mechanisms to serve a warranting role. John Wilkins characterized such a strategy as the repudiation of "singularity" (the imprudent aspiration to authority implicit in most claims to authorship) by the instantiation of "society." It was implicitly modeled on the corporate structure of local governance and trading bodies, such as the Stationers' Company. It also had the effect of defusing suspicions of inappropriate self-elevation. Clopton Havers thus prefaced his book on osteology (1691) by hoping that "no one will think my addresses a rude transgression of the Laws of Decency." The Society had ordered him to publish, so gaining "a Title to it." The unfavorable connotations of authorship were dispersed over the Fellowship, while the favorable remained with Havers.[24]

The other major response related not to mechanisms of production, distribution, and use, but to printed objects themselves. Would-be authors tried to reduce the size, cost, and (in some cases) sheer presumption of the publications being authored. Hitherto, learned authorship had tended to mean the production of enormous, expensive, and aggrandizing folios of an Ulisse Aldrovandi, a Robert Fludd, or an Athanasius Kircher. The Royal Society's choice of Willughby's *Historia* as an ambassador was entirely representative of the genre. The smaller quartos of Boyle (or, for that matter, Hobbes) were a real change from this norm, instantiating modesty in their very typography. Periodicals were yet a further step in the same direction. Journals represented major innovations because they offered both the securing of authorship and, simultaneously, its legitimation as modest. The *Philosophical Transactions* was the first example devoted solely to natural knowledge, and it was intended from the start to extend the Royal Society's conventions into a broader public realm. Arguments that it made authorship safe were repeated virtually every year in the prefaces issued with each collected volume. Moreover, the periodicity, low cost, and regularity of such journals meant that they met with favor from Stationers. The "saving register" of the Society (as Martin Lister called the *Transactions*) proved crucial for establishing the rules of experimental authorship.[25]

There were countless instances in which protocols of this kind came into question. One example was the well-known debate over blood transfusion, in which the Royal Society's printer, John Martyn, took advantage of Henry Oldenburg's incarceration in the Tower to issue a letter by French physician Denis laying claim to priority in the technique. The letter was consecutively paged and signatured with the *Philosophical Transactions*, and was widely taken to be an issue of the journal, thereby calling the *Transactions*' role as a register of discovery and authorship into doubt. But Oldenburg was in a position to use the *Transactions* itself to repudiate this impostor as soon as he was released—something of a first in natural-philosophical publishing. Another case in point is Robert Hooke's conflict with Oldenburg. Here Hooke contrasted his own "publication" of discoveries during oral lectures to the secretary's printed *Transactions*, which, he claimed, subverted authorial propriety and "made a trade of intelligence." He charged that printing and distribution were not enough to make the journal a publication, because its governing interest was, in fact, private. He impugned not only Oldenburg but the role of the *Transactions* as guarantor of authorial propriety. The society responded by endorsing Oldenburg publicly—and from that moment on, unlike in previous years, the *Transactions* always appeared with the society's explicit imprimatur. Yet a third instance was astronomer Royal John Flamsteed's espousal of a different form of authorship altogether. Flamsteed saw the piecemeal production of reports characteristic of the *Transactions* as inimical to his own enterprise, since he needed to collate years of observations before any one of them could truly be accounted knowledge. He identified the society's authorial conventions as "philosophical," and his own as "astronomical," thus carrying over into this realm a distinction prominent in epistemological arguments of the early modern period.[26]

The Rise of the Scientific Author

Changes in conventions of authorship emerged in part from the transition of debates into new social spaces. In 1700 to 1750, disputes over the "propriety" of print ceased to be argued out in the stationers' court and were transferred to the realm of the common and statute law. Hitherto, arguments had rested largely on custom. Now it became far more appropriate to argue at the level of principle. Lawyers, booksellers, and writers alike thus seized upon the latest Lockean notions of the entitlement lent by labor, and applied them to intellectual work as Locke himself had applied them to its physical equivalent. Macclesfield's en passant concession that a translator might count as an author was an early indicator of this trend.

These developments were central to, even definitive of, the invention of copyright. The Lockean character of such arguments is consequently well known. But the modeling of literary propriety on technological and natural philosophical protocols deserves attention too. In fact, technology and natural philosophy became essential resources in these debates. They supplied protagonists with powerful arguments—that much is well known. More important, however, they also supplied both exemplary authors and exemplary practices. Aristotle, Gassendi, Descartes, and Newton were the key authorial archetypes for one combatant; others suggested that, in insisting upon his "copy," a possessive bookseller like Dunton was aspiring to the status of an "Inventor of any small Mechanical Instrument." By the 1760s, after two generations of such exchanges, judges' verdicts were routinely resting on the philosophy of property and precedents like that offered by Harrison's chronometer. It should be the same, remarked one, "whether the Case be *mechanical*, or *Literary*; whether it be an *Epic Poem*, or an *Orrery*."[27]

The same held true of arguments *against* strong copyright. Enlightenment principles could lead one to question whether literary property were possible at all. God's truth ought not to be parceled out. "Invention and labour, be they ever so great, cannot change the nature of things," it was said, "or establish a right, where no private right can possibly exist." Boyle's air pump was not a pump for Boyle's air. In France, Condorcet thus recommended that literature should abolish distinctions of authorship. The result, he said, would be a realm of print in which readers selected works by subject; printed texts would be organized into topical periodicals, without authorial designations. The author would be dead almost before having a chance to be born.[28]

Both standpoints owed much to natural-philosophical people and practices. The academies and societies that had arisen across Europe, with their relatively rigorous authorial regimes and their periodicals instantiating those regimes for outsiders, formed their essential bedrock. The Royal Society has here been adduced, but in this respect it should be taken as only one example of a wider phenomenon. Authorship could not be primarily a legal category, if only because it must be as international as the distribution of printed knowledge itself. Its foundations lay deeper: in cultural practices and representations more constant than the law, and on which the law itself rested. In the practices of the virtuosi originated conventions capable of reconciling learned authorship with modest civility.[29]

The Fall of the Scientific Author?

Almost exactly a century after Burnet, the controversial surgeon William Lawrence (1783–1867) went to court with a similar aim, only to prove even less successful. Lawrence was the most brilliantly successful surgical lecturer of his age. Between 1816 and 1820 he engaged in a fierce feud with his erstwhile mentor, John Abernethy, that captivated polite society in an England possessed by fears of revolutionary upheaval. The resulting case threw into sharp relief the constraints to scientific authorship that still prevailed at the close of the typographical old regime.

The subject of the controversy was life itself. Abernethy had lectured that matter, life, and mind were separate fluids—a notion that found ready support in Anglican theological circles, but a mixed reception more generally. One quarterly dismissed his published lectures as a confused mixture of "bombast about genius, and electricity, and Sir Isaac Newton." Lawrence joined this attack. He preferred the view of Bichat and Cuvier, that life consisted in the very organization of material structures making up the living body. Abernethy replied with an accusation of anti-authorial practice, remarking that Cuvier had appropriated John Hunter's discoveries without due acknowledgment. He further condemned Lawrence's stance as that of a "Party" of "Modern Sceptics" inclined to atheist materialism. Lawrence's response to this appeared in 1819 in his *Lectures on Physiology*. The *Lectures* spurned Abernethy's insinuations about a "party" dedicated to physiological irreligion. Lawrence then proceeded to reject the inspirational nature of scripture, the immaterial soul, and the truth of the Mosaic descriptions on which Burnet's reconstruction had been based a century earlier.[30] At this point theologians, vicars, journalists, and anonymous pamphleteers joined in the fray. Lawrence soon attained notoriety as one pillar of a "radical triumvirate," along with Tom Paine and Lord Byron.

The association led to Lawrence's being forced to recant—in public, at any rate. Conceding a political defeat if not a physiological one, Lawrence, like Burnet a century earlier, arranged to have his book removed from circulation. He also undertook never to reprint it, nor in future to publish texts "on similar subjects." For the rest of his life, he would dedicate himself to surgical practice and instruction alone. Yet this very recantation accommodated itself to romantic views of Galilean persecution, as Lawrence acknowledged in sending a copy of his work secretly to a radical bookseller recently prosecuted for libeling the litany. He told William Hone that its withdrawal had been a matter of expediency, and that the gift testified to Hone's greater courage in what he recognized as a similar situation.

In 1821 and 1822, Lawrence's *Lectures* appeared in an unauthorized edition. He soon gained an injunction to prevent the copies from being sold. Their publisher, one Smith, then moved to have this ban dissolved. His ingenious argument was that since the book contained passages "hostile to natural and revealed religion," and since it "impugned the doctrines of the immateriality and immortality of the soul," it was in violation of the law against blasphemous libel, so it could not be protected by copyright. He had identified a genuine opportunity. Illegitimate works did indeed occupy an ambiguous position, which Macclesfield's successor as Lord Chancellor, Lord Eldon, was currently in the process of settling. Eldon's view was that a libel—that is, a book containing blasphemy, sedition, or licentiousness—could not be accorded the protection of copyright. He had recently determined on this basis that alleged piracies of Byron were permissible, because of the poet's libertinism. When the Lawrence case came before him in March 1822, Eldon now faced the same paradox in a case involving scientific piracy. He obeyed his own principle rigidly. He perused the text and scanned the reviews in the weightier quarterlies. On the basis of this research he decided that Lawrence's book did indeed contain materialist opinions in violation of revealed religion. Therefore, it should be outlawed as libelous. But it could not actually be suppressed. This meant that, being outside the law, it could neither be curtailed nor protected from piracy. The injunction against Smith's edition was therefore unsustainable, and must be dissolved. The piracy, suddenly deemed unstoppable, could proceed to sale—because the book was inherently illegal and hence deserved no protection.[31]

The intervention of a soi-disant modern Galileo then made matters even worse for Lawrence. His admired William Hone was an acquaintance of radical publisher Richard Carlile, already notorious for issuing Paine's and Byron's works. Carlile himself seems to have believed that life was a manifestation of a fluid that caused the organization of matter. But at the same time he wished to see doctrines endorsing the immaterial soul abolished, and priestcraft erased. Newton and Bacon themselves, Carlile remarked, had been blinded by the "kingcraft" to which they had been exposed by living at court. That was why Newton, in particular, had voiced such rubbish about religion: "Rather than be called the author of such trash," Carlile proclaimed, "I would consent to be considered an idiot."[32] This man now decided to add Lawrence definitively to the two members of the radical triumvirate he had already made his own. Carlile printed the *Lectures* in a cheap and popular format. He had already developed an unorthodox, risky, but effective sales strategy, which involved seeking the notoriety of prosecution in an attempt to boost circulation, and may have wished for this to happen again.[33] It did not, so Carlile was robbed of the opportunity to read the entire 600 pages aloud in court—another favored tactic. But his edition reached a new, mass audience. By virtue of this, its notoriety as the source of popular irreligion was rapidly secure.

This was in one sense just an everyday story of early nineteenth-century folk: Harriet Martineau sandwiched it between mechanics' institutes and body-snatching in her own list of phenomena representative of those times.[34] Its significance, however, was twofold. First, it helped inspire Mary Shelley to write the first version of *Frankenstein*—which, as Marilyn Butler argues, displayed a much clearer debt to such physiological controversy than later editions would be permitted to retain. Second, the *Lectures* itself survived, and lurched the length and breadth of Britain. Carlile's version went to nine editions, printed on cheap paper and sold at affordable rates. He purposely addressed artisans and the new class of factory workers. These people, he was convinced, were hungry for strains of knowledge not supplied to them in the expensive formats of more esteemed publishers. In their hands, the reading of physiological lectures could not be predicted. Whether the learned rebuttals of Lawrence himself would even reach their eyes could realistically be doubted. Proletarians might make themselves into materialists by appropriating the pirated work of a recanted author. The result, then, was anxiety about the proper distribution and appropriation of natural knowledge—and about the *kinds* of natural knowledge proper to be distributed. The major material consequences of that anxiety were the so-called Bridgewater Treatises, which aimed to provide a popular literature independent of such sources as Carlile, and to save the people for Anglicanism. The only way to prevent the monstrous materialism in Carlile's piracy from running amok was to let loose such creatures of the steam press.[35]

It is worth reminding ourselves what had *not* changed between Burnet and Lawrence. In neither case, ultimately, were decisions reached on the basis of authorial principles. But anti-authorial practices and, in particular, practices of piracy were central to both. Those practices were so significant primarily because they related to what Burnet called "philosophizing with the multitude." Characterizations of popular reading were more at stake than characterizations of authorship. Authorial identities played roles in these trials, and emerged (or not) out of them, only partly as autonomous principles. At least as consequential were their roles in representing the multitude and its reading practices. It is true, of course, that Eldon's reasoning transited a legal principle of authorship (which he felt necessary to deny Lawrence's lectures), whereas no such principle detained Macclesfield. But that in itself represents a major way in which the two cases contravene the accepted trajectory of authorial identity. In terms of practical outcome, "the author" was recognized in 1720, but not in 1823.

In conclusion, the image of print implicit in most histories of learned authorship needs to be rethought. We tend to assume a priori that Burnet and Lawrence—not to mention Newton and Flamsteed—stood apart from the denizens of Grub Street. Where it appears at all, the world of Edmund Curll, Richard Carlile, Charles Blount, and the mysterious Smith is generally characterized in rather abstracted terms—as "print capitalism" or "print logic." Its less familiar aspects then appear, if at all, as defined by contrast: as elements of a "subculture" or an "underworld." This has undoubted romantic appeal, and is certainly not without evidential warrant. However, for much early modern scientific authorship it rather puts the cart before the horse. Hindsight aside, in that period it was people like Curll and Carlile—and more generally the craft community in which they participated—that made authorship, and unmade it too. The learned adapted to them. "Sub" and "under" are thus inappropriately demeaning terms of analysis. Theirs was the culture; theirs was the world.

Notes

Department of History, University of Chicago, 1126 East 59th Street, Chicago, IL 60637.

1. J. E. Force, *William Whiston: Honest Newtonian* (Cambridge: Cambridge University Press, 1985), 32–40; P. Rossi, *The Dark Abyss of Time: The History of the Earth and the History of Nations from Hooke to Vico*, trans. L. Cochrane (Chicago: University of Chicago Press, 1984), 33–41.
2. T. Burnet, *Archaeologiae Philosophicae* (London, 1692), sigs. a^{v}-a2^{r}, 280–81; *Archaeologiae Philosophicae: Or, The Ancient Doctrine Concerning the Originals of Things* (London, n.d.), vii-viii, 5–7.
3. R. Chartier, "L'ancien regime typographique: Réflexions sur quelques travaux récents," *Annales E.S.C.* 36 (1981): 191–209.
4. C. Blount et al., *The Oracles of Reason* (London, 1693), sig. A2^{r}ff, 1–19, 20–51.
5. T. Burnet, *Of the State of the Dead*, trans. M. Earbery (London, 1727), ttp, sig (π ^{v} T. Burnet, *A Treatise Concerning the State of Departed Souls* (London, 1730), sig.[A4]$^{r-v}$.
6. *Burnet vs. Chetwood*, 2 Mer. 441 (1720). See also M. Rose, *Authors and Owners: The Invention of Copyright* (Cambridge, Mass.: Harvard University Press, 1993), 49–50.
7. For the achievement of facticity in legal documents, see N. Z. Davis, *Fiction in the Archives: Pardon Tales and Their Tellers in Sixteenth-Century France* (Stanford: Stanford University Press, 1987).
8. This is a simplified redaction (an epitome, as it were) of the survey in A. Johns, *The Nature of the Book: Print and Knowledge in the Making* (Chicago: University of Chicago Press, 1998), where a more elaborate version may be found.
9. R. Darnton, *The Business of Enlightenment: A Publishing History of the* Encyclopédie (Cambridge, Mass.: Harvard University Press, 1979), 33, 131; B. Mandeville, *The Fable of the Bees* (Oxford: Oxford University Press, 1924), i, 4; [D. Defoe], *Essay on the Regulation of the Press* (London, 1704), 19; A. Desmond, *The Politics of Evolution: Morphology, Medicine, and Reform in Radical London* (Chicago: Chicago University Press, 1990). passim.

10. See, for example, Johns, *Nature of the Book*, chap. 7, 489–90.
11. Defoe, *Essay on the Regulation of the Press*, 20; Dunton, *Life and Errors*, 56.
12. See, for example, R. Campbell, *The London Tradesman* (London, 1747), 132–33; S. J. Rigaud, ed., *Correspondence of Scientific Men of the Seventeenth Century*, 2 vols. (Oxford, 1841), ii, 552–54.
13. Bodl. MS. Rawl. *Letters* 114, fols. 158^{r}–159^{r}; Johns, *Nature of the Book*, 227–28, 507.
14. Collins, *Authorship*, 26–27; J. M. Levine, *Dr Woodward's Shield: History, Science, and Satire in Augustan England* (Berkeley: University of California Press, 1977), 247*f*; RS MS. Dom. 5, No. 23.
15. [C. Gildon], *The History of the Athenian Society* (London, [1691?]); G. D. McEwen, *The Oracle of the Coffee House: John Dunton's* Athenian Mercury (San Marino, Calif.: Huntington Library, 1972).
16. S. Buckley, *A Short State of the Publick Encouragement Given to Printing and Bookselling in France, Holland, Germany, and at London* (London, n.d.), 3.
17. M. Hunter, "The Reluctant Philanthropist: Robert Boyle and the 'Communication of Secrets and Receits in Physick,'" in *Religio Medici: Religion and Medicine in Seventeenth Century England,* eds. O. Grell and A. Cunningham (Aldershot, England: Scolar Press of Brookfield, Vermont: Ashgate Publishing Company, 1996), 247–72.
18. J. Streater, *A Glympse of that Jewel, Judicial, Just Preserving Libertie* (London, 1653), 15; [Streater], *Observations Historical, Political, and Philosophical, upon Aristotle's First Book of Political Government* (London, 1654), 6–7, 18–20, 37–38; A. Smith, *Lectures on Jurisprudence* (Oxford, 1978), 224, 310, 527–28.
19. This line of argument will be fleshed out in a book I am currently working on.
20. S. Shapin, *A Social History of Truth: Civility and Science in Seventeenth-Century England* (Chicago: University of Chicago Press, 1994), 38–41, 43–52, 156–92; M. Woodmansee, "On the Author Effect: Recovering Collectivity," in *The Construction of Authorship: Textual Appropriation in Law and Literature*, eds.Woodmansee and P. Jaszi (Durham, N.C.: Duke University Press, 1994), 15–28.
21. R. Burton, *The Anatomy of Melancholy,* 2 vols. (Oxford, 1989–90), i, xl–xliv; D. F. McKenzie, "Speech-Manuscript-Print," in *New Directions in Textual Studies*, eds. D. Oliphant and R. Bradford (Austin, Tex.: Harry Ransom Humanities Research Center, 1990), 87–109.
22. H. Love, *Scribal Publication in Seventeenth-Century England* (Oxford: Clarendon Press; Oxford; New York: Oxford University Press, 1993); L. Principe, "Robert Boyle's Alchemical Secrecy: Codes, Ciphers, and Concealments," *Ambix* 39 (1992): 63–74.
23. K. Figala, "Pierre des Maizeaux's View of Newton's Character," *Vistas in Astronomy* 22 (1978): 477–81; L. Principe, *The Aspiring Adept: Robert Boyle and His Alchemical Quest* (Princeton: Princeton University Press, 1998), 138–49.

24. C. Havers, *Osteologia Nova* (London, 1691), sigs. $A3^r$–$A4^v$, $A5^r$–$A6^v$; A. Johns, "Prudence and Pedantry in Early Modern Cosmology: The Trade of Al Ross," *History of Science* 35 (1997): 23–59.
25. For natural history publishing see P. Findlen and T. Nummedal, "Words of Nature: Scientific Books in the Seventeenth Century," in *Scientific Books, Libraries, and Collectors*, 4th ed., ed. A. Hunter (Aldershot: Ashgate, 2000), 164–215.
26. R. C. Iliffe, "'In the Warehouse': Privacy, Property and Priority in the Early Royal Society," *History of Science* 30 (1992): 29–68; S. J. Schaffer, "Regeneration: The Body of Natural Philosophers in Restoration England," in *Science Incarnate: Historical Embodiments of Natural Knowledge*, eds. C. Lawrence and S. Shapin (Chicago: Chicago University Press, 1998), 83–120; Johns, *Nature of the Book*, chap. 8.
27. Johns, *Nature of the Book*, 246–68; [Gildon], *History of the Athenian Society*, 7, 14, 33–35.
28. *Speeches or Arguments of the Judges of the Court of King's Bench* (Leith, 1771), 50; compare Rose, *Authors and Owners*, 87; C. Hesse, *Publishing and Cultural Politics in Revolutionary Paris, 1789–1810* (Berkeley: University of California Press, 1991), 104; Campbell, *London Tradesman*, 136.
29. On the civility of academies, see M. Biagioli, "Scientific Revolution, Social Bricolage, and Etiquette," in *The Scientific Revolution in National Context*, eds. R. Porter and M. Teich (Cambridge: Cambridge University Press, 1992), 11–54.
30. O. Temkin, "Basic Science, Medicine and the Romantic Era," *Bulletin of the History of Medicine* 37 (1963): 97–129; J. Goodfield-Toulmin, "Some Aspects of English Physiology, 1780–1840," *Journal of the History of Biology* 2 (1969): 283–320.
31. E. Jacob, ed., *Reports of Cases Argued and Determined in the High Court of Chancery* (London, 1828), 471–74.
32. R. Carlile, *An Address to Men of Science* (London, 1821), 5, 9–12, 27–28.
33. *The Report of the Court of King's Bench . . . Being the Mock Trials of Richard Carlile* (London, 1822), iii-iv and passim.
34. H. Martineau, *History of the Thirty Years' Peace* (London, 1849–50), ii, 86–88.
35. M. Shelley, *Frankenstein, or The Modern Prometheus: The 1818 Text*, ed. M. Butler (London, 1993); J. Topham, "Science and Popular Education in the 1830s: The Role of the 'Bridgewater Treatises,'" *British Journal for the History of Science* 25 (1992): 397–430.

4.

The Uses of Anonymity in the Age of Reason

MARY TERRALL

What happens to the meaning of authorship when a text appears anonymously? Such texts are certainly not authorless; the author has merely slipped behind a mask that makes his or her identity an open question for the reader. In the early modern period, anonymous publication was often no more than a temporary screening of authorship of contentious or otherwise controversial books. The secrecy afforded by anonymity might well be no more than an ephemeral feature of the text. Most scientific writing appeared in print firmly attached to the name, and often the portrait, of an author. There is, however, a small set of anonymous books on scientific subjects interesting for what they reveal about publication strategies, about the relation of authors to their readers, and more generally about the cultural context of scientific knowledge in this period. Institutions and individuals, in fact, manipulated authorial invisibility for a variety of purposes.

With very few exceptions, anonymous authors of texts on scientific subjects either unmasked themselves or were unmasked by others within a few years, if not months, of publication. Why, then, bother with publishing anonymously at all?[1] The mask of anonymity could of course insure against the risks associated with clandestine publication. In old-regime France, these included the very real possibility of arrest and imprisonment, but also less tangible risks like evoking ridicule or scorn in one's readers. Writers worried endlessly about offending patrons in high places, and could protect themselves from the consequences of offensive remarks by maintaining anonymity. The most familiar anonymous texts of the Enlightenment are those that did put their authors at risk of persecution, whether for libel, sedition, immorality, or irreligion.[2] While anonymity often correlated with vulnerability on the part of the author, this was by no means always the case. Writers chose to keep their identities secret for a variety of reasons related to other aspects of their public status. Ironically enough, anonymity could become a resource for making and defending reputation.

Attention to different modes of authorship leads to recognition of the many genres subsumed under the general category of "scientific writing." More than that, the multiple uses of anonymity illuminate not only the risks attendant on publication, but the integration of wit, humor, satire, and irony into science (and vice versa). Scientific authorship in the eighteenth century was not always readily distinguished from literary authorship, and the use of familiar literary forms for scientific subjects tells us something about the place of science in the wider field of knowledge production, and of publishing.

The Paris Academy of Sciences

As the royally sanctioned arbiter of scientific knowledge, the Paris Academy of Sciences provided its members with the very opposite of anonymity: title, rank, and the privilege of publishing without the approval of state censors.[3] This was a mark of the academy's authority, of course, rather than a license to publish without constraint. In fact, academicians did not always take advantage of this privilege if they were printing works inappropriate to the style or subject matter of the academy. One of its prerogatives, following its reorganization in 1699, was the policing of publication, which can be construed partly as the policing of authorship. The institution used its prerogative to defend and maintain its authority over natural knowledge.[4] Naming and publicity were essential to confirming the legitimacy bestowed by academic membership. However, the academy institutionalized anonymity in the annual prize competition. These occasions allowed the academy to display its authority by conferring approval on texts, and, by extension, on their authors.[5] (Other continental academies sponsored similar prize competitions.) Members of the academy could not compete, in part to guarantee fair play, but also to enforce the tacit notion that once elected, individuals were not subject to institutional judgment as outsiders could be. Anyone outside the academy was eligible to send an anonymous submission for the competition, and essays typically came in from provincial and foreign savants. The process of blind judging culminated in the revelation of the identity of the author of the winning piece (or pieces). At this point, a monetary reward was bestowed on the author, and the academy published the essay with its seal of approval. The judgment conferred a name on the author; the unsuccessful entries were usually not attributed to their authors and faded into obscurity.

The selection of prizewinning essays took place in secret, preserving the fiction of the unanimity of the academy's corporate judgment. The system implied that the truth value of the work had to be rigorously separated from knowledge of the author's identity, and also subsumed the identities of the individual judges under the corporate umbrella of the institution. At the same time, though, the enforced anonymity of the competition reasserted the individuality of the author, whose name and status, if they were known, might interfere with the impartiality of the judges. The authority of the prize commissioners to judge controversial questions was grounded in the academy's formal regulations and privileges. The judgments demonstrated an in-principle consensus, which often eluded the academy in practice, especially behind closed doors.

By setting the prize questions, the academy prompted a flow of learned treatises from the provinces and abroad to the domain of the academy in the French capital. When it determined an essay to be worthy of the prize, the academy transformed the anonymous writer into a renowned author. Once the identities of successful contestants were revealed, they often turned out to be well-known foreign savants, men like Leonhard Euler or any one of the prolific Bernoulli family, but they might also be provincial medical men, or other contenders completely unknown in the Republic of Letters. Eager to display their skills and to compete with each other in such a visible arena, essayists submitted to a temporary anonymity that might ultimately enhance their reputations and assert their authorship more powerfully than publication in some other form, precisely because the author's identity was a surprise to the public and to the judges.

The Marquise du Châtelet

Essayists might welcome the anonymity of the competition for a variety of reasons. A case in point involves Voltaire, more famous for his literary works than for science, and Emilie de Breteuil, the marquise du Châtelet. The Paris prize question for 1738 concerned the nature and propagation of fire. It attracted contributions from a wide field of candidates, including Voltaire, who set up a laboratory at the country estate in Cirey where he was living with Châtelet, and Euler. As an aristocrat with close ties to the French court, the marquise had access to political power, and used her connections to protect her allies and promote their philosophical causes. She became famous in her own time as the ranking femme savante of France, and she particularly strove to make her mark in the masculine realm of physics and mathematics.[6] However, her gender kept her outside the institutional centers of science and colored her choices about whether and how to publish her work. At the time of the prize competition, she was not a published author.

Observing Voltaire's attempts to measure differences in the weight of iron samples at different temperatures, Châtelet found herself disagreeing with his explanations and decided to write a contribution of her own for the competition.[7] She worked in secret and alone, after the rest of the household was asleep. She wrote her essay as a tentative challenge to the academy, to herself, and no doubt to Voltaire as well. "I wanted to test my strength under cover of anonymity, as I imagined that I would never be recognized," she confided subsequently.[8] She was evidently perfectly conscious of the advantages of anonymity for someone with scientific ambitions who had never written a scientific work, someone who was not and could never be a member of the academy, a woman, and an aristocrat. She would be writing for the scientific elite, and at least a few of them would have to read her work. If the commissioners did not crown her work, she ran no risk of condemnation or ridicule. Anonymity earned her work a gender-blind reading; as she was to find out later, this was not always so easy to obtain.

When she learned that neither of them had won the prize, Châtelet disclosed her secret to Voltaire, who decided to lobby his acquaintances in the academy to permit publication of their essays along with the three winners.[9] In doing so, he revealed his own identity to the academicians. Voltaire's prominence in the world of letters meant that he could not be ignored, and Châtelet, although initially reluctant to be named publicly, welcomed the chance to have her work accorded even this modest measure of recognition. At this point, she entered into protracted behind-the-scenes negotiations over whether her name would appear with the text and what the printed text would include.

When she asked to make minor revisions to her text before publication, the prize commissioners refused to allow any alterations whatsoever. She especially wanted to alter a footnote in which she had referred favorably to Dortous de Mairan's attack on *vis viva* (mv^2) as the correct measure of force, an assessment she had come to revise in the course of further study.[10] The footnote was not problematic as long as she remained anonymous; it had been included to flatter a senior academician she suspected of being on the prize committee. In the interim, she had developed a harsh critique of the very paper by Mairan that she had casually praised earlier, and she was formulating ambitious plans to write her own book about physics. Once she was named, she had much at stake, given the uncertainty of her future career as an author. Authorship had taken on intense personal significance because of her marginal position in the community of writers and readers concerned with scientific matters. "It is very sad to see," she complained, "in perhaps the only work of mine that will ever be published, an opinion that is so opposed to my present ideas."[11]

Châtelet tried to exploit her personal connections at the academy to get around the intransigence of the prize commission. She wrote to her old friend and former lover, Maupertuis, asking him to use his access to the academy's papers to remove the footnote surreptitiously. "I asked you," she reminded him, "to erase it incognito, being sure that no one would notice it." But he categorically refused to cooperate.[12] In spite of their long association, he proved unwilling to act anonymously in the academic context where his identity was well established, and where such an act might jeopardize his status with his peers. Châtelet persisted, however, even without his help, and eventually convinced the academy to print a list of errata at the end of the volume, where she could indicate her correction to the offending statement. Her list of mostly trivial changes ran to two printed pages; buried among them was a reformulated version of the problematic footnote.[13] Her dogged persistence over the errata (where she proved powerful enough to win a compromise) reflected her aristocratic hauteur, but her failure to get more substantial concessions reflected her weakness relative to the academy's royal privilege to control its competitions and publications.

She was to have the last word on this score, however. The official volume of prize essays appeared in a very limited edition primarily for members of the Academy, but the public learned of her contribution through a laudatory anonymous review in the monthly *Mercure de France*, which was then reprinted in the *Nouvelle Bibliothèque*.[14] This appreciation was written, in point of fact, by none other than Voltaire. Then in 1744 Châtelet published her own edition of the essay, no longer anonymous, adding revisions to reflect her current commitment to Leibnizian dynamics, and especially to the conservation of vis viva. The publisher's afterword recounted the history of the prize competition, and the essay's previous publication history. Thus she shed the cloak of anonymity and appropriated to herself a text that had earlier been controlled by the academy in Paris.[15]

The academy, under pressure from outside its chambers, had somewhat reluctantly made Châtelet into an author. Shortly after the adventure of the prize competition, however, she pushed ahead with her book project, *Institutions de physique*, which appeared anonymously from a Paris printing house in 1740. This was a treatise on mechanics, with serious pretensions to summarizing the current state of physics from a highly idiosyncratic point of view, synthesizing Newton and Leibniz.[16] Though she took special care to have the manuscript shepherded through the censorship process by a go-between, she also went to considerable lengths to preserve the secret of her authorship. Anonymity veiled her gender and her nobility, as it had in the prize essay competition, but this time for a much grander project. Again, she hoped to gauge reactions from a safe position behind her mask. Her cover was rudely forced, however, by the accusations of her one-time tutor, Samuel König, who claimed that he had written large portions of her book. Once her identity had been so harshly, even slanderously, exposed, she took an aggressive stance to defend her authorship as well as the content of her work. When the academician Mairan printed a pamphlet criticizing her discussion of force, she responded in kind, dropping all pretense of anonymity.[17] In a second edition of *Institutions de physique*, she changed the gender of all the adjectives referring to the author from the masculine to the feminine forms, and replaced an allegorical frontispiece with her own portrait.[18] She carefully managed her authorship as a means of making her name in the Republic of Letters. Control of textual revisions, illustrations, printing, and distribution allowed her to present herself as a woman who could be feminine (evident in her portrait) as well as conversant with the arcane disburser of controversies in the realm of mathematical physics. She only got to this point by starting out behind the screen of anonymous authorship.

Maupertuis

In principle, the academy authorized scientific knowledge for the Crown—this was one of its designated functions. Academic commissions certified manuscripts as worthy of publication, just as they approved inventions and machines. In practice, Enlightenment publishing was neither so simple nor so centralized. Many books appeared without approbation and many were published outside Paris, and hence outside the jurisdiction of the academy.[19] Apart from prize essays and book reviews in periodicals, anonymity was usually associated with texts that were "dangerous," or outside the bounds of propriety in some way—or that aspired to be so.

Madame du Châtelet's friend and correspondent Maupertuis was a successful academician who occasionally used anonymous publication to reach a constituency beyond the academies. He had a productive career in the Paris Academy of Sciences starting in the late 1720s and moved to Berlin to take over the presidency of the Prussian Academy of Sciences in 1746. He cultivated a reputation for defending adventurous, even unorthodox, positions on scientific questions, and he maintained social connections in the elite world of salons and coffeehouses, as well as at the Prussian court.[20]

At just the time that Châtelet was begging Maupertuis to use his privileged access to Academy papers to erase her regretted footnote, he was embroiled in his own dispute with Jacques Cassini, the preeminent astronomer in Paris. Cassini had challenged the accuracy of observations Maupertuis had made on a recent academy-sponsored expedition to determine the shape of the earth. As fellow academicians (*confrères*), they were enjoined from attacking each other publicly, but each devised his own strategies for continuing the battle. Maupertuis impugned Cassini's honor and credibility in *Examen désintéressé des differentes ouvrages qui ont été faits pour déterminer la figure de la terre*, an ironical anonymous pamphlet that posed as an evenhanded summary of both sides of the dispute. I have written elsewhere about Maupertuis's use of the purported neutrality of numbers as a polemical device in this text.[21] He also fanned the flames of public interest by encouraging rumors that it had actually been written by one of his enemies, as a defense of Cassini. The confusion about the text's point of view was played out in the press in speculation about the identity of the author. Anonymity, and stubborn denial of his authorship even to intimate friends, allowed Maupertuis to create a minor literary phenomenon. He intensified the hoax by publishing a second edition with a preface about the book's mysterious provenance, boasting that "There were people in Paris who sought the place of publication of the *Examen* with as much care as others had employed to discover the shape of the earth."[22]

The fact that people conversant with the issues and claims at stake could be misled about the polemical stance of the book attests to the subtlety of its critiques. The detailed "Histoire du livre" printed in the second edition hinted broadly at the author's true identity, but went on to repeat rumors that claimed the text for the other side of the dispute. The false attributions then served ironically as evidence that the book could not possibly be hostile to Cassini. Quoting positive responses to the first edition by Cassini himself, Fontenelle, and Mairan, the preface went on, "Surely no one will believe that [these men] lack sufficient knowledge [*science*] and intelligence [*lumières*] to be mistaken about what favors or destroys their own position. Even if the common man may be an Oedipus in judging that which concerns him personally, are we to believe that the most enlightened men in the kingdom of France are deluded and take for praise what are only Ironies?"[23] This commentary justified the second edition and played on readers' interest in stories of authorship and publication of controversial texts. It gave readers clues about how to read the book, and how to identify the author and his victims. The meaning of anonymity was tied up with publishing practices (multiplying editions, for example, and masking the place of publication to pique reader interest) and with the complex, and often masked, interaction between publisher, author, and reader.

The manipulation of anonymity allowed Maupertuis to display his cleverness to an audience that appreciated (and knew how to read) such authorial contortions. Tricking not just Cassini, who had no facility for wit, but the elegant and eloquent Fontenelle as well was a subversive claim to fashionable literary territory. Maupertuis pursued this line even after the question of the shape of the earth had been settled in the academy, which some of his friends thought to be in bad taste. But this shows that there was more at stake than the approbation of the official scientific elite. He was playing a game with the conventions of satirical character assassination, but it was not a trivial game. He was out for personal revenge on Cassini, certainly, but he shifted the ground of debate from the academy to the press, the wider Republic of Letters, and conversations in elite drawing rooms. This move made the scientific quarrel into a literary one, drawing on a long tradition of such exchanges (often anonymous) going back to the dispute between self-styled ancients and moderns at the end of the seventeenth century.[24] Maupertuis could not have written a satirical attack on a fellow academician under his own name without risking his position in the academy. He also knew that the full impact of the satire would only be apparent once the identity of the author was known definitively. Although he never acknowledged his authorship in print, it is clear from correspondence that he was widely recognized by the time the second edition appeared in 1743, so that it would have been read with this knowledge in mind.[25] There were, then, multiple kinds of authorship in play. Without his academic credentials, which could not be displayed openly in this exchange, Maupertuis could not have been convincing as a witty author of "Ironies" (as he called them). He wished to defend his credibility as an astronomer and mathematician, but he also set out to establish credibility in yet another social context, where literary cleverness might grab his audience's attention.

Anonymous authorship was one way to bend the unwritten code of comportment that came with the mantle of academic membership. Publishing anonymously might mean putting aside that mantle for a more playful style, but the two identities did not necessarily remain mutually exclusive. When Maupertuis wrote a short tract occasioned by the comet of 1742, for example, he suppressed his identity as academician to play the role of elegant wit, this time with no nastiness. He gestured at the conventions of gallantry made famous by Fontenelle, addressing himself to a nameless lady fascinated by the new comet. "You wished, Madame," the text begins, "that I should speak to you of the comet that is today the subject of all the conversations of Paris; and I take all your desires as my commands."[26] Comets crossed the boundaries dividing polite conversation and gossip from the discourse of academy and observatory. They could be seen as wonders or portents by "the common man," as curiosities by elegant ladies, as intractable calculational challenges by mathematicians, or as poorly understood celestial phenomena by *physiciens*. Comets represented both the triumph and the limitations of Newtonian methods, since calculating exact orbits for particular comets from the theory of central forces was a tricky business. Neither can cometary science in this period be separated from public fears about the dangers associated with comets, as Simon Schaffer has shown.[27]

Maupertuis undertook to show his gentle reader what enlightened modern astronomy, authorized by the best analytical mathematics and the best telescopes, had to offer a public anxious to understand the appearance and possible consequences of comets. The author's voice is both authoritative (he commands the technical knowledge to understand the erratic behavior of comets) and entertaining (he unrolls fanciful scenarios for cometary encounters with the earth). No longer the province of the astrologer, comets are subject to physical law, and any dangers must be assessed rationally. This requires the support of mathematics, for which even the enlightened elite reader needs the assistance of the expert.

Although it is framed in the discourse of flirtatious didacticism, most of Maupertuis's little book is devoted to a clear summary of the current state of knowledge about comets, including an overview of the dynamics of the Newtonian solar system and Halley's heroic calculations of cometary orbits. This becomes a kind of set piece on the analytical power of Newtonian cosmology. But the reader finds that nothing in Newton's physics precludes the earth from crossing paths with an errant comet, and then learns about the possible consequences of such an encounter. Far from leaving the reader in mortal terror, the text teaches her to temper speculation about possible futures with a calculation of their likelihood.

> Everything shows us that comets could bring deadly changes to our earth and to the whole economy of the heavens. . . . But we are right to feel safe. . . . The tiny place that we occupy in the immense expanse where these great events happen, annihilates the risk for us, although it does not change the nature of the danger.[28]

In other words, the intelligent reader will recognize that comet-induced catastrophe, though physically possible, is also improbable enough to become interesting and amusing rather than terrifying.

Why should such a text be published anonymously? It was not particularly risky, morally or scientifically, though publishing such a work as an academician would have stretched the parameters of the academic identity. Even so, the academy would have had no grounds for any kind of official condemnation of such a work. But that still leaves the question of why this alternative identity was appealing to a successful academician. Writing such a book claimed a role for men of science as the agents of an enlightened understanding of nature. The conventions of gallant literature allowed the author to present a version of rationality, transmitted by a male authority to his feminized readers. The anonymous author created an individual persona for himself as a particular kind of man of science, who can operate equally well in polite society and in the academy. Authorship enabled him to move from the world of equations and telescopes into a parallel social space where literary and conversational style marked the man of letters. His text flaunts his own place in that elite society through the intimate terms of his reference to his female companion and reader. This is the anonymity of a fancy-dress ball, where the relation of the mask to the identity of its wearer is exposed at the moment of unmasking.[29] Reading implicates the reader in the unmasking and brings author and reader together through a kind of inside joke.

In this case, we have some evidence for the reception of the text. The author did not remain masked for long. At least two reviewers identified Maupertuis by name, describing him as "a Philosopher who is known no less as a man of wit [*homme d'esprit*] than as a profound mathematician [*géometre*]."[30] These reviews noted the gallant style of the book, and applauded it as an authoritative compendium of current knowledge about comets.[31] In the contentious terrain of the literary marketplace, the text also provoked an outright attack. This pamphlet, *Critique de la Lettre sur la comète, ou lettre d'un philosophe à une demoiselle âgée de 9 ans*, objected violently to a review comparing Maupertuis favorably to Fontenelle. The critique's author, as it turned out, was Gilles Basset des Rosiers, a Jesuit professor of philosophy. He did not put his name on the title page, but he did sign a poem appended to the end of the text, addressed to the young girl of the title. He viewed Maupertuis's book as an almost sacrilegious perversion and co-optation of Fontenelle's style. The *Critique* is interesting for the parallelism of its rabid attacks on the literary form and the Newtonian content of the *Lettre*, while it also participates ironically in the same discourse by addressing a girl who is not yet part of the adult world of irony and sexual intrigue.

> The address [to the lady] is the only thing that links it to [Fontenelle's] *Mondes*. . . . In the new work, there are only English names, that you could not pronounce without giving yourself a sore throat, and that your ears could not hear without being wounded. It is nothing but cones, ovals, parabolas, conic sections. . . . I tell you, Mademoiselle, it is frightening. I don't believe there is in France a single woman tough enough to read this *Lettre* and learned enough to understand it, unless it be Madame the Marquise du Châtelet.[32]

Maupertuis's choice of style and genre for his discussion of comets once again enabled him to make a splash on the literary scene. The *Critique* only made it more visible. One of his aims, as with his satire of Cassini, was to provoke people to talk about him. If the book were appreciated by its target audience, the authority of the *savant astronome* would be enhanced, because he would have succeeded at displaying his virtuosity.

Among academicians, Maupertuis made the most extensive use of the conventions of anonymity, hiding his identity again for two books on heredity and organic generation.[33] Though speculative, these texts made use of substantial experimental and observational data. The first of them also took the gallant conventions that frame the *Letter on the Comet* to much greater lengths. When Maupertuis cloaked his contentions about the inherent activity of matter in authorial anonymity, he was appealing to a fashionable audience by invoking literary and philosophical conventions incompatible with his academic status. Once again, he provoked Basset des Rosiers to publish a vitriolic anonymous critique, impugning the morality of the author while broadly hinting at his identity.[34] Critical attacks of this kind exploited the potential of anonymity to make libelous claims about other authors. Anonymous authorship, then, was a common strategy for articulating controversy and criticism, across a broad spectrum of issues, from literary style to religious heresy, and including scientific method or results. Personal animosity cannot be separated from these controversies either, since the world of letters was small and filled with contentious and prickly personalities, ready to attack or take offense at the smallest perceived slight. Science was part of this world, as evidenced by its appearance in ironic and gallant, and perhaps heretical, works. In the latter category we might put Buffon, whose monumental work was not anonymous, but who provoked the censure of the Sorbonne and an extensive anonymous attack by the Jesuit scholar Lelarge de Lignac. Like Maupertuis's anonymous critic, Lignac objected vociferously to Buffon's literary style, as well as to his scientific claims and their religious implications.[35] The link between style and content was clearly perceived by readers and writers alike. Style and genre were deployed quite self-consciously by scientific authors anxious to cross the boundaries marking off the corporate academy from the more free-ranging world of the salon, the journalist, and the bookseller.

La Mettrie

My final example of an author who exploited anonymity is the French materialist Julien Offroy de la Mettrie. He was more willing than either Maupertuis or Châtelet to risk position and reputation, although it might be argued that he had less at stake than either of them. As a provincial physician newly arrived in the capital, La Mettrie began his publishing career by antagonizing the Paris medical establishment with a series of vicious pamphlets attacking current practices.[36] Banned in Paris, and then in Holland, he found refuge in tolerant Prussia, where Frederick II welcomed him to the circle of philosophers and wits at Potsdam. La Mettrie's occasionally scurrilous sense of humor, and his willingness to play the buffoon, were probably more responsible for his success at court than his unorthodox views. He is remembered for the scandal associated with the radical materialism of his books, but these books also belong in a culture of irony, wit, and eroticism where anonymity was more than a protection from the civil and religious authorities. La Mettrie's books, like those of Diderot, brought the subject matter of science into this literary culture.

In *Natural History of the Soul*, published in 1745 when he was still living in Paris, La Mettrie gave a materialist account of the soul's functions based on the inherent activity and irritability of organic matter. Writing about the "nature and properties of the soul," La Mettrie made a risky move from physiology into the territory of the theologians, using the language of natural history, physiology, and mechanics, tempered by irony. The title page declared the work to be "translated from the English of M. Charp by the late M. H. of the Academy of Sciences." The book was instantly seized by the police. Under the circumstances, Maupertuis cannot have been pleased to see that La Mettrie had dedicated the book to him as his friend and as "the philosopher most capable of judging it."[37] By the time the Parliament of Paris condemned it to the flames, along with Diderot's anonymous *Pensées philosophiques*, the identity of the author was well enough known that La Mettrie fled to Holland. Here again, anonymity provided only limited protection from persecution. It did, however, mark La Mettrie's book as scandalous and hence as interesting to readers seeking "philosophical" books.

The next edition (1747) of the offending work appeared once La Mettrie had left Holland for the Prussian court. It bore Oxford as its fictitious imprint, retained the dedication to Maupertuis, and added a "Critical Letter from M. de L. M. to M. la Marquise du Châtelet." This bizarre text mounts a tongue-in-cheek attack on La Mettrie himself, in the guise of the pseudonymous M. Charp, as a way of reiterating arguments for the purposive activity and sensitivity of matter. "My research," he says, "has directed me to evident truths entirely opposed to this writer's [his own] doctrine, and I was agreeably surprised to find that Faith has forestalled all the discoveries on this delicate subject by the most penetrating Philosophers." He goes on to berate himself (or his alter ego) point by point, remarking that "the whole work could pass for a chaos of explanations, as obscure as they are dangerous."[38] Given that *Natural History of the Soul* had been burned in Paris the previous year, La Mettrie was somewhat presumptuous and certainly indiscreet in allying himself with Châtelet, and especially in naming her. Her position as a proponent of Leibniz in France, and as a noblewoman treading on the toes of certain Parisian academicians, was tenuous at best, and she could not have welcomed La Mettrie's brazen declaration that he "would seize with pleasure the opportunity to have metaphysical conversations" with her.[39] He had nothing to lose by this pretense of familiarity, and made it into an elaborate joke at her expense. Manufacturing false identities, and mixing them with actual—and controversial—authors, made the game more complicated. Evidently, writing and presenting texts—the public and private aspects of authorship—were crucially intertwined for La Mettrie.

His most notorious book, *L'homme machine*, was published anonymously in Leiden in 1747.[40] Its materialism remained grounded in medical and scientific evidence, but also deployed the language of eroticism and irony. This time, La Mettrie dedicated his work to an enemy, the anatomy professor Albrecht von Haller. The dedication played with Haller's dual identity as poet and physician, in a salacious description of the parallel pleasures of the flesh and the mind.

> Sensual pleasure . . . has only a single climax which is its tomb. . . . How different are the mind's resources of pleasure! The more one approaches Truth, the more one finds her charming. Not only does enjoyment of her enhance desire; but one climaxes here as soon as one wishes to. . . . Study has its ecstasies, just like love.[41]

Haller disavowed any connection to the text or its author, though he did not initially realize that the perpetrator was in fact La Mettrie. Again, anonymity functioned as more than a screen for gossip and innuendo. The "machine man" of the title, self-moving if not self-fashioning, was appropriately described by an anonymous author. The materialist world, running on its own powers according to no one's plan, had no author in particular. Readers, themselves no more or less than the mechanisms described by the scandalous anonymous author, were meant to take pleasure in the text, as in the world of active matter that was their home.

> To be a machine and to feel, to think and to be able to distinguish right from wrong . . . are thus things which are no more contradictory than to be an ape or a parrot and to be able to give oneself pleasure. For since here we have an opportunity to say so, who would ever have guessed a priori that a drop of liquid ejaculated in mating would provoke such divine pleasure and that from it would be born a little creature that one day, given certain laws, would be able to enjoy the same delights? I believe thought to be so little incompatible with organized matter that it seems to be one of its properties, like electricity, motive power, impenetrability, extension, etc."[42]

La Mettrie may have taken both the content and the style of his books beyond the bounds of propriety, but not far beyond his readers' tolerance levels. His books were bought, reprinted, bought again, and presumably read.[43]

There was, of course, no place for La Mettrie's extreme materialist account of the soul and its functions in the Academy of Sciences, even in Berlin.[44] But there was no place for irony, eroticism, or galanterie in the academies either. A character like La Mettrie, with a quick wit and a tendency to be offensive, turned to the literary marketplace where conventions could be stretched, and where readers knew how to read ironies and to identify anonymous authors. Anonymity marked his books as clandestine, but it also marked him as a participant in a literary culture that linked natural knowledge, eroticism, and wit—the same culture that Maupertuis tapped into with his anonymous books. This was a realm where, at least on scientific matters, men wrote for women as well as for other men. As a woman trying to force her way into the male world of science, Châtelet did not write in an ironic mode, though she was certainly part of the culture that recognized and appreciated such style. For her, anonymity served a different purpose, a way out of irony and into the high seriousness of the academy. Her position in the fashionable literary world was assured by her aristocratic status and her willingness to play hostess and muse to men of science and men of wit. She had to struggle to earn a reputation as a thinker who was *not* frivolous, whereas Maupertuis and La Mettrie both strove to temper their scientific claims with a measure of frivolity.

Châtelet used anonymous authorship in a bid to be taken seriously by the officially sanctioned arbiters of science. For Maupertuis, it served as a technique for multiplying his audiences, while preserving his formal status in royally sponsored institutions. For La Mettrie, as an intellectual gadfly, it was a necessity. In all three cases, anonymity focused the attention of readers on the question of authorial identity. One measure of the success of a book (and of an author) was the extent of the talk it provoked, so that anonymity contributed to the success of these authors, all of whom courted a broad spectrum of readers. Maupertuis could afford to diversify his published personas in a way that Châtelet could not. She published only texts that would support her self-presentation as a learned woman, and only after testing her audience anonymously. She certainly played other roles as well, but they were not manifested to the public in her books. La Mettrie's case presents yet other features, since he had no official position or noble rank to protect. Anonymity or pseudonymity suited his clandestine pose, though his mask was by no means impenetrable.

Boundaries between audiences were no more sharply drawn than boundaries between disciplines or between genres in this period. Anonymous publication was one tool for navigating through this contentious terrain while participating in the fashion for literary games. Making science fashionable, amusing, or risqué did not mean making it trivial, however. None of the texts I have mentioned are negligible in content, even if some of them adopt a lightness of tone. The scientific author in the Enlightenment was normally a named individual, who placed himself on the institutional and cultural map in various ways by listing affiliations and titles. Then, as now, readers used authorial identity as one key to interpreting a text, based on what was known about the person of its author. Leaving off the author's name gave books a kind of life of their own, allowing for multiple readings depending on speculations and guesses about authorship. In practice, for scientific texts this anonymity was temporary, and somewhat flexible. It had the effect of making a book provocative, as authors maneuvered to hide their identities, to have their books produced outside the official censorship system, and to distribute them anonymously. This realm of scientific practice existed alongside the elaborate system of credit and credibility based on naming authors and witnesses, familiar from recent historiography.[45]

Anonymous authors rarely achieved complete invisibility. The most sensational example of long-term anonymity of a scientific text comes from the nineteenth century, when *Vestiges of the Natural History of Creation* kept readers guessing about its author through many editions. Robert Chambers, the masked author, noted (in an anonymous essay) that anonymity allowed him "to be absolutely nobody and to live absolutely nowhere; . . . [to be] everything and nothing; every sex and no sex. . . ."[46] In short, the writer enjoyed a freedom that bordered on the absurd, or the fantastic. But Chambers knew as well as any author that the view from nowhere was illusory, and that his books would be read in the real world of literary, scientific, and theological controversy.

Notes

1. Joan DeJean, "Lafayette's Ellipses: The Privileges of Anonymity," *PMLA* 99 (1984): 884–902. DeJean discusses "transparent anonymity" in seventeenth-century French fiction, where readers often knew authors' names even when they did not appear in print.
2. See, for example, Diderot, *Lettre sur les aveugles* (1749); Diderot, *Les bijoux indiscrets* (1748); La Mettrie, *L'homme machine* (1748); Helvetius, *De l'esprit* (1757); d'Holbach, *Système de la nature* (1770).
3. Roger Hahn, *The Anatomy of a Scientific Institution: The Paris Academy of Sciences, 1666–1803* (Berkeley: University of California Press, 1971), 60.
4. See academy regulations, in Ernest Maindron, *L'Académie des sciences* (Paris: F. Alcan, 1888). On the significance of the academy's right to judge scientific work, see Daniel Roche, "Sciences et pouvoirs dans la France au XVIIIe siècle," *Annales. E,S,P* 29 (1974): 738–48.
5. On prize essays, with a complete list of questions and winners, see Ernest Maindron, *Les fondations de prix à l'Académie des sciences* (Paris: Gauthier-Villars, 1881).
6. Mary Terrall, "Emilie du Châtelet and the Gendering of Science," *History of Science* 33 (1995): 283–310; Elisabeth Badinter, *Emilie, Emilie: L'ambition féminine au XVIIIième siècle* (Paris, 1983); Esther Ehrman,

Mme du Châtelet: Scientist, Philosopher and Feminist of the Enlightenment (Leamington Spa, 1986).

7. Voltaire followed Musschenbroek in assigning weight to the matter of fire; Châtelet decided that fire must be an imponderable, expansible fluid. See W. A. Smeaton and Robert L. Walters, Introduction to *Essai sur la nature du feu, et sur sa propagation*, in *Oeuvres complètes de Voltaire* by Voltaire (Oxford: Voltaire Foundation, 1991), 17: 1–89. See also Robert L. Walters, "Chemistry at Cirey," *Studies on Voltaire and the 18th Century* 58 (1967): 1807–27.
8. Châtelet to Maupertuis, 21 June 1738, in *Les lettres de la Marquise du Châtelet*, ed. Theodore Besterman (Geneva: Institut et Musée Voltaire, 1958), 1: 236. On the connection between gender and anonymity, see Erica Harth, *Cartesian Women: Versions and Subversions of Rational Discourse in the Old Regime* (Ithaca, N.Y.: Cornell University Press, 1992), 26–27; Joan DeJean, "Lafayette's Ellipses," n. 1.
9. Though not entirely unprecedented, this kind of publication was unusual. See Smeaton and Walters, "Introduction," 15–18, n. 7. Leonhard Euler, Louis-Antoine Lozeran Du Fech, and Jean-Antoine de Créquy shared the prize.
10. Châtelet to Maupertuis, 10 February and 1 September 1738, *Lettres*, ed. Besterman 1: 215–18, 252–26. The original footnote stated, "as we would still believe [in living forces] if it hadn't been for the admirable way M. de Mairan has proved the contrary." *Receuil des pièces qui ont remporté le prix de l'Académie Royale des Sciences en 1738, avec les pièces qui ont concourus* (Paris, 1739), 87–168; 105 n. On the hotly contested question of force and its measure, see Thomas Hankins, "Eighteenth-Century Attempts to Resolve the *Vis Viva* Controversy," *Isis* 56 (1965): 281–97.
11. Châtelet to Maupertuis, 19 November 1738, in *Lettres*, ed. Besterman, 2: 266.
12. Châtelet to Maupertuis, 24 October, 5 November, 19 November, 1 December 1738, in *Lettres*, ed. Besterman, 2: 266–67, 268, 270, quotation on 272.
13. In place of "as we would still believe if it hadn't been for the admirable way Mairan proved the contrary," she substituted "as a large portion of the *monde sçavant* still believes, in spite of the admirable way Mairan established the contrary . . . " *Receuil des pièces* n.p., n. 10.
14. Académie des Sciences, *Pièces qui ont remporté les prix de l'Académie Royale des Sciences en 1738* (Paris: Imprimerie royale, 1739). This was an edition of three hundred copies; a larger edition came out years later in the multivolume series *Receuil des pièces qui ont remporté les prix de l'Académie Royale des Sciences . . . avec les pièces qui ont concourus (1738–1740)* 4 (Paris: Imprimerie royale, 1752). Voltaire, "Mémoire sur un ouvrage de physique de Madame la Marquise du Châtelet, lequel a concouru pour le prix de l'Académie des Sciences en 1738," *Oeuvres de Voltaire* 38, ed. M. Beuchot (Paris: Lefèvre, 1830), 353. Originally published *Mercure de France*, June 1739, 1274–1310; reprinted *Nouvelle bibliothèque, ou histoire littéraire*, July 1739, 414–22.
15. du Châtelet, *Dissertation sur la nature et la propagation du feu* (Paris:

Prault, 1744). On the revisions, see Walters, "Chemistry at Cirey," 1819–20.

16. [du Châtelet], *Institutions du physique* (Paris: Prault, 1740); another anonymous edition, Amsterdam, 1741. See Carolyn Iltis, "Madame du Châtelet's Metaphysics and Mechanics," *Studies in the History and Philosophy of Science* 8 (1977): 29–48; Linda Gardiner Janik, "Searching for the Metaphysics of Science: The Structure and Composition of Mme. du Châtelet's *Institutions de physique, 1737–1740*," *Studies on Voltaire and the 18th Century* 201 (1982): 85–113.
17. J. J. Dortous de Mairan, *Lettre à Mme *** sur la question des forces vives* (Paris, 1741); du Châtelet, *Réponse de Mme la Marquise du Châtelet à la lettre que M. de Mairan, secretaire perpetuel de l'Académie Royale des Sciences, lui à écrite le 18 février, 1741, sur la question des forces vives* (Brussels, 1741). She appended this exchange to her 1744 edition of *Dissertation sur la nature et la propagation du feu*, n. 15.
18. du Châtelet, *Institutions du physique* (Amsterdam: Au dépens de la compagnie, 1742). On the dispute with König, see Mary Terrall, "Emilie du Châtelet and the Gendering of Science," *History of Science* 33 (1995): 283–310. On the revisions, see Keiko Kawashima, "Les idées scientifiques de Madame du Châtelet dans ses *Institutions de physique*: Un rêve de femme de la haute société dans la culture scientifique au Siècle des Lumières," *Historia Scientiarum* 3 (1993): 63–82, 137–55.
19. On censorship and the clandestine book trade, see David Pottinger, *The French Book Trade in the Ancien Régime, 1500–1791* (Cambridge, Mass.: Harvard University Press, 1958); Daniel Roche, "Censorship and the Publishing Industry," in *Revolution in Print*, ed. D. Roche and R. Darnton (Berkeley and Los Angeles: University of California Press, 1989); Ira Wade, *The Clandestine Organization and Diffusion of Philosophical Ideas in France from 1700 to 1750* (New York: Octagon, 1967); Robert Darnton, *The Literary Underground of the Old Regime* (Cambridge, Mass.: Harvard University Press, 1982).
20. On Maupertuis, see Mary Terrall, *The Man Who Flattened the Earth: Maupertuis and the Sciences in the Enlightenment* (Chicago: University of Chicago Press, 2002).
21. Mary Terrall, "Representing the Earth's Shape: The Polemics Surrounding Maupertuis's Expedition to Lapland," *Isis* 83 (1992): 218-37.
22. "Histoire du livre," *Examen désintéressé des différentes ouvrages qui ont été faits pour déterminer la figure de la terre*, 2d ed., 1741 [1743?], n.p.
23. Ibid.
24. Joan DeJean, *Ancients against Moderns: Culture Wars and the Making of a Fin de Siècle* (Chicago: University of Chicago Press, 1997).
25. This edition was falsely dated 1741; the first edition, printed in 1740, was falsely dated 1738.
26. [Maupertuis], *Lettre sur la comète qui paroissoit en 1742* (Paris, 1742). English translation by Esther Burney was published in *An Essay towards a History of the Principal Comets That Have Appeared Since the Year 1742*, ed. Charles Burney (Glasgow, 1770). The title page of this edition claims

that Maupertuis's letter was "written to the marchioness Du Châtelet," but there is no basis for this in the original.

27. Simon Schaffer, "Authorized Prophets: Comets and Astronomers after 1759," *Studies in 18th-Century Culture* 17 (1989): 45–74. On popular and learned perceptions of comets, see Sara Schechner Genuth, *Comets, Popular Culture, and the Birth of Modern Cosmology* (Princeton: Princeton University Press, 1997).
28. *Lettre sur la comète*, in Maupertuis, *Oeuvres*, 4 vols. (Lyon: Bruyset, 1756) 3: 247–48.
29. On the cultural meanings of masquerade, albeit in a different national context, see Terry Castle, *Masquerade and Civilization: The Carnivalesque in Eighteenth-Century English Culture and Fiction* (Stanford: Stanford University Press, 1986).
30. *Journal des savants*, June 1742, 351.
31. Desfontaines called Maupertuis "the ingenious rival of the author of *The Plurality of Worlds*." *Observations sur les écrits modernes* 31 (1742): 135.
32. [Gilles Basset des Rosiers], *Critique de la Lettre sur la comète, ou lettre d'un philosophe à une demoiselle âgée de 9 ans* (Paris, 1742), 30–31.
33. [Maupertuis], *Dissertation physique à l'occasion du nègre blanc* (Leyden, 1744), republished as part of *Vénus physique* (Paris, 1745); [Maupertuis], *Essai sur la formation des corps organisés* (Berlin, 1754); reprinted as *Systême de la nature* in *Oeuvres*, vol. 2. On the connection between gallantry and theoretical content, see Mary Terrall, "Salon, Academy, and Boudoir: Generation and Desire in Maupertuis's Science of Life," *Isis* 87 (1996): 217–29.
34. [Gilles Basset des Rosiers], *L'Anti-Vénus physique* ([Paris?], 1746), n.p.
35. On Buffon, see Jacques Roger, *Buffon: A Life in Natural History*, trans. Sarah Bonnefoi (Ithaca, N.Y.: Cornell University Press, 1997). Lelarge de Lignac, *Lettres à un Amériquain sur l'Histoire naturelle . . . de M. de Buffon*, 5 vols. (Hambourg, 1751–56). On critiques of Buffon, see Jeff Loveland, *Rhetoric and Natural History: Buffon in Polemical and Literary Context* (Oxford: Voltaire Foundation, 2001).
36. Kathleen Wellman, *La Mettrie: Medicine, Philosophy, and the Enlightenment* (Durham: Duke University Press, 1992).
37. [Julien Offroy de la Mettrie], *Histoire naturelle de l'âme, traduite de l'anglois de M. Charp, par M. H de l'Académie des sciences* (La Haye: J. Neulme, 1745).
38. [La mettrie], "Lettre critique," in *Histoire naturelle de l'âme, traduite de l'anglois de M. Charp . . . Nouvelle edition . . .* (Oxford: Aux depens de l'auteur, 1747), quotations on 2, 12.
39. Ibid., 7.
40. [La Mettrie], *L'homme machine* (Leiden: Elie Luzac, 1748).
41. La Mettrie, *L'homme machine; A Study in the Origins of an Idea*, ed. A. Vartanian, trans. Mary Terrall (Princeton: Princeton University Press, 1960), 143–45.
42. La Mettrie, *Machine Man and OtherWritings*. Trans. and ed. Ann Thomson (Cambridge: Cambridge University Press, 1996), 35.

43. Vartanian, introduction to *L'homme machine*, by La Mettrie, n. 41, cites ten posthumous editions of the collected *Oeuvres philosophiques* between 1751 and 1775.
44. La Mettrie was a member of the Berlin Academy, but did not publish in its *Mémoires*.
45. Steven Shapin, *The Social History of Truth* (Chicago: University of Chicago Press, 1994); Christian Licoppe, *La formation de la pratique scientifique: le discours de l'expérience en France et en Angleterre (1630–1820)* (Paris: Editions La Découverte, 1996).
46. Cited in James Second, *Victorian Sensation: The Extraordinary Publication, Reception and Secret Authorship of* Vestiges of the Natural History of Creation (Chicago: University of Chicago Press, 2000), 366–67. Second discusses Victorian uses of anonymity, and Chambers's fascination with it; ibid., 364–400.

5.

Can Artisans Be Scientific Authors?

The Unique Case of Fraunhofer's Artisanal Optics and the German Republic of Letters

MYLES W. JACKSON

> The class that is the ruling *material* force of society is at the same time its ruling *intellectual* force. The class that has the means of material production at its disposal simultaneously has control over the means of mental production; therefore, generally speaking, the ideas of those who lack the means of mental production are subject to it. The ruling ideas are nothing more than the ideal expressions of the dominant material relationships, the dominant material relationships grasped as ideas.
>
> —Karl Marx, 1845–46

Joseph von Fraunhofer (1787–1826) was a rather remarkable figure in the history of science and technology. He was a working-class optician whose work on physical optics revolutionized the production of achromatic glass, telescopes, heliometers, and ordnance surveying instruments. He served as a bridge that spanned two distinctive, yet critically linked communities: artisans and savants, or scientific instrument makers and *Naturwissenschaftler*. Although instrument makers had been crucial to the scientific enterprise since the Scientific Revolution, by the early nineteenth century, experimental natural philosophers generally could not do without these artisans, as few savants possessed the necessary manual skills to build their instruments. Yet artisans were rarely granted the status of experimental philosopher for three reasons. First, as I have argued elsewhere, the importance of secrecy to the artisanal trade was seen as anathema to the Republic of Letters, whose members prided themselves on the openness of scientific knowledge.[1] Second, savants were reluctant to accept artisans as their intellectual equals, as craftsmen were members of a commercial nexus and financial interests tainted their work.[2] And finally, members of the Republic of Letters argued that instrument makers merely manipulated preexisting materials; they did not create anything. This slavish "following of craft rules" was deemed as the antithesis of creative, scientific knowledge.

Because Fraunhofer was so precariously perched between these two groups, he is such a fascinating and historically informative character. Although he undoubtedly belonged to the artisan population, he strove for scientific recognition. He clearly contributed to the corpus of scientific knowledge. His work on the dark lines of the solar spectrum, which now bear his name, as well as his work on diffraction gratings, which was to support the nascent undulatory theory of light so eloquently proposed by Thomas Young and Augustin Fresnel, formed the cornerstone of an impressive spectrum of disciplinary research during the nineteenth century, including spectroscopy, photochemistry, and of course stellar and planetary astronomy. But Fraunhofer also offered the scientific community first-rate optical instruments, particularly his superior achromatic lenses and prisms. And herein lies the tension that this paper explores. The craft processes necessary for the construction of those optical lenses and prisms were a company secret. Because Fraunhofer was employed by a profit-seeking company, the Optical Institute, he was never permitted to divulge the processes of manufacture. Indeed, because a portion of his annual salary was based on profits of sold merchandise, it was in his financial interest not to make his artisanal practices public. Here, the commercial and clandestine practices of craftsmen, which savants so deplored, were intimately related. On the one hand, he was the creator of superior scientific devices. On the other hand, Fraunhofer authored important scientific articles. But, as I shall argue below, some savants questioned Fraunhofer's status as a scientific author.

This doubt was a very painful and devastating critique of Fraunhofer, as it denied him scientific recognition. Normally, artisans were rewarded with patents for their inventions. Questions of authorship seldom arose in practice, because instrument makers had no interest in rendering commercial secrets public. Patents, however, were never an issue for Fraunhofer, as his employer, Joseph von Utzschneider, the co-owner of the Optical Institute, was convinced that no one could reproduce his institute's products. Hence, Fraunhofer never applied for patents. And Fraunhofer was also unique as he, unlike other craftsmen, was able to publish papers in journals highly regarded by *Naturwissenschaftler*. Although artisans often published articles about their instruments, often to publicize their products, this entrepreneurial connection disqualified them from being considered scientific authors.

Fraunhofer's position, and the attempt by several savants to thwart his status as a scientific author, must be understood in its proper historical context. The German Republic of Letters (*Deutsche Gelehrtenrepublik*) was defining the rights of authors, including scientific authors. They attempted to wrestle the ownership of their printed works away from the book dealers. In so doing, they drew upon the works of Edward Young and others in order to argue that an author was someone who was not a craftsman, but an inspired genius. Creativity was a product of the mind, not the hands. Manual skills were merely ways of following certain rules laid down by the master. Authors transcended those rules, which were intended for those less intellectually gifted. Also, the German Republic of Letters began to police itself during the late eighteenth and early nineteenth centuries by discouraging its members from authoring books for the masses. Such commercialism, they argued, harmed literature and corrupted the Republic. These condemnations greatly affected Fraunhofer's attempt to ascend into the Republic of Letters. In short, Fraunhofer's role as scientific author was inextricably bound to debates on social class, secrecy, market forces, and creativity.

I. The Secret of Success: The Market

During the early nineteenth century, two different sets of patent laws existed in Bavaria. The first, which was applicable to the Bavarian Palatinate on the Rhine, was France's patent law of January 7, 1791, resulting from Napoleon's occupation of that region. The second, which was practiced in the "seven older districts" of Bavaria, including where Fraunhofer labored, did not come into existence until September 11,

1825, or less than a year before Fraunhofer's death.[3] Previous to that date, Bavaria had not enjoyed an official patent policy. Rather, the *Strafgesetzbuch Bayerns* of 1813 was used in cases where individuals profited from others' inventions. But the paragraph dealing with the "rights of the inventor" was obscure and not quite relevant to Fraunhofer's optical lenses.[4]

Patents never seemed to concern Utzschneider; there is no archival evidence of patenting the work done at his Optical Institute, even before 1825. Lenses for microscopes and telescopes did not lend themselves to patenting. Achromatic lens manufacture is an extremely sophisticated procedure that cannot be easily replicated. The percentage of ingredients needs to be tinkered with until the most efficient combination results. Stirring the molten glass proves to be a highly coveted skill. Because lead oxide, which is a very dense material, is a key ingredient in flint optical glass, complex stirring techniques are required to ensure homogeneity throughout the glass. And, the glassmaker must ensure that oxygen bubbles do not form as a result of an aggressive stirring of the mixture. After a highly regulated cooling procedure, a glass slab is formed. This slab needs to be cut and polished into the correct dimensions for the requisite telescope. Fraunhofer performed all of these diverse practices with astonishing precision.

Because of this complexity, Utzschneider felt that the chance of someone successfully copying this procedure was minimal. Indeed, to ensure the market's fidelity, he did not permit Fraunhofer to either publicize the recipe or techniques for achromatic glass production, or to demonstrate the procedure to anyone, save a few chosen apprentices of the Optical Institute. Because of the company's strict policy of disclosure, the only way one could successfully replicate these optical devices was to ferret out the skilled techniques solely from the final products. Both Utzschneider and Fraunhofer, however, were convinced that such attempts would be futile. Indeed, Fraunhofer often offered his final products free of charge to visiting experimental natural philosophers to convince them of the superior quality of his craftsmanship. Also, both patent systems existing in Bavaria and many of the German territories in the early nineteenth century required that the applicant provide detailed descriptions of the inventions as well as the process procuring them. Neither Utzschneider nor Fraunhofer wished to disclose such information. And in the case of achromatic lenses, each lens was unique, depending upon the specifications of the refracting telescope. In addition to these reasons for not patenting the fruits of his labors, the design of Fraunhofer's optical lenses was not original, as he used the typical construction of the period: a convex/concave doublet of crown and flint glass first implemented by John Dollond and Chester Moor Hall in mid-eighteenth-century England.[5] Fraunhofer's contribution to his profession lay in his skilled manipulations of optical glass production, changes in the recipes, and his calibration technique for determining the refractive and dispersive indices of glass with amazing precision and accuracy. Generally, changes in the chemical composition of lenses were not sufficient to warrant a patent.[6]

As was the case with all forms of artisanal knowledge, secrecy was paramount to Fraunhofer's manufacture of optical glass. The importance of secrecy to the success of the Optical Institute is evidenced by the contracts between Utzschneider and the Swiss watch and bell maker, Pierre Louis Guinand, who was the Optical Institute's first achromatic glassmaker. In the first contract dated May 10, 1806, the first paragraph states that

> The [optical glass] work must be done by him [Guinand] and his wife, Rosalie Bourverot, with their own hands only to ensure that the secret of their glass production . . . will be guarded in the utmost and will, under no circumstances, be told to a third party.[7]

Utzschneider reminded Guinand of his obligation in paragraph 5 of the contract by declaring that the artisan's pay was contingent upon him strictly adhering to this policy.[8] In case of death, his wife would continue to work and be compensated as long as she kept to the agreement signed by her husband.

In the second contract between Utzschneider and Guinand signed nearly a year later, secrecy is once again underscored. Utzschneider informed him that he was neither permitted to dismiss anyone, nor to permit anyone to see the glass hut without Utzschneider's consent.[9] Also, Utzschneider warned Guinand that should the Swiss craftsman decide to leave Benediktbeuern, he would not be allowed to teach anyone the method of producing optical glass.[10] After numerous quarrels with his young assistant, Fraunhofer, Guinand decided to leave Benediktbeuern and return to his native canton of Neufchatel, Switzerland. Before he and his wife departed, Guinand needed to sign an agreement with Utzschneider that neither he nor his spouse would divulge to anyone any information regarding the production of optical glass, and that neither one would work on optical glass ever again. Otherwise, Utzschneider would suspend further payments.[11]

Unfortunately, we do not know what the original agreement between Fraunhofer and Utzschneider entailed when the young apprentice was hired in 1806, but since Guinand's contracts were rather typical for the period, one would assume Fraunhofer's original contract would have emphasized the importance of secrecy as well. In a contract dated February 7, 1809, when Fraunhofer was promoted to assist Guinand in optical glass manufacture, Utzschneider made it perfectly clear that he alone was the owner of the flint and crown glass goods, all of the buildings in Benediktbeuern, and the optical glass machines, tools, and various other forms of equipment.[12] And Utzschneider also reminded Fraunhofer that visitors to Benediktbeuern were not permitted access to the optical glass hut.[13]

The important point about secrecy here is that Utzschneider felt that he, as owner of the Optical Institute, was also the owner of the practical knowledge of optical glass production at Benediktbeuern, even though he himself conducted only a few experiments in the manufacture of optical lenses, all of which were failures. His role was predominantly, and nearly exclusively, managerial. He owned the materials of production: the glass hut, the raw materials, the optical equipment, and he owned any practical, artisanal knowledge produced by those employed by him, particularly Guinand and Fraunhofer. This was clear from contractual agreements with his employees. Any breech of these contracts was punishable under Bavarian law, ranging from fines to imprisonment. But enforcement was difficult, as many glassmakers, such as Guinand, would leave Bavaria (and indeed the German territories altogether) in order to resume their careers. In the case of the Optical Institute, ownership was coextensive with authorship.

But one need not conclude from this that Utzschneider was a diabolical owner exploiting his workers. Early nineteenth-century Bavaria was certainly not early nineteenth-century Britain, and the secularized cloister of Benediktbeuern was neither a dark, satanic mill, nor a Dickensian workhouse. Fraunhofer was rewarded handsomely for his labors. He undoubtedly belonged to the group of artisans who financially benefited from their association with science and technology. The earliest record of Fraunhofer's salary dates from May 7, 1808, two years after entering the Optical Institute as an apprentice. His monthly wage was a meager 40 guilders per month.[14] Such an amount was slightly less than the average income of a glassmaker at the time. From 1809, the year of his second contract, until 1813, he received 67 guilders a month.[15] From 1814, the year of his third contract, to 1819, his monthly income was 125 guilders a month, plus profits earned, usually totaling 700 guilders a year. He also received an astonishing 10,000 guilders in 1814 as he now shared in the Optical Institute's ownership with Utzschneider, but as a junior partner.[16] Fraunhofer was now set for life. And, although the company's secrets were still the property of Utzschneider, he encouraged Fraunhofer to publish scientific articles on the theory of his work—in the optician's own name, without disclosing any information on production—so as to increase his reputation, as well as the reputation of the institute. From 1819, when the Optical Institute returned to Munich (although the optical glass manufacture remained at Benediktbeuern) until his death in 1826, Fraunhofer earned 150 guilders per month, plus approximately 700 to 800 guilders per year from profits. In 1823, after being named *Konservator* of the Royal Bavarian Academy of Sciences' Mathematical and Physical Instrument Collection, he received an additional 800 guilders per year.[17] This was a very good wage indeed, certainly permitting Fraunhofer to enjoy a lifestyle far more lavish than he experienced as a child.

In short, secrecy was necessary to ensure the Optical Institute's monopoly of the optical-glass market. Even after Fraunhofer was appointed co-owner, the veil of secrecy was never lifted. But secrecy also proved deleterious to Fraunhofer, as it thwarted his attempts to be recognized as a *Naturwissenschaftler*. As will be argued in the next section, among nineteenth-century circles of German *Naturforscher*, secretive knowledge stood starkly opposed to scientific knowledge. Savants claimed that scientific knowledge was knowledge made accessible to other savants for the common good. This knowledge, however, necessarily discriminated against any form of artisanal knowledge, which by its very nature was secretive. Guild secrets, and later trade secrets, were standard practices throughout the German territories until the late nineteenth century. But the German Republic of Letters also discriminated against the possessors of such knowledge, the skilled artisans. Most skilled craftsmen were not interested in scientific authorship. They certainly expected rewards for their inventions, but in the form of patents. Fraunhofer, who could not obtain patents for his work, sought the reward of scientific authorship. It is when the craftsmen desired the creative status of the savant that tensions arose.

II. The Academy Must Not Become a "Corporation of Artists, Factory Owners, and Artisans"

Although Utzschneider argued that Fraunhofer was a gifted *Naturforscher* who should receive all the benefits and privileges of any other savant, including scientific authorship, would an article published by an artisan that dealt with the theory of science, rather than being a mere description of his newly invented scientific instrument, be considered to proffer the same philosophical and creative results as the essays published by members of the Republic of Letters, such as Carl Friedrich Gauss or John Herschel? While it is certainly clear that well before the 1810s instrument makers were claimed to be crucial to the scientific enterprise, whether the knowledge those skilled artisans generated was creative, like the intellectual labor of experimental natural philosophers, was a point of contention.

Before discussing the attitude of several savants to Fraunhofer's work, it is necessary to see how the German Republic of Letters defined the role of authorship during the late eighteenth and early nineteenth centuries. During the late eighteenth century, a literary culture was finally blossoming in the German territories, lagging far behind either Britain or France. As a result of this rapid transformation of the literate middle class, a bifurcation resulted among German literary circles: those high-brow authors forming around Friedrich Gottlob Klopstock's *Deutsche Gelehrtenrepublik* (German Republic of Letters) of 1772, and those authors who catered to the predilections of "the masses." Not surprisingly, many of the latter low-brow authors needed to sustain their rather humble existence; hence, the move to pen simplistic tales and to *self-plagiarize*—the process of regenerating themes in order to write more numerous works—was rather common.[18] Those authors enjoying less challenging financial circumstances, such as Johann Wolfgang von Goethe, who earned his keep as the privy councilor to Duke Carl August of Weimar, frowned upon those authors who capitulated to the base whims of the public, thereby squashing the author's creativity.

The sentiment that succumbing to the predilections of the lower classes destroys creativity was a powerful one by the turn of the eighteenth into the nineteenth century in the German territories. Artisanal knowledge had historically been considered the antithesis of inspired genius. During the Renaissance, for example, being a craftsman, or a "master of a body of rules or techniques," had been deemed to be one of the two necessary components of authorship.[19] Craftsmanship had been defined in contrast to the other component of authorship—inspiration, or genius. Inspiration was seen as being creative and intellectual—as opposed to manual, a higher form of knowledge not shamefully following the rules or techniques required of the craftsman. And as time went on the role of the author as craftsman began to wane, until the late eighteenth century when, as theorists of the period exclaimed, it had been totally eclipsed by inspirational genius. Following Edward Young's claim that "imitations are often a sort of manufacture wrought by those mechanics, art and labor, out of preexistent materials not their own,"[20] the German literary intelligentsia saw the author as the transcender of rules. Indeed, between 1773 and 1794 a debate over the ownership of intellectual property flared throughout the German territories sparked by Klopstock's announcement in *Deutsche Gelehrtenrepublik* that authors should circumvent publishers and present their work directly to the public via subscriptions.[21] The critical shift in defining the author as a creative, inspired genius was accompanied by the belief that the author had rights; the message of the book was his or hers, rather than the audience's. When authors had been seen as being mere craftsmen, the book dealers and publishers had been the owners of the knowledge presented in the text. The debate culminated with a lengthy treatise by Ernst Martin Gräf, *Forschungsbericht: Versuch einer einleuchtenden Darstellung des Eigenthums des Schriftstellers und Verlegers und ihrer gegenseitigen Rechte und*

Verbindlichkeiten. Mit vier Beylagen. Nebst einem kritischen Verzeichnisse aller deutschen besonderen Schriften und in periodischen und anderen Werken stehenden Aufsätze über das Bücherwesen überhaupt und den Büchernachdruck insbesondere (*A Research Report: An Attempt toward a Classification of the Property and Property Rights of Writers and Publishers and Their Mutual Rights and Obligations. With Four Appendices. Including a Critical Inventory of All Separate Publications and Essays in Periodical and Other Works in German Which Concern Matters of the Book as Such and Especially Reprinting*). Gräf's *Forschungsbericht* sided with Klopstock's *Deutsche Gelehrtenrepublik* that authors should be the owners of their work. He wanted

> to ascertain whether it might be possible by arranging such subscriptions for scholars to become the owners of their writings. For at present they are so only in appearance; book dealers are the real proprietors, because scholars must turn their writings over to them if they want to have these writings printed. This occasion will show whether or not one might hope that the public, and the scholars among themselves, . . . will be instrumental in helping scholars achieve actual ownership of their property.[22]

As obvious as Klopstock's plea might sound, before this period throughout the German territories, a book had been seen as a collaborative enterprise, each group of artisans receiving the same amount of credit as the others. For example, in 1753, the *Allgemeine Oeconomisches Lexicon* listed all of the artisans responsible for producing a book in its entry for "Book": the writer, the paper maker, the type founder, the typesetter and the printer, the proofreaders, the publisher, and the book binder. They had all been equally deserving of credit of the manufacture, authorship, and ownership of the book's contents. Klopstock's intervention attempted to thwart the egalitarianism of the mid-eighteenth century by granting ownership exclusively to the author, who was no longer considered to be part artisan.

Klopstock's trials were to be rewarded, but not until 1810, after the Napoleonic occupation of the German territories. Baden jurists added laws covering literary property to the *Code Civile*:

> ¶ 577.da. Every written transaction is originally the property of the person who composed it, as long as he did not write it on the commission of another and for the advantage of another, in which it would be the property of the person who commissioned it.[23]

Fraunhofer's Bavaria defined the object of the author's proprietary rights in 1813 by drawing upon Johann Gottlieb Fichte's work and Article 397 of the Bavarian Penal Code:

> Anyone who publicizes a work of science or art without the permission of its creator, his heirs, or others who have obtained the rights of the creator by reproducing it in print or in some other way without having reworked it into an original form will be punished.[24]

As detailed below, these debates about authorship spilled over into debates about scientific authorship during the 1820s.

Within certain circles of Bavaria, whether artisanal knowledge counted as scientific knowledge and whether instrument makers should be considered scientific authors were topics of debate, particularly within the confines of the Royal Academy of Sciences in Munich. Fraunhofer's nomination to the rank of ordinary member of the mathematics and physics section of the academy can be used to trace the contours of the debate over the status of artisanal labor and its relationship to scientific authorship. The debate centered on the argument of whether Fraunhofer was a *Naturforscher* (investigator of nature) or a *geschickte Handwerker* (skilled artisan).

In 1820, Fraunhofer was proposed for a promotion from corresponding member, which he had been since February 15, 1817, to ordinary visiting member.[25] The recommendation stated that

> Herr Professor Fraunhofer has become famous among physicists over the past several years for his direction of the Optical Institute formally in Benediktbeuern, now in Munich. His secret of the manufacture of flint and crown glass and the production of optical glass of a size hitherto unheard of have secured an everlasting name for him in the history of science.[26]

The letter continued by praising Fraunhofer's sharp sense of observation, which had led to discoveries in the field of optics.[27]

This recommendation elicited an immediate protest from the director of machinery of the prince's Coin and Mining Office (*Maschinendirektor beim kurfürstlichen Münz- und Bergmeisteramt*), Joseph von Baader, an ordinary member of the mathematics and physics section. In his letter to the academy dated March 31, 1820, Baader complained that Fraunhofer's reputation was insufficient for an ordinary membership. He quoted the academy's constitution paragraph XIII, title 1, which stated that ordinary members may be accepted only if the world of scholars has been convinced of the merit of the potential member's published works, or if the academy has been privy to important discoveries made by the potential member in lectures.[28] Baader continued by emphasizing that Fraunhofer was not university educated, and indeed never even attended *Gymnasium* (high school). Although Fraunhofer was admittedly well versed in the *Kunstfach* (art) of practical optics as a result of his training in optical glassmaking, this knowledge was insufficient for Fraunhofer to be called a mathematician or physicist.[29] Baader's mean-spirited attack became most vitriolic when he warned that the academy must not become a "corporation of artists (*Künstler*), factory owners (*Fabrikanten*), and artisans (*Handwerker*)."[30]

Baader's diatribe, however, did not stop there. He proceeded to attack Fraunhofer's article "On the Determination of the Refractive and Dispersive Indices of Differing Types of Glass" by asserting that although the article was very interesting and useful for artisans working on the perfection of optical instruments, it lacked any form of a scientific discovery (*wissenschaftliche Entdeckung*). To Baader, an artisan, by his very nature, could not make any theoretical, philosophical discoveries relevant to science, the product of creative genius. He even questioned whether Fraunhofer had written it himself. Since his essay was published in the Academy's *Denkschriften*, Fraunhofer had fulfilled one of the necessary (and sufficient) conditions for appointment as a regular member of the mathematics and physics section. His essay offered a detailed account of how one could observe the dark lines of the solar spectrum, and how one could use those lines as a calibration technique for the precise determination of the refractive and dispersive indices of various types of optical glass samples. But, he never divulged how he procured his superior glass samples, which he would have needed to do if applying for a patent under Bavarian law. In order to disqualify Fraunhofer's candidacy, Baader publicly questioned whether Fraunhofer actually authored the piece himself, strongly suggesting that the true author was his employer, Utzschneider. Clearly, questions of authorship were deeply rooted in issues of social class. Baader concluded his letter by arguing that the secretive nature of glassmaking is "not of a scientific nature, but of an artistic, artisanal one."[31] Fraunhofer's private and entrepreneurial knowledge was, in Baader's eyes, the antithesis of science. For Baader, being a scientific author and *Naturwissenschaftler* were coextensive; therefore, social, historical, and epistemological issues were inextricably intertwined.

Baader's protest was joined by an attack on Fraunhofer's character by Julius Konrad Ritter von Yelin, Munich's chief financial advisor (*Oberfinanzrat*). Ritter von Yelin was a physicist and chemist. His assault was brief, but just as harsh as Baader's. Yelin echoed Baader's concern that Fraunhofer was self-educated. Such a lack of formal education, Yelin argued, would result in an inability to follow the complex lectures that periodically took place in the mathematics and physics section of the academy. Yelin's anger was most evident in his concluding remark: he found it personally insulting that Fraunhofer would join the same section, at the same rank, as he.

It should be noted that Yelin was not opposed to the application of technology in the service of the state. Quite the contrary, he extolled the progress made by Bavarians in science and technology and how such progress had strengthened Bavaria. Indeed, Yelin played a critical role in Bavaria's *polytechnischer Verein*, which had the express goal of applying scientific and technological advances to the fledgling Bavarian economy. What Yelin objected to was that those responsible for such progress should necessarily be considered *Naturforscher*.[32]

Baader's and Yelin's resistance sparked Fraunhofer's supporters in the academy to take a concerted action. They argued that there had been historical precedents whereby men without a university education had become ordinary members of the academy.[33] Indeed, the section's secretary and botanist Franz von Paula Schrank composed a memorandum listing members of other academies who did not possess a university education. Johann Georg von Soldner, Bavarian Court astronomer and major contributor to the theory and practice of geodesy, defended his friend's work. He simply could not agree with Baader's claim that Fraunhofer's work did not contain a single scientific discovery. The dark lines of the solar spectrum, for Soldner, was an example of such a discovery.[34] At the actual vote, he continued his plea: "Through these lines *exact measurements* of the solar spectrum are now possible, and the possibility of exact measurements and their implementation is the goal of what one considers to be *exact science*. I consider this discovery of Fraunhofer's to be the most important one in the area of light and colors since Newton."[35] Hence, the academy should not bar Fraunhofer based on the grounds that his work did not belong to a recognized scientific canon. By placing Fraunhofer's name in the same sentence as Newton's, Soldner was indeed considering Fraunhofer to be a great physicist and mathematician.

In the end, it was decided that appointing Fraunhofer to an ordinary membership was too controversial. On June 27, 1821, then, he was promoted from corresponding member to an extraordinary visiting member of the academy after a vote on the previous May 30 of 19 to 1 in his favor.[36] With an extraordinary visiting membership, Fraunhofer was at least permitted to attend sectional meetings. In that same year, Fraunhofer's most theoretical piece hitherto was accepted for publication in the Academy's journal, *Denkschriften der königlichen Akademie*, entitled: "Neue Modification des Lichtes durch gegenseitige Einwirkung und Beugung der Strahlen" ("New Modification of Light through Reciprocal Effects and Diffraction of the Rays"). Soldner took the lead in proposing the paper to the *Denkschriften*. He argued that it marked "a new epoch in the physical theory of light."[37] Other academy members agreed, including Reichenbach; the physicist, Benedictine monk, and later professor of the University of Munich, Thaddäus Siber; Schrank; and even Baader. This article was not simply a description of a scientific instrument, such as Fraunhofer's amazing diffraction gratings. Rather, it sought scientific status as it offered a compelling account of the undulatory theory of light proposed by Thomas Young and Augustin Fresnel. Nothing in the paper was based on secret knowledge. Yelin protested again, arguing that the work of someone with such little formal education should not be included in such a prestigious journal.[38] This time, his protest was to no avail. In 1823 Fraunhofer was appointed *Konservator* and professor of the academy's collection of mathematical and physical instruments with a stipend compensating him for lectures periodically delivered at the academy until his death in 1826.[39] He was also elected member of the *Gesellschaft für Naturwissenschaften und Heilkunde* of Heidelberg in 1825 and was even knighted by King Maximilian I in 1824.[40] It is clear, then, that the issue of whether the skilled labor of artisans such as Fraunhofer was sufficiently elevated and creative to be granted the status of scientific authorship was hotly contested during the third decade of the nineteenth century in Bavaria.

III. Conclusion

In short, during the late eighteenth and early nineteenth centuries, the German Republic of Letters was defining the role of author and actively seeking a change in the archaic laws. They successfully lobbied for the linkage of the author to ownership by severing the previous link between the market (book dealers), authorship, and ownership. The commercial interests of the market place thwarted their role as authors. This was one of their objections to Fraunhofer's work, as it too arose out of market interests. And because they were keen to distinguish themselves from the craftsmen-like elements of the profession, they were not interested in including artisans such as scientific instrument makers in their "republic." Craft secrecy challenged their commitment to the openness of scientific knowledge. Finally, echoing the concern of Young, German intellectuals questioned the creativity of artisans, arguing that they merely manipulated preexisting materials rather than creating something truly novel. Allowing artisans the status of scientific authorship would mean returning to mid-eighteenth-century obscurity and irrelevance.

This story has raised an important question concerning the status of artisanal knowledge vis à vis scientific authorship. The craftsman seeking scientific credibility faced a serious dilemma: because artisans worked in guilds, their knowledge was necessarily shrouded in secrecy and connected to a commercial network. And, their labor was seen as being uncreative, unlike the savant. Members of the Republic of

Letters, on the other hand, could take individual credit for their intellectual labors by becoming scientific authors. Craftsmen affiliated with the scientific enterprise were often denied such status. Questions of authorship during the late eighteenth and early nineteenth centuries were part of a larger attempt by experimental natural philosophers to distinguish clearly between intellectual and mechanical labor and skill. During this period of intense mechanization, what counted as skill, creativity, or authorship was being redefined. And the politics of labor sheds light on how certain individuals created and maintained these categories.

Notes

1. See Myles W. Jackson, "Illuminating the Opacity of Achromatic Lens Manufacture: Joseph von Fraunhofer and the Architecture of his Monastic Laboratory," in *Architecture in Science*, eds. Galison and Thompson (Cambridge, Mass. and London: MIT Press, 1999). See also Myles W. Jackson, *Spectrum of Belief: Joseph von Fraunhofer and the Craft of Precision of Optics* (Cambridge, Mass., and London: MIT Press, 2000); and Myles W. Jackson, "Buying the Dark Lines of the Solar Spectrum: Joseph Fraunhofer and His Optical Institute," vol. 1 of *Archimedes*, ed. Jed Z. Buchwald, 1996, 1–22.
2. Such logic also prevailed in seventeenth-century England. See Steven Shapin, *A Social History of Truth. Civility and Science in Seventeenth-Century England* (Chicago and London: University of Chicago Press, 1994), 355–407.
3. R. Klostermann, *Die Patentgesetzgebung aller Länder nebst den Gesetzen über Meisterschutz und Waarenbezeichnungen systematisch und vergleichend* (Berlin: I. Guttengerb, 1869), 217.
4. Ludwig Gieseke, *Die geschichtliche Entwicklung des deutschen Urheberrechts* (Göttingen: Otto Schwarz & Co., 1957), 122.
5. Interestingly, there was a priority dispute over the patenting of optical lenses between Dollond and Hall during the 1730s.
6. Klostermann, *Die Patentgesetzgebung*, 40.
7. As reprinted in Adolf Seitz, *Joseph Fraunhofer und sein Optisches Institut* (Berlin, 1926), 15.
8. Ibid., 17.
9. Ibid., 21.
10. Ibid., 23.
11. Ibid., 50.
12. Fraunhofer Nachlaß, Archiv 6079, Deutsches Museum (Munich), 1.
13. See Myles W. Jackson, "Artisanal Knowledge and Experimental Natural Philosophers: The British Response to Joseph Fraunhofer and the Bavarian Usurpation of Their Optical Empire," *Studies in History and*

Philosophy of Science, 25 (1994): 549–75, particularly 567–72; and Jackson, *Spectrum of Beliefs*, 74–77.

14. Staatsbibliothek Preußischer Kulturbesitz Berlin (hereafter cited as SPKB), Utzschneiders Nachlaß, Box 1, 7 May 1808.
15. Joseph von Utzschneider, *Kurzer Umriß der Lebens-Geschichte des Herrn Dr. Joseph von Fraunhofer, königlich bayerischen Professors und Akademikers, Ritters des königlichen bayerischen Civil-Verdienst, und des königlich dänischen Dannebrog-Ordens, Mitgliedes mehrerer gelehrten Gesellschaften, etc.* (Munich: Rösl'schen Schriften, 1826), 19.
16. Ibid.; and SPKB, Utzschneiders Nachlaß, Box 1, 21 June 1817. Twice a month from 1809 to 1814, Utzschneider sent Fraunhofer 500 guilders for expenses incurred during the glass-making procedure. It should be noted that 10,000 guilders were 40 to 50 times the average monthly salary of a worker at that time. See Deutsches Musuem, Archiv 7323; and Hans-Peter Sang, *Joseph von Fraunhofer* (1987), 73.
17. Von Rohr, *Leben, Leistung, und Wirken* (1929), 196; and Seitz, 91. As a comparison, his assistants were generally earning between 18 and 24 guilders per month in 1820. SPKB, Fraunhofer Nachlaß, Box 5, "Lohne der Arbeiter."
18. We are told that Karl Philipp Moritz (1756–1793) was a shameless recycler of his ideas. This portion of my paper owes much to the outstanding work of Martha Woodmansee, *The Author, Art, and the Market: Rereading the History of Aesthetics* (New York: Columbia University Press, 1994), 29–53.
19. Ibid., 36.
20. Edward Young, *Conjectures on Original Composition in a Letter to the Author of Sir Charles Grandison*, in *English Critical Essays: Sixteenth, Seventeenth, and Eighteenth Centuries*, ed. Edmund D. Jones (London: Oxford University Press, 1975), 274.
21. Woodmansee, *The Author*, 48.
22. As quoted in Woodmansee, 48, and Helmut Pape, "Klopstocks Autorenhonorare und Selbstverlagsgewinne," *Archiv für Geschichte des Buchwesens*, 10 (1970): cols 1–268, here columns 103–4. See also Philipp Erasmus Reich, *Zufällige Gedanken eines Buchhändlers über Herrn Klopstocks Anzeige einer gelehrten Republik* (Leipzig, 1773).
23. As quoted in Woodmansee, *The Author*, 53. Original: Ch. F. M. Eisenlohr, ed., *Sammlung der Gesetze und internationalen Verträge zum Schutze des literarischen-artistischen Eigenthums in Deutschland, Frankreich und England* (Heidelberg: Bangel and Schmitt, 1856), 11.
24. As quoted in Woodmansee, *The Author*, 53. Original: Ludwig Gieseke, *Die geschichtliche Entwicklung des deutschen Urheberrechts* (Göttingen: Otto Schwartz, 1957), 122.
25. From the time he worked on the dark lines of the solar spectrum in 1813 and 1814, Fraunhofer wanted to be regarded as a *Naturwissenschaftler* for personal reasons. This is quite clear from his private correspondence with Utzschneider.
26. Bayerische Akademie der Wissenschaften, Akten der königlichen

Akademie der Wissenschaften, Personal Akten: Herr Joseph Fraunhofer, 1821–26 (hereafter cited as BAW:JF), 1.

27. Ibid.
28. Ibid., 8–9. See also Sang, *Joseph von Fraunhofer*, 96–97.
29. BAW:JF, 8.
30. Ibid. It should be noted that the German word *Künstler* meant something like the French word *artiste*, someone well versed in the mechanical (as well as fine) arts.
31. Ibid.
32. von Yelin, "Rede gehalten" (1818), 601–8.
33. BAW:JF, 16–17.
34. Ibid., Akt V, 12, no. 25, 4 April 1820.
35. Ibid. Emphasis in the original.
36. This might sound confusing to the modern reader. In German states, ordinary memberships and ordinary professorships were higher in stature and salary than extraordinary members and extraordinary professorships; ibid., 27; 28.
37. Ibid., 28.
38. Ibid., 26.
39. DMA, Fraunhofer Nachlaß, N14/27, 5417.
40. See SPKB, Haus II, Fraunhofer Kasten 3, letter no. 7; and DMA, Fraunhofer Nachlass, 14/9, 5415; and 14/7, 5420 a/c.

Bibliography

Eisenlohr, Ch. F. M., ed., *Sammlung der Gesetze und internationalen Verträge zum Schutze des literarischen-artistischen Eigenthums in Deutschland, Frankreich und England.* Heidelberg: Bangel and Schmitt, 1856.

Gieseke, Ludwig. *Die geschichtliche Entwicklung des deutschen Urheberrechts.* Göttingen: Otto Schwarz & Co., 1957.

Jackson, Myles W. "Artisanal Knowledge and Experimental Natural Philosophers: The British Response to Joseph Fraunhofer and the Bavarian Usurpation of Their Optical Empire." *Studies in History and Philosophy of Science.* Vol. 25. 1994, 549–75.

———. "Buying the Dark Lines of the Solar Spectrum: Joseph Fraunhofer and His Optical Institute." In *Archimedes*, ed. Jed Z. Buchwald, Vol. 1. 1996, 1–22.

———. "Illuminating the Opacity of Achromatic Lens Manufacture: Joseph von Fraunhofer and the Architecture of His Monastic Laboratory." In *Architecture in Science*, edited by Peter Galison and Emily Thompson. Cambridge, Mass., and London: MIT Press, 1999.

———. *Spectrum of Belief: Joseph von Fraunhofer and the Craft of Precision Optics.* Cambridge, Mass., and London: MIT Press, 2000.

Klostermann, R. *Die Patentgesetzgebung aller Länder nebst den Gesetzen über Meisterschutz und Waarenbezeichnungen systematisch und vergleichend.* Berlin: I. Guttengerb, 1869.

Pape, Helmut. "Klopstocks Autorenhonorare und Selbstverlagsgewinne." *Archiv für Geschichte des Buchwesens* 10 (1970): cols 1–268.

Reich, Philipp Erasmus. *Zufällige Gedanken eines Buchhändlers über Herrn Klopstocks Anzeige einer gelehrten Republik.* Leipzig, 1773.

Rohr, Moritz von. *Joseph Fraunhofers Leben, Leistungen und Wirksamkeit.* Leipzig: Akademische Verlagsgesellschaft, 1929.

Sang, Hans-Peter. *Joseph von Fraunhofer: Forscher, Erfinder, Unternehmer.* Munich: Dr. Peter Glas, 1987.

Seitz, Adolf. *Joseph Fraunhofer und sein Optisches Institut.* Berlin: Julius Springer Verlag, 1926.

Shapin, Steven. *A Social History of Truth: Civility and Science in Seventeenth-Century England.* Chicago and London: University of Chicago Press, 1994.

Utzschneider, Joseph von. *Kurzer Umriß der Lebens-Geschichte des Herrn Dr. Joseph von Fraunhofer, königlich bayerischen Professors und Akademikers, Ritters des königlichen bayerischen Civil-Verdienst, und des königlich dänischen Dannebrog-Ordens, Mitgliedes mehrerer gelehrten Gesellschaften, etc.* Munich: Rösl'schen Schriften, 1826.

Woodmansee, Martha. *The Author, Art, and the Market: Rereading the History of Aesthetics.* New York: Columbia University Press, 1994.

Yelin, Ritter Julius von. "Rede gehalten in der Versammlung des Verwaltungs-Ausschusses des polytechnischen Vereins in Bayern zu München am 26. August 1818 zur Feyer des Stiftungsfestes, der neuen Beamtenwohl und der Niederlegung seiner Stelle als diesjährigen Vorstands des Vereins." *Kunst- und Gewerb-Blatt des polytechnischen Vereins im König-Reiche Bayern* 4 (1818): 601–8.

Young, Edward. *Conjectures on Original Composition in a Letter to the Author of Sir Charles Grandison.* In *English Critical Essays: Sixteenth, Seventeenth and Eighteenth Centuries*, edited by Edmund D. Jones. London: Oxford University Press, 1975.

6.

"A Very Hard Nut to Crack"

or Making Sense of Maxwell's *Treatise on Electricity and Magnetism* in Mid-Victorian Cambridge

ANDREW WARWICK

> The self that speaks in the preface to a treatise on mathematics . . . is identical neither in its position nor in its functioning to the self that speaks in the course of a demonstration. . . . In the first case, the "I" refers to an individual without an equivalent who, in a determined place and time, completed a certain task; in the second, the "I" indicates an instance and a level of demonstration which any individual could perform provided that he accepted the same system of symbols, play of axioms, and set of previous demonstrations.[1]
>
> —Michel Foucault

> The difficulties experienced by a good mathematician in Maxwell's treatise arise more from what it omits than from what it contains. The difficulty lies in following Maxwell's train of thought, and in seeing what exactly it is he is trying to prove.[2]
>
> —Charles Chree

I. A Treatise for the Over-Educated?

In his seminal essay "What Is an Author?" Michel Foucault at one point draws an interesting distinction between the authorial self that speaks in the preface to a mathematical treatise, and that which speaks in the technical demonstrations in the text that follows. Foucault's purpose in making this distinction is to persuade the reader that even in what might be regarded as one of the driest and most formal genres of writing, the voice in which the author speaks can change from page to page. The voice in the preface, Foucault argues, lays claim to the whole work by speaking in the first person singular and from a given location in space and time; in a mathematical demonstration, by contrast, the "I" of the author and the "I" of the reader are conflated in a realm beyond time and space, defined only through a shared technical expertise. Foucault goes on to identify yet another voice in the treatise, this one speaking of the work's "meaning, the obstacles encountered, the results obtained, and the remaining problems," but the general point he wishes to establish is that in acknowledging the existence of more than one authorial self in a work, we undermine the notion that any one of these selves "refer[s] purely and simply to a real individual."[3] On Foucault's showing, a published work is a compilation of voices speaking in different modes with different purposes, neither the sum total nor any subset of which necessarily constitutes a consistent and unified whole. Foucault concludes that it is the notion of a transcendental author that lends unity to such polyphonic texts.[4] The text becomes a token of the author's thought, which, in turn, assumes the role of an explanatory principle—referred to by Foucault as the "author function"—guaranteeing the existence of a unique and consistent meaning in the work.

Foucault's brief comments on the narrative style of mathematical treatises are intriguing partly because it is unusual for literary analysts to evoke books of this kind alongside works in theology, history, and literature, but also because they provoke some interesting questions concerning the relationship between the authors, texts, and readers of highly technical writings. Consider, for example, James Clerk Maxwell's *Treatise on Electricity and Magnetism* (hereafter *Treatise*), a book filled with discussion of electrical apparatus, advanced electromagnetic theory, and higher mathematics. Part of the narrative structure of this book neatly exemplifies the first two voices alluded to above. In the preface, dated February 11, 1873, Maxwell describes the origins, purpose, and plan of the work, referring to himself as "I" no less than thirty-seven times in just eight pages of text. In the almost one thousand pages of technical exposition that follow, the personal pronoun "I" gives way almost entirely to the passive voice and to the inclusive "we." Thus in the preface Maxwell asserts that it is *his* accomplishment to have mathematized Michael Faraday's electromagnetic theory and to have shown its equivalence to more traditional theories: "When *I* translated what *I* considered to be Faraday's ideas into mathematical form, *I* found that in general the results of the two methods coincided." In the main body of the text, however, the demonstration of technical proofs and theorems becomes the joint accomplishment of author and reader: "*We* shall afterwards prove that if *we* have obtained a value of ψ which satisfies these conditions, it is the only value of ψ which satisfies them."[5]

Maxwell's narrative is subject to precisely the shifting modalities noted by Foucault, but these shifts signal far more than the introduction of new voices; they also announce profound changes in the implied relationship between the reader and the text. This point can be illustrated by consideration of the reading experiences produced respectively by the *Treatise*'s preface and the first substantial chapter. Almost any literate person in mid-Victorian Britain would not only have found the preface comprehensible but would have constructed an acceptable meaning coextensively with the real-time experience of reading. It is unlikely that many of Maxwell's contemporaries spent long hours poring over the preface wondering whether they had fully grasped the import of this or that statement—they would generally have read and understood it as they would a novel or a column in a newspaper. But if this same reader strayed just a few paragraphs into what Maxwell termed the "Preliminary" chapter, the reading experience would have changed dramatically. Suddenly the reader is confronted by specialized vocabularies and turns of phrase from electrical engineering, electrical theory, metrology, and higher mathematics. Nor is it simply terminology that bars general access to the body of the text. The narrative assumes that the reader knows how to build apparatus, make measurements, understand physical theory, and solve differential equations. In this sense it points far beyond the knowledge and experience of a literate reader to skills mastered through protracted training in the laboratory, the classroom, and the study. As a disgruntled bookseller grumbled to Oliver Lodge in the mid-1870s, Maxwell's book was "a product of the over-educated" and, as this remark suggests, readers who found Maxwell's work at all accessible beyond the preface formed exclusive groups.[6]

These observations suggest that in analyzing highly technical works it might be useful to acknowledge the presence of authorial selves beyond those which Foucault identifies. Foucault locates voices according to their narrative style and to the kinds of knowledge to which they lay claim, but these are in the main the attributes of literary convention. The forms of address adopted by Maxwell, for example, are fairly typical of technical expositions in mid-Victorian Britain and can be traced both forward into the twentieth century and back until at least the early modern period. What is more unusual about the authorial selves presented in the *Treatise* is the way in which he speaks sometimes as a physical theorist, sometimes as a mathematician, and sometimes as an electrical engineer. These selves refer not only to the kinds of knowledge of which Maxwell speaks, but, implicitly, to his own biography and to the skills a reader has to possess in order to make some sense of various sections of the work. As we shall see, Maxwell the theoretician, the mathematician, and the experimenter can in large measure be understood historically as, respectively, Maxwell the Scottish natural philosopher, the Cambridge graduate, and member of the British Association Committee on Electrical Standards. It was from these traditions that Maxwell's selves were made, and those who hoped to find something akin to his overall understanding of the *Treatise* needed to possess at least a comparable range of skills.

Yet although specialized training was a necessary condition of those who sought access to more technical sections of the *Treatise*, it was by no means a sufficient one. In the first epigraph at the beginning of the chapter, Foucault characterizes the authorial voice in the body of a technical work as one that narrates demonstrations that "any individual could perform provided that he accepted the same system of symbols, play of axioms, and set of previous demonstrations." The important point to notice about this claim is that it refers to the author's voice rather than to the experience of the reader. Maxwell's adoption of the inclusive "we" certainly implies that the reader ought immediately to share the author's sense of conviction at each demonstration, but no such reader actually existed. The second epigraph, written by a Cambridge graduate who struggled with Maxwell's text in the early 1880s, informs us that even a "good mathematician" often found it difficult not only to see how a particular line of argument led legitimately to the conclusion claimed by Maxwell, but why such a conclusion was of any special significance.

Chree's comments capture both the strengths and limitations of Foucault's analysis. On the one hand they illustrate an important aspect of the author-function in operation: Chree depicts the narrative in the *Treatise* as a transcription of "Maxwell's train of thought" and refers his difficulties in making sense of the text to what he assumes Maxwell has either concealed or else failed to clearly convey to the reader. As Foucault suggests, belief in the existence of a profound and consistent meaning in the work is preserved by referring that meaning to the thought of the author. On the other hand, Chree's remarks indicate that the inclusive tone of the voice that narrates technical demonstrations needs to be treated with caution. The expression "we conclude" is certainly an incitement to the reader to find in himself the conviction professed by the author, yet it cannot compel the reader's assent and offers no additional resources to those who remain unconvinced by the demonstration given.

In this sense the meaning of a technical demonstration is often not imminent in the way it is in many other narrative forms. To understand a difficult demonstration a reader must expect to draw heavily upon one or more kinds of technical expertise, to puzzle over and to experiment with difficult or opaque steps in the argument, to return to difficult passages having mastered subsequent demonstrations, to consult works referred to by the author, and to consult other readers (perhaps including the author himself). Given that these processes can take days, months, or even years, it is clear that the production of meaning around one or more technical demonstrations can be a highly protracted and collective activity. Put another way, the production of the ideal reader, who can not only work through the technical demonstrations without hesitation but also explain their meaning to others, is by no means a straightforward process. Moreover, the preservation of the greater meaning of a work will typically require readers, and even the author himself, to eventually modify or discredit some demonstrations as misprinted, incomplete, or erroneous. The editors (Maxwell included) of the second and third editions of the *Treatise* made numerous decisions of this kind, most of which emerged from the struggles of Cambridge mathematicians to understand the book and to teach its contents to undergraduates.[7]

In this essay I use the writing of Maxwell's *Treatise* and its reception in mid-Victorian Cambridge as a means of exploring the interaction between a scientific author, his technical text, and one geographically-bound faction of his readership. My analysis will address all of the issues raised above under the rubric of one or other of the following two general themes. The first concerns the process by which ideal readers of Maxwell's *Treatise* were gradually produced in Cambridge in the 1870s and early 1880s. When the book first appeared in 1873, Cambridge mathematicians found it extremely hard going. Very few of them were used to dealing with novel physical theory, and virtually none had sufficient knowledge of experimental electricity to tackle Maxwell's discussion of electrical apparatus and metrology. Even the sections that ought to have been most accessible to Cambridge men—those on the application of higher mathematics to the solution of problems in electrostatics—often proved impenetrable because Maxwell had solved problems on a case-by-case basis without fully explaining the methods he had employed. During the 1870s, different teachers working at respectively different sites in the university began slowly to master specific sections of the text. Historical analysis of the process by which this mastery was accomplished will enable us to cast considerable light on the resources necessary to bring consistent meaning to a highly technical work.

The second general theme addressed below concerns the nature of the interpretation of the *Treatise* that emerged in Cambridge in the 1870s. I emphasized above that the construction of meaning around a technical work depends upon the relationship between the skills and assumptions written into the text by the author and those brought to it by the reader. In Berlin, for example, Hermann von Helmholtz, who came from a distinctly German tradition in electromagnetic theory, attributed a rather different meaning to Maxwell's theory from that divined by Maxwell's Cambridge readers.[8] An interesting peculiarity of the Cambridge interpretation is that it was in many respects very similar to the one Maxwell himself would probably have ascribed to his book. This might not at first sight appear especially surprising since, as I have already noted, Maxwell himself was teaching electromagnetic theory at Cambridge in the 1870s. But, as we shall see, Maxwell's lectures were actually of very little direct significance to the Cambridge interpretation of the most mathematical and theoretical aspects of his work. In order to explain how some Cambridge students were able, by the late 1870s, to master virtually the whole of the *Treatise* in Maxwell's style, we must explore the similarities between the skills written into the text by Maxwell and the pedagogical resources available in mid-Victorian Cambridge.

II. The Making of the Treatise in Pedagogical Context

The publication of the *Treatise* in 1873 marked the culmination of almost twenty years of work by Maxwell on theoretical and experimental electricity. Born in Edinburgh in 1831, his youthful interests in geometry, natural philosophy, and experimental science were nurtured through the 1840s during his time as a schoolboy at the Edinburgh Academy and undergraduate at Edinburgh University. Visits to meetings of the Royal Society of Edinburgh and William Nicol's optical laboratory helped to shape the young Maxwell's approach to science, as did attendance at J. D. Forbes undergraduate lectures on natural philosophy.[9] Maxwell's precocious ability in geometry and mathematical physics took him next to Cambridge, where he was placed second in the Mathematical Tripos of 1854. After graduation, Maxwell devoted much of his research effort to recasting Michael Faraday's novel conception of electric and magnetic lines of force in the form of a new, mathematical, field theory of electromagnetism, based on a dynamical ether and the principle of the conservation of energy.[10] In contrast to earlier action-at-a-distance theories of electromagnetism, Maxwell's theory attributed all electromagnetic effects to changing dynamical states in an all-pervading ether. He also showed that the equations he had developed to describe electromagnetic phenomena predicted that electric and magnetic forces were transmitted through the ether at the speed of light and, famously, that light itself was a transverse electromagnetic wave in the ether. The change in physicists' conception of physical reality eventually wrought by Maxwell's writings on electromagnetism was later claimed by Albert Einstein to constitute "the most profound and the most fruitful that physics has experienced since Newton."[11]

During the 1860s, Maxwell also served as a member of the British

Association Committee on electrical standards, an experience that familiarized him with the instruments and techniques of experimental electricity, the definition and measurement of fundamental electrical units, and the theory and operation of the electric telegraph.[12] But, despite his participation in these important projects, by the late 1860s Maxwell's highly mathematical and conceptually difficult papers on electromagnetic theory had generated little informed response even from the handful of professional mathematical physicists capable of following his work.[13] The *Treatise*, written between 1868 and 1872, was intended to address this problem by providing an advanced textbook on the mathematical theory and experimental practice of electromagnetism, couched in terms of Maxwell's field-theoretic approach. By the time the book was published in March 1873, Maxwell had also been appointed to a new professorship of experimental physics in Cambridge, a job that expressly required him both to offer public lectures on electromagnetic theory to students of the Mathematical Tripos and to direct experimental research at the new Cavendish Laboratory to be opened in 1874.[14]

The establishment of the new chair and laboratory reflected the changing status of physics in undergraduate studies in mid-Victorian Cambridge. A number of important physical subjects, including electricity and magnetism, had been excluded from the Mathematical Tripos in 1848 on the grounds that their theoretical foundations were neither secure nor exemplary of the more fundamental science of dynamics. From the mid-1860s, however, a number of reformers, including Maxwell, argued that these progressive and commercially important subjects ought to be studied at Britain's leading mathematical center. Maxwell's new theory strengthened this argument by showing that the theoretical science of electromagnetism could be built upon dynamical foundations. The Cambridge Board of Mathematical Studies decided accordingly in 1867 that electricity and magnetism would be reintroduced to undergraduate studies from the early 1870s.

This decision raised two pedagogical difficulties for the university. First, the most able undergraduate mathematicians in Cambridge were taught not by the university professors or college lecturers, but by private tutors, or "coaches." However, these coaches, the most famous of whom was Edward Routh, had little or no expertise in electromagnetism, and, since there were no good textbooks on the subject, had no way of acquiring such expertise. Second, electricity and magnetism were profoundly experimental subjects, yet the university had no facilities for offering instruction in experimental physics. It was partly in anticipation of these difficulties that Maxwell decided to write the *Treatise*, and the university decided to build a physical laboratory and to appoint a new professor of experimental physics. The professor's job would be to teach experimental physics at the laboratory and to hold public lectures on the mathematical theories of thermodynamics, electricity, and magnetism. A group of fellows drawn from some of the larger colleges also hoped to reform college teaching, so that the most able among their number could attract students from the whole university. Colleges had traditionally restricted their lectures to their own students, but, from the early 1870s, a number of intercollegiate courses were offered, which were open to students of all colleges for a fee. As we shall see, it was through a course of lectures of this kind at Trinity College that novel aspects of Maxwell's *Treatise* were first taught to undergraduates.

By the mid-1870s, then, Cambridge University would appear to have been well placed not only to teach the new physical subjects but to build a research school on the original field theory of electromagnetism propounded in the *Treatise* by the new professor of experimental physics. But in practice, students and teachers in the university found it extremely difficult to make sense of the work as a whole. Maxwell wrote the *Treatise* for a broad audience, and, to this end, he tried as far as was possible to keep advanced mathematical methods, novel physical theory, and electrical apparatus in respectively separate chapters. This separation meant that, say, an electrical engineer could read about a piece of electrical apparatus without running into higher mathematics, while an undergraduate mathematician could study the application of harmonic analysis to problems in electrostatics without encountering unfamiliar physical theory or electrical instruments. For Maxwell himself, who had spent much of his professional life mastering mathematical, theoretical, and experimental electricity, this division would have seemed artificial. He almost certainly hoped that his book would help to usher in a new era in university physics in which undergraduates would feel equally at home with higher mathematics, advanced physical theory, and electrical instruments. But in Cambridge in the early 1870s there were few if any teachers, apart from Maxwell himself, who possessed the skills required to deal effectively with such a broad range of material. Far from being reunited in undergraduate training, the three strands of electrical studies mentioned above were torn further apart as the *Treatise*'s contents were fragmented among the three major sites at which physics was taught—the coaching room, the intercollegiate classroom, and the Cavendish Laboratory. Teachers at these sites offered respectively different accounts of the book's contents, each emphasizing those aspects that suited their technical competencies and pedagogical responsibilities. The coherent field of theoretical and experimental study envisaged by Maxwell was thus fragmented through pedagogical expediency into three separate projects in, respectively, applied mathematics, novel physical theory, and electrical metrology.

A related problem to be overcome before the *Treatise* could be used effectively as a textbook on mathematical electrical theory concerned the difficulties inherent in translating what might be termed Maxwell's "personal knowledge" into one or more collectively comprehensible disciplines that could be taught to undergraduates and young mathematicians in a matter of months. Much of Maxwell's presentation in the *Treatise* relied upon the physical intuition and case-by-case problem solutions he had developed through long years of experience. Yet Maxwell did not engage in the kind of private tutoring or small-class teaching through which he might have nurtured similar physical and technical sensibilities in a new generation of undergraduates. It was the coaches, intercollegiate lecturers, Tripos examiners, and, to some extent, the students themselves, who were left in the mid-1870s to provide a consistent and comprehensible account of the theory contained in Maxwell's book. In the following sections I discuss the production of this account, paying special attention to the different readings of the *Treatise* provided at different teaching sites. I shall argue that the collective understanding of electromagnetic field theory that emerged in Cambridge circa 1880 relied mainly upon two resources: first, the mathematical expertise generated by the top coaches; and, second, communal discussion of the meaning of Maxwell's work in intercollegiate lectures held at Trinity College by W. D. Niven.

III. "A Very Hard Nut to Crack"

One of the points I want to emphasize strongly is that making sense of a theoretically novel and technically difficult book such as Maxwell's *Treatise* was a collective enterprise and one that relied heavily on the whole pedagogical economy of undergraduate mathematical training. The years spent by Maxwell mastering and refashioning electromagnetic theory had led him to take for granted both the purpose and inherent difficulties of his project as well as the idiosyncratic techniques by which he solved specific problems. As we saw in the introduction, Charles Chree claimed that it was the absence not only of those steps in an argument that seemed too self-evident to Maxwell to be worthy of inclusion, but sometimes of a clear statement of the point of the argument itself, that made the *Treatise* so difficult for other mathematicians to follow. The technical expertise of Cambridge mathematicians was generated and preserved through a pedagogical economy based on small-class teaching, long hours of carefully supervised study, and the use of locally written textbooks and past examination questions. When it came to an original book like the *Treatise*, the only person to whom a baffled reader could turn for enlightenment was the author himself. But inquiries of this kind were as likely to end in a two-way discussion over the meaning of the text as in a simple statement of clarification by Maxwell.

When the *Treatise* was published in March 1873, the only lectures on electromagnetic theory being held in Cambridge were those offered by Maxwell himself. It is clear, however, that these lectures were singularly ineffective in preparing undergraduates to master even the more straightforward sections of the book. One such student, Edward Nanson, attended the lectures in the academic year 1872–73, but he quickly ran into difficulties when he began to work through the *Treatise* in the autumn of 1873. Evidently frustrated by his inability to follow the argument in several sections, this outstanding young mathematician wrote to Maxwell, drawing his attention to several errors of mathematical reasoning and to a number of passages where the meaning was opaque:

> Pardon me for remarking that the whole of Art[icle] 165 is very hard to make out; and I speak from knowledge of the experience of others in reading the same passage—this is the first case of successive images in the book and the principle of the method is not sufficiently explained. I have asked several men who have read the passage and not one of them could tell me *how* or why the images would determine the electrification.

Nanson went on to draw attention to incorrect definitions, the use of inconsistent nomenclature, significant mathematical and typographical errors, inaccurate or inappropriate citations, and a lack of consistency in what a mathematically literate British reader might reasonably be expected to know. In conclusion, Nanson begged Maxwell's pardon if he had "made any mistakes," and added, in mitigation, that the *Treatise* was "considered by all I know to be a very hard nut to crack."[15] These comments confirm that even the most able mathematicians had great difficulty following the application of advanced mathematical methods to new and unfamiliar physical problems, even when no novel physical theory was involved.

A letter to Maxwell by George Chrystal reveals that the introduction of new and difficult physical concepts provided further problems for the student. Chrystal made a careful study of the *Treatise* during the summer of 1874, some six months before he sat his Tripos examination. His private tutor, Edward Routh, would almost certainly by this point have introduced the theory of electrical potentials covered in the first volume, but when Chrystal strayed into the less familiar territory of the second volume he quickly ran into difficulties. He was troubled to find that an expression derived by Maxwell (for the force acting on an inductively magnetized body in a magnetic field) did not agree with the expression given elsewhere by William Thomson following a similar derivation. Unable to resolve the problem himself, Chrystal wrote to Maxwell asking for clarification. In his reply, Maxwell acknowledged that it was he, rather than Thomson, who had given the wrong expression for the force acting on the body, but he insisted that the error was no "mere slip." It sprang rather from the conceptual difficulties inherent in calculating the components of the total magnetic energy, though Maxwell did concede that such errors had taught him 'not to be miserly in using symbols."[16]

These examples illustrate the physical and mathematical difficulties experienced by young mathematicians reading Maxwell's book, but they also highlight the interactive relationship that quickly emerged between the *Treatise* and its Cambridge readers. Many of these enquiries prompted Maxwell to amend, rearrange, simplify, and, in some cases, entirely rewrite sections of the first nine chapters while preparing the second edition of the *Treatise* for publication. When the new edition appeared in 1882, most of the errors pointed out by Nanson had been corrected, and Maxwell's detailed response to Chrystal had become the basis of a whole new appendix explaining how the energy of induced magnetic fields was correctly calculated.[17] The *Treatise* was becoming, in effect, a collective product whose contents were based partly on Maxwell's original design but increasingly on the interactive struggles of other Cambridge mathematicians to understand and clarify his work.

IV. Electricity in the Coaching Room

> There is a substantial substratum of truth in the remark once made to the writer that it would have been an immense improvement to Maxwell's "Electricity" to have been written by Routh.[18]
>
> —Charles Chree

The site at which the vast majority of Cambridge undergraduates were formally introduced to technical aspects of the *Treatise* during the 1870s and 1880s was the coaching room. As soon as the *Treatise* was published, the leading coach in the university, Edward Routh, set to work to master the sections he thought most relevant to undergraduate studies and to incorporate them in his coaching regime. Routh's thorough courses already covered most of the important mathematical methods used in the *Treatise*, and within months of reading the book he had written a new course in which examples from electrostatics, magnetostatics, and current electricity were used as illustrations of these methods. At the beginning of February 1874, one of Routh's most brilliant pupils, Joseph

Ward, noted in his diary that he was "reading Electricity with Routh this term (using Maxwell's book)" and that Routh had recommended that he and his peers attend Maxwell's first course of lectures on experimental electricity at the Cavendish, as it would "be as well to see some experiments on the subject."[19] These remarks confirm the speed with which Routh incorporated the new subjects into his third-year coaching schedule and highlight his awareness that examination questions in these subjects were likely to refer to instruments and practical devices with which he could not familiarize his pupils.

Routh's teaching of electricity and magnetism was almost certainly the most thorough given by any coach in Cambridge. It is nevertheless a very striking aspect of his teaching that he introduced these new subjects, at least implicitly, in the form of an action-at-a-distance theory of electricity. This is nicely illustrated by the fact that Routh began his teaching of electromagnetic theory by establishing the inverse square law of attraction and repulsion for electrostatic force, an exercise calculated to drive home the formal mathematical analogy between gravitational and electrostatic phenomena.[20] Routh in fact made no reference at all in his lecture notes to the field-theoretic approach adopted by Maxwell in the latter sections of the *Treatise*, nor did he discuss the electromagnetic theory of light. The directions that he gave his pupils concerning which sections of the *Treatise* were to be read and which were to be ignored almost certainly reinforced this same conservative interpretation of electromagnetic theory on the book itself.

The complex relationship that existed in the mid-1870s between coaching, the contents of the *Treatise*, Maxwell's lectures at the Cavendish Laboratory, and the questions being set by Tripos examiners is succinctly illustrated by some reminiscences of the Tripos of 1876. The examiner in mathematical physics for that year, Lord Rayleigh, thought he could make attendance at Maxwell's lectures pay in the examination by setting a question on the use of Wheatstone's Bridge, an instrument widely used by telegraphic engineers and in physical laboratories to measure electrical resistance. The first part of the question required the student to explain both the principle of the instrument and the way it should be used to locate the position of a fault in the insulation of a cable submerged in a tank of water (both ends of the cable being accessible).[21] The content of this question is particularly interesting because the required technique was neither discussed in any of the books likely to have been read by Cambridge undergraduates nor taught even by Routh. The likely explanation is that Rayleigh had deliberately set a question that could only be answered by students who had witnessed Maxwell's demonstrations at the Cavendish. Maxwell certainly covered Wheatstone's Bridge and the electric telegraph in his lectures, and almost certainly had the apparatus to demonstrate exactly the problem Rayleigh had set.[22]

The deep impression made by Rayleigh's question on one of the candidates, Richard Glazebrook, illustrates both the student's general lack of familiarity with common electrical apparatus and the coach's role in shaping a student's reading of the *Treatise*. Glazebrook, who subsequently worked with Maxwell at the Cavendish, was one of the few undergraduates tutored by Thomas Dale, a minor coach at Trinity, who, unlike Routh, had neither required his students to read the relevant chapter of the *Treatise* nor recommended that they attend Maxwell's Cavendish lectures. On two occasions when Glazebrook was invited to reminisce about Cambridge physics in the 1870s, he recalled Rayleigh's question and his own irritation at not being able to attempt a mathematically simple problem. Half a century after his examination he lamented:

> I had read carefully much of Maxwell's "Treatise on Electricity and Magnetism," published in 1873, but, alas, had paid no attention to chapter xi on "The Measurement of Electrical Resistance." My coach had originally marked it with a large O—omit. It is true he had corrected this later, but I had no idea of what was meant by Wheatstone bridge. My annoyance was great when, on my first visit to the Cavendish, Maxwell himself explained it and I realised the simplicity of the question I had passed over.[23]

Glazebrook's comments reveal both his lasting irritation at having dropped marks for a mathematically simple question and the way a coach would dictate which sections of Maxwell's book a student should read or ignore. In another account of the same events, Glazebrook alluded to the difficulties experienced by teachers and students alike in coming to terms with the new physical subjects and, on this occasion, made an even more explicit reference to Dale's role in shaping his understanding of Maxwell's work.[24] Glazebrook recalled how the chapters of the first volume recommended by Dale were the most pedestrian and that he had been told to omit those that later seemed the most important. The second volume of the *Treatise* was almost totally ignored by Glazebrook, and it was this volume that contained Maxwell's dynamical theory of the electromagnetic field, his equations of electromagnetism, and the electromagnetic theory of light. Dale's selective reading of the *Treatise* was probably typical of that offered by other mathematics coaches.

Rayleigh's questions of 1876 represented the last attempt by an examiner to make success in the mathematical Tripos dependent upon attending lectures at the Cavendish Laboratory. Within a year or two it had become clear that mathematics students could grasp the principles and applications of instruments such as Wheatstone's Bridge without actually witnessing the instrument in action. As coaches like Routh mastered the basic technical content of the first three quarters of the *Treatise* and learned to solve the kind of questions that examiners were likely to set, electricity and magnetism were effectively incorporated within the standard repertoire of undergraduate mathematical studies. Indeed, Routh in particular became a powerful gatekeeper to the mathematical methods necessary to make sense of Maxwell's book. This point was neatly illustrated in 1883 when Michael Pupin, a mathematics student from Columbia College in New York, arrived in Cambridge in the hope of meeting Maxwell and learning his new electromagnetic theory. W. D. Niven examined Pupin in mathematics and then offered him the following advice:

> Niven pointed out that a prospective physicist who wished to master some day Maxwell's new electrical theory must first master a good part of the mathematical work prescribed for students preparing for the Cambridge mathematical tripos examinations. "Doctor Routh could fix you up in a quicker time than anybody," said Niven with a smile, and then he added cautiously, "that is, if Routh consents to your joining his private classes, and if you can keep up the pace of the youngsters who are under his training."

After two terms of intense training with Routh, Pupin not only found himself able to "handle the mathematics of Maxwell's theory of electricity with considerable ease," but was able to follow the professorial lectures in electromagnetic theory and optics given by Rayleigh and George Stokes, respectively.[25]

V. Teaching the *Treatise* at Trinity College

The site at which the novel, field-theoretic aspects of Maxwell's work were first taught and discussed in Cambridge was William Davidson Niven's intercollegiate class in electricity and magnetism at Trinity College.[26] Having taken a high place in the Mathematical Tripos of 1867, Niven had left Cambridge to teach mathematical physics in London. He kept in touch with Cambridge mathematics during his absence by acting regularly as a Tripos examiner, and, in 1875, he returned to his old college as a mathematics lecturer.[27] At Cambridge he struck up a close friendship with Maxwell and became heir to his scientific writings following the latter's untimely death in 1879.[28] Niven subsequently completed the revisions for the second edition of the *Treatise* and edited the two posthumously published volumes of Maxwell's collected scientific papers.[29]

Niven's lectures were the first in Cambridge to treat the second volume of the *Treatise* as an advanced textbook on electromagnetic theory, and they soon made Trinity College by far the most important site in the university for the discussion of mathematically difficult and physically obscure aspects of Maxwell's work.[30] One of Niven's most distinguished students, Joseph Larmor, recalled not only that Niven's "class-room was a focus for [Maxwell's] theory, in which congregated practically all the mathematicians of the University," but also that Niven's influence as a teacher of the theory was so pervasive that "in later years he could count *nearly all* the active developers of electrical science on the mathematical side as his friends and former pupils."[31] As Larmor also noted, the great strength of Niven's teaching was not simply that he offered a systematic treatment of the most technically demanding sections of the book, but that he drew attention to, and attempted to clarify, the novel physical hypotheses upon which the book was based at a time "when they were largely misunderstood or not understood at all elsewhere."[32]

An equally distinguished contemporary of Larmor's, J. J. Thomson, who attended the lectures during the academic year 1877 to 1878, corroborates Larmor's assessment of Niven's teaching and provides further insight into his style. According to Thomson, Niven was "not a fluent lecturer nor was his meaning always clear," but he was a great enthusiast for Maxwell's "views" and managed to "impart his enthusiasm to the class." Thomson and his peers recognized Niven's ability to inspire enthusiasm for Maxwell's work as his most important quality as a teacher, and one which, when combined with his apparently enigmatic attempts to explicate Maxwell's theory, was extremely productive.[33]

Niven's lectures helped Cambridge students to master Maxwell's text in three ways. First, he convinced Cambridge mathematicians of the importance of Maxwell's new theory and inspired them to investigate it through the pages of the *Treatise*. His weekly lectures sent his students back again and again to search for meaning in Maxwell's enigmatic equations and prose. Second, Niven's classroom provided a meeting place where those engaged in this enterprise could share their insights and difficulties. This collective activity enabled Cambridge mathematicians to pool their skills in puzzling out opaque passages and difficult derivations, and, most importantly, to discriminate with confidence among problems that could be solved by adopting a particular interpretation, those due both to Maxwell's errors and to the fact that some aspects of the theory had simply been left unfinished. Finally, Niven's actual exegesis of the last quarter of the *Treatise* impressed a particular reading of the book upon his class. Unlike Routh, Niven discussed the fact that, according to Maxwell's theory, electromagnetic effects were due to the flow of electric and magnetic energy in the space surrounding charged bodies, and that Maxwell's equations produced quite new solutions to such seemingly straightforward problems as the currents generated during the discharge of a conductor.

We cannot know for certain exactly what kinds of issues were discussed in Niven's lectures, but we can make some reasonable inferences using the notes he appended to the second edition of the *Treatise*. Niven explained in his preface to this edition that he had felt it appropriate to make "the insertion here and there of a step in the mathematical reasoning" and to add a "few foot-notes on parts of the subject which [his] own experience or that of pupils attending [his] classes shewed to require further elucidation."[34] These steps and notes provide a useful guide to the sections with which Niven and his students had difficulties and reveal two interesting characteristics of Niven's reading of the *Treatise*. The first concerns the specific corrections that Niven made to the text. When Maxwell died in 1879, he had revised only the first nine of the twenty-three chapters in the book. Niven's editing of the remaining chapters was relatively light, but he did correct numerous errors and amend various derivations along the lines recommended by those of his Cambridge peers who had struggled with the book.[35] The second characteristic revealed by Niven's notes is that his technical interest in the *Treatise* focused mainly on the advanced mathematical methods employed by Maxwell to find the potential distribution surrounding charged conductors, and on the application of the general equations of the electromagnetic field to the solution of problems in current electricity.[36] These were the topics on which Niven offered points of clarification and alternative derivations, and it is reasonable to infer that it was on these issues that advanced discussion in his classroom was focused. Conversely, Niven had no points of explanation, clarification, or alternative derivations to add either to the sections on the physical and dynamical foundations of the theory or to those on the electromagnetic theory of light. Those who studied the *Treatise* in Niven's classroom would therefore have gained the impression that Maxwell's theory was primarily a collection of equations and mathematical techniques that could be applied to produce novel solutions to standard electrical problems.

Learning electromagnetism with Routh and Niven and mastering difficult Tripos problems based on the contents of the *Treatise* prepared the outstanding mathematics graduates of the late 1870s to make their own contribution to the subject after graduation. Only two of the students who studied experimental physics with Maxwell at the Cavendish Laboratory, R. T. Glazebrook and J. H. Poynting, subsequently published important papers in theoretical electricity, but their work owed virtually nothing to their personal relationship with Maxwell. Niven, by contrast, trained virtually the whole of the generation of Cambridge Maxwellians who took Maxwell's work forward in the late-Victorian period.[37] Thanks to his lectures, men like Richard Glazebrook, J. H. Poynting, J. J. Thomson, and Joseph Larmor had an excellent working knowledge of the most mathematical and theoretical sections of the *Treatise*. They felt confident that they understood not only the strengths of Maxwell's new theory, but also its present limitations, areas where it might be developed further, and aspects of Maxwell's exposition that were unclear, incomplete, inconsistent, or erroneous. Those students, such as Glazebrook and Thomson, who also studied experimental electricity at the Cavendish Laboratory, were truly ideal readers, equipped to master not only most theoretical and mathematical parts of the *Treatise* but also the sections on instrumentation and metrology.

VI. Conclusion: Training the Ideal Reader

> I have endeavoured to add something in explanation of the argument in those passages in which I have found from my experience as a teacher that nearly all students find considerable difficulties; to have added an explanation of all passages in which I have known students find difficulties would have required more volumes than were at my disposal.[38]
>
> —J. J. Thomson

The *Treatise* is remembered by physicists today almost entirely as the book that contained Maxwell's mature and most widely read account both of his new electromagnetic field equations and of the electromagnetic theory of light. Yet these topics, important as they are, constitute just four of some fifty-six chapters in the book. In Cambridge in the 1870s and 1880s, by contrast, the *Treatise* was studied as a whole and placed electrical science at the heart of mathematical and experimental physics in the university. The thorough assimilation of electrical science in Cambridge during the 1870s derived partly from the interpretive work done by Maxwell's Cambridge readers and partly from the rich pedagogical resources through which different aspects of the book were taught. It also relied upon the intimate relationship that existed between the wide range of skills written into the *Treatise* by its author and those possessed by its Cambridge teachers. As I suggested in the introduction to this chapter, three authorial voices, or selves, are clearly discernible in Maxwell's electrical writings, each of which can be traced to specific periods and experiences in his life. What enabled much of Maxwell's project in electromagnetism to be reconstructed so effectively in Cambridge in the 1870s was the distributed presence of very similar selves among coaches, intercollegiate lecturers, and, in time, demonstrators at the Cavendish Laboratory. Consider first the man who bore the greatest responsibility for teaching undergraduates the mathematical methods that underpinned the most technical sections of Maxwell's book, Edward Routh. Maxwell and Routh had been undergraduate contemporaries at Cambridge in the early 1850s, and were trained together by William Hopkins, the first of the great mathematical coaches. It was Routh who beat Maxwell into second place in the Mathematical Tripos of 1854, the two brilliant young mathematicians being indistinguishable at the top of the Smith's Prize examination a few weeks later.[39] Routh was ideally suited to pass on the mathematical tradition in which he and Maxwell had been trained, and to fathom the clever techniques by which Maxwell had solved those problems in electrostatics, magnetostatics, and current electricity that permeated the first three quarters of the *Treatise*.

What Routh would have found far less penetrable, and, indeed, far less interesting in Maxwell's book, was the discussion of electrical instruments, metrology, and novel physical theory. One of the things that sharply distinguished Maxwell from most of his Cambridge contemporaries was his broad interest in natural philosophy (including physical theory beyond mechanics and dynamics) and experimental science. This he owed to his early training in Scotland and to his ongoing relationship with other Cambridge-trained, Scottish natural philosophers, such as William Thomson and P. G. Tait.[40] Among Cambridge graduates it was the Niven brothers who had first explored novel aspects of Maxwell's new electromagnetic theory, and W. D. Niven who taught the theory to Cambridge undergraduates. The Nivens too were Scotsmen who were initially trained as undergraduates in Scottish natural philosophy at King's College, Aberdeen. Moreover, the undergraduate career of W. D. Niven (and possibly that of his brother) would have overlapped with Maxwell's brief tenure as professor of natural philosophy at Marischal College in the same town in the late 1850s.[41] At Cambridge, both W. D. Niven and Charles Niven were coached in mathematics by Routh, and were examined by Maxwell in the Triposes of 1866 and 1867, respectively. W. D. Niven later studied experimental physics with Maxwell at the Cavendish Laboratory, and became his close acquaintance and literary executor. Niven was thus better placed than any other Cambridge teacher—apart from Maxwell himself—to try to interpret and to teach the most novel aspects of Maxwell's new field theory.

Finally, there was Maxwell's own teaching at the Cavendish Laboratory. The other of Maxwell's selves that would have distinguished him from *all* Cambridge teachers in the early 1870s was that formed by his experience as a member of the British Association Committee on Electrical Standards in the 1860s. It was as a leading member of this committee that Maxwell had the opportunity to build and to undertake large electrical experiments, to work with the electrical apparatus commonly employed by electrical engineers, and to investigate the theory and practice of electrical metrology.[42] In 1871 the British Association agreed that all of the apparatus acquired by the committee would be deposited on permanent loan at the Cavendish Laboratory. When the Cavendish opened in the spring of 1874, this collection not only formed an important part of the apparatus available in the Laboratory, but was central to Maxwell's teaching and research throughout the 1870s.[43] Recall, for example, Richard Glazebrook's first visit in 1876, when Maxwell initiated him and his peers in experimental physics by explaining the theory and operation of Wheatstone's Bridge. Many young Cambridge mathematicians who worked in the Cavendish in the 1870s first learned to use instruments and make measurements using electrical apparatus, and several subsequently worked with Maxwell in his ongoing program on electrical standards. In the 1880s, Glazebrook and Willam Shaw, now demonstrators at the Cavendish, developed and implemented regimes of practical training in experimental physics at the laboratory that drew directly upon the metrological doctrine expounded by Maxwell in the *Treatise*.[44]

The work done by Maxwell's Cambridge readers in making the *Treatise* fit to teach in the university was also written into subsequent editions by the Cambridge editors. As we have seen, when Maxwell died in 1879 he had revised less than a quarter of the book, much of which he had rewritten in the light of comments by the likes of Nanson and Chrystal. With the new editors came new voices in the text. The editing of the second edition was completed by W. D. Niven, who added an appendix based on Chrystal's exchange with Maxwell, a section on self-induction taken from Maxwell's last course of lectures at the Cavendish Laboratory, and numerous footnotes clarifying difficult proofs over which he and his students had struggled.[45] Niven also alluded to the collective nature of his editorial work by acknowledging the help he had received from his brother, Charles, and his young protégé at Trinity College, J. J. Thomson.[46]

By the time Niven left Cambridge in 1882, Thomson had established himself as the university's leading authority on Maxwell's electromagnetic theory: Thomson replaced Niven as a lecturer at Trinity College in 1882; was appointed to one of five new university lectureships (with special responsibility for electromagnetic theory) in June 1884; and was elected at the end of that year to succeed Rayleigh as professor of experimental physics and director of the Cavendish Laboratory.[47] The rapid development of electromagnetic field theory through the 1880s, and, especially, the production of electromagnetic waves by Heinrich Hertz in 1888, prompted the Clarendon Press to commission a third edition of Maxwell's increasingly famous work. When Niven declined the task due to the "pressure of his official duties," Thomson willingly stepped into the breach.[48] Thomson's initial plan was to give an account of recent advances in electrical science in a series of new footnotes, but he quickly realized that the number of notes required would "disfigure the book."[49] He therefore opted to confine these advances to a supplementary volume and to preserve Maxwell's *Treatise* as the "source from which [students] learn[ed] the great principles of the science."[50] For Thomson, the *Treatise* was no longer a book at the cutting edge of electrical theory and practice, but a textbook from which students could begin to master the new approach to electromagnetism pioneered by Maxwell.

In order to assist the learning process, Thomson not only added a large number of explanatory footnotes based on his experience as a reader and teacher of the *Treatise* (see epigraph on p. 152), but attempted to verify the results that had been quoted in the book without mathematical proof. He had to confess that in the last of these enterprises he and his Cambridge colleagues had "not in all instances succeeded in arriving at the result given by Maxwell."[51] After eighteen years of communal exploration in Britain's leading center of mathematical and experimental physics, some results in the *Treatise* remained inexplicable even to Maxwell's most ardent Cambridge disciples. Thomson's prefatory voice further modulated the author-function by manipulating the reader's expectations with respect to such anomalies. The profundity of the author's original insights was buttressed by the observation that "all" electrical research over the previous fifteen years had "tended to confirm in the most remarkable way the views advanced by Maxwell," while the results that Thomson had been unable to reproduce were marginalized as irrelevant.[52] Niven and Thomson's emendations and footnotes smoothed and accelerated the reader's journey through the text, facilitating agreement with Maxwell's conclusions whenever possible and flagging points where agreement should not be sought.

This is not to say that these notes made the *Treatise* easily accessible to any reader. Rather, their existence points to the broader pedagogical economy in Cambridge within which teachers like Routh, Niven, Thomson, Rayleigh, and Glazebrook passed on their own understanding of the work. The holistic grasp of the book propagated by their combined efforts was almost certainly peculiar to Cambridge in the 1880s. At no other academic site in Britain was the *Treatise* made anywhere near as central to undergraduate studies, and in continental Europe the field equations and electromagnetic theory of light were slowly and partially incorporated in quite different traditions. What is especially interesting about the use of the *Treatise* as a textbook in Cambridge in the 1870s and 1880s is the extent to which it reveals the amount of labor required not only to build consistent meaning around the work as a whole, but also to pass on that meaning to a new generation of students. The nearest thing the *Treatise* ever had to an ideal readership (as envisioned by Maxwell) was that composed of those Cambridge students who successively coached with Routh, attended Niven's lectures at Trinity College, and worked with Maxwell or, later, with Rayleigh at the Cavendish Laboratory. The production of the ideal reading experience thus depended not just on the enthusiasm and ability of the reader, but on the knowledge and pedagogical expertise of highly skilled guides in applied mathematics, physical theory, and experimental physics.

Notes

1. Foucault (1984), 112.
2. Chree (1908), 537.
3. Foucault (1984), 112–13.
4. Ibid., 111.
5. *Treatise*, vol. 1, ix, 133. All references are to the third edition of the *Treatise* unless otherwise stated.
6. Quoted in Hunt (1991), 26.
7. See J. J. Thomson's footnotes in *Treatise*, vol. 2, 85, 258, 391.
8. On Helmholtz's interpretation of Maxwell's theory, see Buchwald (1985), chap. 21; and (1994), chaps. 2 and 3.
9. On Maxwell's years in Edinburgh, see Harman (1990), 2–7; and Smith (1998), 212–18.
10. For an overview of Maxwell's electromagnetic field theory, see Smith (1998), 218–32.
11. Torrance (1982), 31.
12. On Maxwell's work on electrical standards, see Schaffer (1995).
13. J. J. Thomson noted that, even in the early 1870s, Maxwell's scientific work "was known to very few" and his reputation was "based mainly on his work on the kinetic theory of gases." Thomson (1936), 100–101.
14. Maxwell's work standards, work, and professorial appointment in Cambridge are discussed in Sviedrys (1976) and Schaffer (1992).
15. Nanson to Maxwell, 5 December 1873 (JCM-CUL).
16. Chrystal to Maxwell, 7 July 1874; Maxwell to Chrystal, 22 July 1874 (JCM-CUL).
17. *Treatise*, vol. 2, 284.
18. Chree (1908), 537.
19. (JTW-SJC), 4 February 1874.
20. (EJR-PC), Vol. 4, 377.
21. Rayleigh, *Scientific Papers*, vol. 1, question ix, 282.
22. On apparatus at the Cavendish Laboratory, see Maxwell, *Scientific Letters*, 868–75.
23. Glazebrook (1926), 53. The story is first referred to in Glazebrook (1896), 77.
24. Glazebrook (1931), 132–33.
25. Pupin (1923), 174, 197.
26. Thomson (1936), 42; Rayleigh (1942), 8. Thomson recalled that Niven's lectures were "on mathematical physics, mainly on Maxwell's treatise on *Electricity and Magnetism* which had then lately been published."
27. Niven was a Tripos examiner in 1871, 1873, and 1874.
28. Thomson et al. (1910), 331; Glazebrook (1896), 78.
29. Larmor (1917), xxxix; Maxwell (1890).
30. Thomson et al. (1931), 21.
31. Larmor (1917), xxxix. My italics.
32. Ibid.

33. Thomson (1936), 42–43.
34. Maxwell (1873), 2d ed. (1881), xiii.
35. George Chrystal noted that only those "perfectly familiar" with the *Treatise* would appreciate the "conscientious labour" expended by Niven in preparing the second edition. Chrystal (1882), 239.
36. See, for example, Niven's notes to Articles 160, 200, 603, 659, 701, and 705.
37. Niven's role in forming a research school in electromagnetic theory in Cambridge circa 1880 is discussed in Warwick (forthcoming).
38. From Thomson's preface to the third edition of the *Treatise*, xvi.
39. The Smith's Prize examination was held shortly after the examinations for the Mathematical Tripos and offered the most able mathematicians a final mathematical contest.
40. On the differences between science education in Cambridge and Scottish universities at this time, see Wilson (1985); and Smith (1998), chap. 1.
41. Harman (1990), 18–20; Larmor (1917), xxxviii; Macdonald (1923), xxvii.
42. The reports published by the committee are collected in Jenkin (1873).
43. Campbell and Garnett (1882), 353.
44. Schaffer (1992), 38.
45. The section on self-induction was based on notes taken by J. A. Fleming at Maxwell's lectures in spring 1979. *Treatise*, vol. 2, Article 755; Fleming (1934), 66.
46. Thomson discussed his close relationship with Niven in Thomson (1936), 42–43.
47. Rayleigh (1942), 14; *Cambridge University Reporter*, 3 June 1884, 791.
48. *Treatise*, vol. 1, xv.
49. Ibid, xvi.
50. Thomson (1893).
51. *Treatise*, xvi. Thomson acknowledged the help of Joseph Larmor, Charles Chree, L. R. Wilberforce, and G. T. Walker, all of whom had graduated from the Mathematical Tripos in the 1880s.
52. Ibid, xv.

Bibliography

Buchwald, J. Z. *The Creation of Scientific Effects: Heinrich Hertz and Electric Waves*. Chicago: Chicago University Press, 1994.

———. *From Maxwell to Microphysics: Aspects of Electromagnetic Theory in the Last Quarter of the Nineteenth Century*. Chicago: Chicago University Press, 1985.

Campbell, L., and W. Garnett. *The Life of James Clerk Maxwell*. London: Macmillan and Co., 1882.

Chree, C. "Mathematical Aspects of Electricity and Magnetism." *Nature* 78 (1908): 537–38.

Chrystal, G. "Clerk Maxwell's 'Electricity and Magnetism.'" *Nature* (12 January 1882): 237–40.

Fleming, J. A. *Memoirs of a Scientific Life*. London and Edinburgh: Marshall, Morgan & Scott, 1934.

———. Collection. (MS Add 122/34), University College, London (JAF-UCL).

Forsyth, A. R. "Old Tripos Days at Cambridge." *Mathematical Gazette* 19 (1935): 162–79.

Foucault, M. "What Is an Author?" in *The Foucault Reader*, ed. P. Rabinow, 101–20. New York: Viking Press, 1984.

Glaisher, J. W. L. Ed. *Solutions of the Cambridge Senate-House Problems and Riders for the Year 1878*. London: Macmillan and Co., 1879.

Glazebrook, R. T. "The Cavendish Laboratory: 1876–1900," *Nature* 118, supplement (18 December 1926): 52-?.

———. "Early Days at the Cavendish Laboratory." In James Clerk Maxwell: *A Commemoration Volume 1831–1931*, by J. J. Thomson et al., 130–41. Cambridge University Press, 1931.

———. *James Clerk Maxwell and Modern Physics*. London: Cassell and Company, Ltd., 1896.

———. "Sir Horace Lamb 1849–1934." *Obituary Notices of Fellows of the Royal Society* 1 (1932/35): 376–92.

Greenhill, A. G. *Solutions of the Cambridge Senate-House Problems and Riders for the Year 1875*. London: Macmillan and Co., 1876.

Harman, P. M. Introduction to Maxwell, vol. 2, 1–37, 1990.

Harman, P. M., ed. *Wranglers and Physicists: Studies on Cambridge Physics in the Nineteenth Century*. Manchester: Manchester University Press, 1985.

Hunt, B. *The Maxwellians*. Ithaca and London: Cornell University Press, 1991.

Jenkin, F., ed. *Reports of the Committee on Electrical Standards*. London: E. & F. N. Spon, 1873.

Lamb, H. "Clerk Maxwell as Lecturer." In *James Clark Maxwell: A Commemorative Volume 1831–1931*, by J. J. Thomson et al., 142–46. Cambridge: Cambridge University Press, 1931.

———. "On Electrical Motions in a Spherical Conductor." *Philosophical Transaction of the Royal Society* 174 (1883): 519–49.

Larmor, J. "Sir William Davidson Niven." *Proceedings of the London Mathematical Society* 16 (1917): xxviii-xliii.

Macdonald, H. M. "Charles Niven, 1845–1923." *Proceedings of the Royal Society* 104 (1923): xxvii-xxviii.

Maxwell, J. C. *An Elementary Treatise on Electricity*. Edited by William Garnett. Oxford: Clarendon Press, 1881.

———. *The Scientific Letters and Papers of James Clerk Maxwell*. Edited by P. M. Harman, 3 vols. Cambridge: Cambridge University Press, 1990–.

———. *Scientific Papers of James Clerk Maxwell*. Edited by W. D. Niven. 2 vols. Cambridge: Cambridge University Press, 1890.

———. *A Treatise on Electricity and Magnetism*. 2d ed. 1881; 3d ed. 1891. 2 vols. Oxford: Clarendon Press, 1873.

———. Collection. (MSS Add 7655), Cambridge University Library (JCM-CUL).

Newall, H. F. "1885–1894." In *History of the Cavendish Library 1871–1910*, by J. J. Thomson et al., 102–58. London: Longman, Green and Co., 1910.
Niven, C. "On the Induction of Electric Currents in Infinite Plates and Spherical Shells." *Philosophical Transactions of the Royal Society* 172 (1881): 307–53.
Niven, W. D. "On the Theory of Electric Images, and Its Application to the Case of Two Charged Spherical Conductors." *Proceedings of the London Mathematical Society*, 8 (1876/77): 64–83.
Pupin, M. *From Immigrant to Inventor*. New York: Charles Scribner's Sons, 1923.
Rabinow, P. *The Foucault Reader*. New York: Viking Penguin, 1984.
Rayleigh, Lord. *The Life of Sir J. J. Thomson*. Cambridge: Cambridge University Press, 1942.
———. *Scientific Papers*. 6 vols. Cambridge: Cambridge University Press, 1899/20.
Rouse Ball, W. W. "Edward John Routh." *The Cambridge Review*, 13 June 1907, 480–81.
Routh, Edward. Papers. Peterhouse College, Cambridge (EJR-PC).
Schaffer, S. J. "Accurate Measurement Is an English Science." In *The Values of Precision*, edited by M. N. Wise, chap. 6. Princeton: Princeton University Press, 1995.
———. "A Manufactory of Ohms: Late Victorian Metrology and Its Instrumentation." In *Invisible Connexions*, edited by S. Cozzens and R. Bud, 23–56. Bellingham: SPIE, 1992.
Schuster, A. "The Clerk Maxwell Period." In *History of the Cavendish Laboratory 1871–1910*, edited by J. J. Thomson et al., 14–39. London: Longmans, Green and Co., 1910.
Smith, C. *The Science of Energy: A Cultural History of Energy Physics in Victorian Britain*. London: The Athlone Press, 1998.
Sviedrys, R. "The Rise of Physics Laboratories in Britain." *Historical Studies in Physical Sciences* 7 (1976): 405–36.
Thomson, J. J. *Notes on Recent Researches in Electricity and Magnetism*. Oxford: Clarendon Press, 1893.
———. *Recollections and Reflections*. London: G. Bell & Sons Ltd., 1936.
Thomson, J. J., et al. *History of the Cavendish Laboratory 1871–1910*. London: Longmans, Green and Co., 1910.
———. *James Clerk Maxwell: A Commemoration Volume 1831–1931*. Cambridge: Cambridge University Press, 1931.
Torrance, T. F., ed. *A Dynamical Theory of the Electromagnetic Field*. Edinburgh: Scottish Academic Press, 1982.
Turner, H. H. "Edward John Routh." *Monthly Notices of the Royal Astronomical Society* 68 (1907/8): 239–41.
Ward, Joseph Timmis. Diary (W2). St. John's College Library, Cambridge (JTW-SJC).
Warwick, A. C. *Masters of Theory: A Pedagogical History of Mathematical Physics in Cambridge from the Enlightenment to World War I*. Cambridge: Cambridge University Press, forthcoming.

Whittaker, E. T. "Andrew Russell Forsyth 1858–1942." *Obituary Notices of Fellows of the Royal Society* 4 (1942/44): 209–27.

Wilson, D.B. "The Educational Matrix: Physics Education at Early-Victorian, Cambridge, Edinburgh and Glasgow Universities." In *Wranglers and Physicists: Studies on Cambridge Physics in the Nineteenth Century*, edited by P. M. Harman, 12–48. Manchester: Manchester University Press, 1985.

———. "Experimentalists among the mathematicians: Physics in the Cambridge Natural Sciences Tripos, 1851–1900." *Historical Studies in Physical Sciences* 12 (1982): 325–71.

Wise, M. N., ed. *The Values of Precision*. Princeton: Princeton University Press, 1995.

PART II
Limits of Authorship

7.

Emergent Relations

MARILYN STRATHERN

Anthropologists often find themselves gravitating toward debate and litigation as telling moments in cultural life. For what may be as interesting as the positions being defended are the cultural resources people bring to their aid—analogies and tropes to make the persuasive point, new properties forced onto old concepts. Debate and litigation offer present-day materials for my own exposition.[1] However, I also have a question about emergent properties and new claims from the early modern English-speaking world. The question is what made the English at that time endow the words *relation* and *relative* with the property of kinship—kinship by blood and marriage, that is. At the least, I hope to show why it might be of interest to ask. The reasons for that begin and end in the present, and I sandwich the historical issue between recent ones. This tracking back and forth will mimic the way in which kinship appears and disappears as a cultural resource for thinking about relations.

I

Multiple Origins

Janet Dolgin (2000), anthropologist, lawyer, and observer of the family as it fares in U.S. litigation, notes how lawmakers concern themselves simultaneously with enduring values *and* with changing conventions. The families constructed by the law may either be traditional and solidary communities or are understood as modern collections of autonomous individuals choosing relationships through negotiation. These can be conjoined, as in the opinion of a lawyer who claimed that his clients, advertising an offer of $50,000 for female gametes chosen for certain characteristics, would always love any child born to them. The ingredients for creating a child may be obtained in the marketplace, then, although once a baby is part of the family, traditional moral views prevail. A new location for individual choice is also a location for expressing enduring values of family solidarity.

Dolgin writes about determinations of parenthood where gamete donation and surrogacy have led to dispute. Although it is possible to track a path through lawsuits, which shows the value Americans put on genetic ties (Dolgin 1990), it is equally possible to show relationships taken as paramount. U.S. courts have been known to refuse evidence about biological paternity, as in the case of one man who tried to sever ties from his son when he discovered that he was not the biological father, and was brought back to the relationship that already existed: if "a parent-child bond" had been formed, then "a relationship still exists at law" (2000, 531).

So what creates a relation? Recent legal decisions have given weight to certain prenatal determinations focused on birth yet abstracted from the birth process. Here claims are based on a mental condition: the parents' intention. Such a case was brought in 1998 to the California Court of Appeal.[2] A child had been born from an embryo created from anonymous donors and gestated for a fee by a surrogate; the original couple were now divorced and the woman sought parental status, arguing that she and her former husband were the legal parents. Despite there being six potential parents (the couple, the surrogate and her husband, and two gamete donors), the trial court concluded that in law the child had no parentage. The appellate court upturned this; intention is sufficient cause when a "child is procreated because a medical procedure was initiated and consented to by intended parents" (72 Cal. Rptr. 2d at 282).[3]

Recognition of intent is consonant with an ideology of family that prizes autonomous choice "in terms once reserved for life in the marketplace" (Dolgin 2000, 542). At the same time, establishing legal parentage would set up obligations, including the child support that the father was trying to avoid. Regardless of the ways in which parenthood may be created, someone must be accountable, and the child looked after.

Numerous arguments are going on at the same time, including the place of the market in the making of families and the fact that relationships entail responsibilities. Taken for granted is the role of medical technology, which (after Latour) has lengthened the chains of circumstances and personnel it takes to produce parents and children. It feeds people's interests in attaching persons to or detaching them from one another. Indeed, technology would not multiply the number of claimants if it were not for the way people seize on new openings. The legal decision drew on a further possibility. Creativity lies in mental acts.

At this point I jump to another arena altogether, from debate in the law courts to debates among practitioners of science, specifically those in biomedicine.

Mario Biagioli (in press)[4] asks how authors of scientific papers become attached to their works. "Guidelines of the International Committee of Medical Journal Editors" require that each name listed in an article's byline refer to a person who is fully responsible for the entire article. In the background lies technological magnification. The "increasingly large-scale, collaborative and capital-intensive contexts of [biomedical] research" (1998, 6) and the sheer amount of work entailed by Big Science projects that bring together different skills mean that (in Biagioli's phrase) multiauthorship has become a fact of scientific life. The result has been an explosion in authorial naming. But people have protested at the idea that authors should vouch for one another. A 1997 letter to *Science* invokes what could almost be Dolgin's modern family of autonomous subjects: "If marriage partners are not held liable for the actions of spouses, why should we assume that scientific collaborators are liable?" (quoted by Biagioli 1998, 10). Others point to preexisting relationships. Indeed, at the further extreme, one organization has adopted a no-choice model: publications automatically contain the names of everyone contributing to the enterprise as a whole.

What Biagioli brings to light is that side-by-side with a model of individual authorship are others that (perhaps like the traditional family) stress solidarity between all those involved in creating knowledge: advocates of a corporate model would include a diversity of scientific workers, although the term *corporate* here carries resonances with commercial corporations rather than community. A different solution is the proposal to replace authorship altogether—for example, by dividing contributor from guarantor.[5] The *British Medical Journal*, which has been interested in this last proposal, has considered another plan—offering copyright ownership to authors while the journal would secure a "license to publish."[6] This move departs from U.K. provisions introduced some years ago, which separated copyright from moral right.[7] Moral rights protect certain relationships between a work and its creator, as in the right to be identified as author; a creator, with a claim on the work's integrity, is thus technically distinguished from an owner claiming economic benefit, usually the publisher. These most recent proposals would universalize the author—literary or scientific—as copyright holder. The division between creator of the work and owner of the economic rights remains, but the term *copyright* would shift—now to rest in the author, while the journal publisher becomes a license holder. Truly, as Biagioli says "[t]he kinship between authors and works is a tricky two-way street."

If claims to scientific authorship are at a tangent to rights created by intellectual property law, literary credit is another matter. Here the very notion of creativity is partly the result of historical struggles over intellectual property where (Biagioli opines) "the focus on the individual author as the holder of . . . property rights misrepresented the long chain of human agency that produced a literary work" (Biagioli 1998, 11).[8] Perhaps for literary producers at least, the emergent figure of the new copyright holder will keep two dimensions in tandem: a new location for individual originality is at the same time a location for a new sense of community.

Why this leap from one arena to another—from parental suits to scientific authorship? In each, debate turns on the implications of multiplicity. Yet surely we could not sustain an analogy long enough to think usefully about the former (parenthood) in terms of the latter (authorship)? The potential parallels can, therefore, be interesting for one reason only: because they bring to mind a possibility already realized, an occasion when someone has proferred connections of just this kind. I have not presented a worked-out analogy between multiple parenthood and multiple authorship, but rather the kinds of raw materials from which such analogies are made and the cultural possibilities these contain: my pretend analogy sets the stage for one that was no pretense at all.

An Analogy

Behind the 1998 appeal was a much cited case brought to the California Supreme Court in 1993.[9] One of the judges, in dissent, analyzed the court's clinching argument: it rested on a hidden comparison between reproductive and intellectual creativity. She exposed the analogy in order to dispose of it.

Anna Johnson had undertaken to act as a gestational surrogate on behalf of Crispina Calvert and her husband; the embryo came from their own gametes. In the dispute that followed, each woman laid claim to motherhood. The Supreme Court found that the Calverts were the "genetic, biological and natural" parents. That "and natural" was determined by one crucial factor, procreative intent.[10] The majority argued: "But for [the Calverts'] acted-on intention, the child would not exist." They quote a commentator, who proceeds to make a dreadful pun: "The mental concept of the child is a controlling factor of its creation, and the originators of that concept merit full credit *as conceivers*" (my emphasis). The pun I return to. The commentator means the conceivers of the mental concept, valuable for fixing in "the initiating parents of a child," a sense of their obligations (cf. Morgan 1994, 392).[11] The efforts of the intended parents, wrote another commentator, meant they were "the first cause, or the prime movers, of the procreative relationship."

Justice Kennard, dissenting, seized on this formula. She pointed out that the originator-of-the-concept rationale is frequently advanced when justifying protection of intellectual property based on the supposition that an idea belongs to its creator as a manifestation of the creator's personality. The majority were implying that "just as a song or invention is protected as the property of the 'originator of the concept,' so too a child should be regarded as belonging to the originator of the concept of the child" (851 P. 2d 776). But, she argued, there is a problem in comparing rights to property: the marketplace. Unlike songs or inventions, rights in children cannot be sold for a consideration or made freely available—no one can have a property right of any kind (intellectual or otherwise) in a child because children are not property in the first place.

Now the comparison is not just with property; it is also with the kind of connections that exist between parent and child and between the originator of a concept and its realization. It is perhaps just as well that the majority of judges did not pursue the analogy with intellectual property further. If the parallel is to patenting an invention, it cannot be the idea of a child to which claim is made—that is already in the public domain; while if it is a particular expression of the idea, as in a song subject to copyright, then claim can only be laid to the unique features of the child itself, and one might have to argue about how much was intended by the parents' intention and what in any case was copiable about it. Yet vague as the claims obviously were, they made cultural sense—the analogy between reproductive and intellectual creativity was not pulled out of thin air. Although the cases are American, I would claim the analogy for the English-speaking world more generally.

People are culturally at home when they can jump across different domains of experience without feeling they have left sense behind.[12] What links the two domains in this case—reproductive and mental creativity—is an entirely commonsensical (though not uncontested) view about the originators of things claiming benefit or having responsibility attributed to them; the language of intellectual property rights emphasizes the naturalness of an identification between conceiver and conceived. Another link is the warning against confusing identification with economic possession when persons are at stake, as in the idea of owning children as property.

Not out of thin air: something is being sustained here that might hold our attention—the fact that, in the same breath, English speakers find it possible to talk about practices to do with making kinship and practices to do with making knowledge. As in the comparison of spouses and scientific collaborators, one might have supposed that kinship relations would be the source of figurative language for the production of knowledge, not the other way around. For instance, the term *paternity* has slipped into regular usage to designate one of the new moral rights in English law (protecting an identification between author and work). However, I gave Justice Kennard's opinions space precisely because of the direction of her analogy. She asserted that the arguments being put forward about parental claims were derived from arguments familiar from the law's protection of authorship. This too is not out of thin air; as we shall see, this directionality has a history of its own. I take up some already much discussed materials in order to thicken the air further. This sets the stage for my historical question.

II

Offspring into Property

If one were not alert to the way in which idioms appear and disappear, one might think that paternity was an old established trope. The truth is that only recently has it been incorporated into English copyright law. It is therefore fascinating to consider its fate at the time, in eighteenth-century England, when authorial rights in literary works were becoming an arena for debate (e.g., Coombe 1998, 219–20; cf. Franklin 1996).

Daniel Defoe's celebrated protest in 1710, "[A] book is the Author's Property, 'tis the Child of his Inventions, the Brat of his Brain," casts back, Rose (1993, 39) suggests, to sixteenth- and seventeenth-century metaphors: "The most common figure [of speech] in the early modern period is paternity: the author as begetter and the book as child." Defoe is not talking about an enduring proprietorship, but complaining of piracy through unacknowledged printing, which he likens to child-stealing. Of those sixteenth- and seventeenth-century usages, Rose comments:

> Inscribed with the notion of likeness more than of property, the paternity metaphor is consonant with the emergence of the individual author in the patriarchal patronage society concerned with blood, lineage, and the dynastic principle that like engenders like. (1993, 39)

Full authorial *property* rights, by contrast, emerged in a liberal society, and with other arguments (Rose 1993, 41, 58). Over the eighteenth century grew the notion that property could refer not just to the material but to the immaterial, not just to the book as a physical body but to a more abstract entity, the composition as a text.[13] Here, far from assisting the new ideas that were developing about authorship, the idiom of paternity seems to have gotten in the way.

At the very moment when a creational concept of author was taking shape, that particular kinship idiom seems to have disappeared. Works might have continued to be referred to as "offspring," but the vivid vision of paternity faded. Was the image of the book as a father's child altogether too concrete? Rose observes that the metaphor would run into trouble if the idea of begetter and offspring was extended to the marketplace. Who would sell their children for profit?[14] He does not claim that this absurdity was the reason for the figure's demise, simply observing that it would present rhetorical difficulties, but he gives a clue as to what else might have been going on. Creeping up on new ways of thinking about property were new ways of linking writers to their writing: the emergent owner was not the bookseller but the author, and the emergent book not the volume but the text. And could it also be that creeping up on paternal begetting as a figure of speech were fresh possibilities in ideas of conception[15] and creation?[16] If so, they offered somewhat different grounds for identifying the author with his work.

Conception and *creation* had long established double connotations, at once procreative and intellectual, and they are still in place—witness the dreadful pun brought into the surrogacy case. By the end of the eighteenth century, the view had taken hold that it was the particular form in which literary authors gave expression to ideas that belonged to them as the mark of their unique work. Woodmansee (1994, 36–37) describes how this notion of the author inspired from within took over from earlier views of the writer as a vehicle inspired by external agencies, human or divine. Recapitulating that earlier relationship in a father-child idiom, the writer fathering his book, just as God fathered the world, would reinforce the writer's perception of dependency. Did those too-vivid images of dependency need to disappear? Was authorial creativity best separated from enmeshment in relationships?

I can only extrapolate. Perhaps the concreteness of the father-child image had lain partly in the relation it presupposed. The imagery had been used to claim the kind of possessiveness that parents felt toward their children.[17] Did a new rhetoric of conception and creation instead allow one to take the child's view? The author's text was now to embody the author's own genius. It was the child's view insofar as the father becomes superfluous: the omnipotent heir can create his own world. If there is pride in creating what never existed before, then the author does not want the preexistence of fathers either, for he must be as original as his work.[18] The relationship between author and text could be imagined as one of correspondence, a kind of nongenerational generation. Evidence of authorial identity would lie not in lineage or genealogy but in an informational matrix, in the way in which the work encoded information about the producer of it.

Sometimes not making connections may be as enabling as making them.[19] We might see dropping an inappropriate kinship metaphor as *part* of a nexus of ideas and concepts that link kinship and knowledge, not *apart* from it. Can one suggest, then, that the metaphor of paternity was actually edged out by new notions of creativity, which were powerful precisely to the extent that the resonances with kinship could be held at a remove? For if *conception* and *creation* retain kinship echoes, they seemingly displace the idea of an interpersonal relationship with more immediate but also more abstract evidence of connection: the work itself informs you about the author. Does creation become a kind of procreation without parenthood? If so, this would be at once consonant with both the emerging originality of the author and the emerging uniqueness of the literary text.

Information into Knowledge

What was happening with the text did not take quite the same route in science. Booksellers originally had authors' names printed in order to point to the person responsible for the contents, should they prove seditious or libelous (cf. Biagioli 1998, 3). Accountability continued to be important in scientific writing. It was not the form of the presention over which claims were made but the *quality of information* being communicated; its value came from how it stood up to other kinds of information. The author was actually abstracted from it in that sense.[20]

However, the author abstracted from the text was made present elsewhere, as one of an assembly of authors. If today there are many names associated with a scientific paper, this is all part of an informational process that places the author within an arena of social relations.

Writing about the problems of trust engendered through the collective character of empirical knowledge-making, Shapin (1994, 359) observes of the seventeenth century that "scientific knowledge is produced by and in a network of actors" (emphasis removed). He asks how verifiability was ascertained, and answers that "knowledge about people was constitutively implicated in knowledge of things" (1994, 302). What counted as knowledge depended on what people were willing to attest, and the value of their testimony rested in turn on the kind of people they were. So texts that circulated with a presumed equality between them were also circulating between persons who could vouch for one another. It was the relations that turned a multiplicity of persons into a social arena of authority.[21]

Relations were also doing something else. It was relations that also produced knowledge out of information. If items of information were judged against one another, any fit was simultaneously a relation between them. Knowledge became understood as accountable information, and it was by virtue of being relational that it was accountable. Indeed I wonder if the concepts of relation and its partner, connection, were not to be directly enabling of secular inquiry fuelled by the Enlightenment conviction that the world (nature) is open to scrutiny. For relations are produced through the very activity of understanding when that understanding has to be produced from within,[22] that is, from within the compass of the human mind, without reference to divinity, and thus when things in the world can only be compared with other things on the same earthly plane. What validates one fact are other facts—always provided the connections can be made to hold. And Shapin's seventeenth-century experimenters were looking for connections everywhere—always provided the facts could be made to hold.

Let me speak for a moment from the perspective of contemporary English speakers. One grasps (a piece of) information as knowledge by being aware of its context or ground, that is, of how it sustains a relationship to other (pieces of) information. Conceptual relations have two significant properties.[23] First, the notion of relation can be applied to any order of connection. For in describing phenomena, the fact of relation instantiates connections in such a way as also to produce instances of itself. The demonstration of a relationship, whether through resemblance, cause and effect, or contiguity, reinforces the understanding that through relational practices—classification, analysis, comparison—relations can be demonstrated. We could call the relation a self-similar or self-organizing construct, a figure whose organizational power is not affected by scale. To return to the seventeenth and eighteenth centuries, perhaps the capacity for making conceptual relations was itself being *conceptualized* (concepts being formulated about concepts) under the pressure of systematic inquiry into practices of knowledge-making. The relation also has a second and quite distinct property: it requires other elements to complete it—relations between what? This makes its connecting functions complex, for the relation always summons entities other than itself, whether these entities are preexisting (the relation is *between* them) or are brought into existence by the relationship (and thus exist *within* it).[24] One does not only see relations between things but things as relations.[25] Yet it is because things (the terms bound by or containing the relation) are routinely conceptualized apart from the relation that the relation can model complex phenomena: it has the power to bring together dissimilar orders of knowledge while conserving their difference. This is the perception that makes Latour's (1986) two-dimensional inscriptions—the diagrams, charts, and tables that have long enabled scientists to superimpose images of different scales and origins—work.[26]

Indeed, working as one might say technology works, conceptual relations are part of the machinery of exposition. One cannot point to a relation without bringing about its effect. The very concept (relation) participates in the way we give expression to what we know about it. So relations themselves can appear at once concrete and abstract. They can produce a sense of an embedded or embodied knowledge out of information that would, otherwise, abstracted from context, float around weightlessly. Or they can seem ethereal or disembodied, hypothetical linkages hovering over the brute facts and realities of information on the ground. We shall see some of the possibilities that lie in this duplex formation. But either way, it has to be said, conceptual relations seem at a remove from the arena of human social relations, including those with which this section began. As dreadful as the double entendre in *conceive*, have I simply conjured more punning? Not if I can articulate my question properly.

Relations into Relations

I have no idea what conceptual relations once connotated,[27] or how to differentiate the eighteenth from the seventeeth century in this regard. So I am not really certain when, or in what social milieu, to locate the question. But this is it. We can imagine the part that the concept of relation played in the unfolding of understandings about the nature of knowledge. *How then did it come to be applied to kin?* For it would seem that *relation*, already in English a combination of Latin roots and variously a narrative reference back to something, or comparison, became in the sixteenth and seventeenth centuries applied to ties, whether by blood or marriage, through kinship. It was not alone—several terms to do with both knowledge practices and kinship practices were seemingly in flux.[28] In many instances it was a case of adding new properties to old, so that existing terms acquired double meanings. I point to two such clusters.

One cluster refers to propagation, and the oldest candidate here is the very term *conceive* and its correlates, *concept* and *conception*. To create offspring and to form an idea: this double sense of *conceive* had been recorded since the 1300s.[29] But there is also *generate*, *reproduce*, *create*, *issue*, and some of these only doubled their reference much later. Thus *creation* was used in the fourteenth century for *begetting*, and with divine connotations of *causing to come into being*; it was first recorded as applying to an intellectual product or form in the late-sixteenth/early-seventeenth century. Other doubles also emerged in the early modern period. Consider the second cluster, dominated in my own mind by the term *relation*, which includes *connection* and *affinity*. *Affinity* seems to have been a relationship by marriage or an alliance between consociates before it became in the sixteenth century a term for structural resemblance or causal connection. Conversely, *connection*, which appears in the seventeenth century, seems to have referred to the joining of words and ideas by logical process before it came in the eighteenth to designate the joining of persons through marriage or consanguinity.[30] The two clusters are connected. One elides mental conceptions and procreative acts; the other elides the kinds of connections these produce. Elucidating the nature of mental conceptions was among philosophy's contributions to new knowledge, while the relationship between procreation and kinship fed into emergent formulations of nature and culture. But that is in prospect; there is something to be explained in retrospect.

If these were originally puns and conjunctions allowed by the English language and the way it created verbal connections, then *they must also have been allowed by English kinship* in the way it set up connections between persons. Was the attention to knowledge-making that we associate with the new sciences also refashioning the way people represented their relations to one another? What was entailed in having *relation* introduce into thinking about kin an intellectualized sense of connection? And embedded there, did it acquire further properties? For, once introduced into kinship, the relation could be borrowed back again.

Listen now to this deliberate analogy, addressed to the elucidation of knowledge processes. How we know kinsfolk and how we know things are drawn together in a parallel with all the force of serious explication. In his *Essay Concerning Human Understanding* (1690), John Locke conjures up the image of two cassowaries on display in St. James's Park, London. (*Cassowaries* are large, flightless birds from Papua New Guinea and Southeast Asia.) The philosopher wanted to illustrate the logical circumstance whereby a relation could be perceived clearly even though the precise nature of the entities themselves might be in doubt. He contrasted the strangeness of this bird with the clearly perceived relationship between the pair: they were dam and chick. The one was the offspring of the other. The parent-child relation, a matter of kinship, illustrated how one could, as a matter of knowledge, conceive of relations between entities.

One might argue that all at issue here were relations between concepts, viz., those of parent and offspring. But not only did Locke draw on the concrete act of propagation,[31] the avian connection had been preceded by several references to human kinship. Thus, in talking about the way in which comparison (bringing items into relationship) is a clarifying exercise, Locke argued that "in comparing two men, in reference to one common parent, it is very easy to frame the idea of brothers, without yet having the perfect idea of a man" (Locke n.d, 236). What he himself is comparing, of course, are the two kinds of relations. Throughout his disquisition, he takes kin relationships as immediately accessible exemplars of logical relations. He gives as examples of correlative terms obvious to everyone: *father* and *son*, *husband* and *wife*.

In making the comparisons, Locke linked a conceptual relation between entities to a procreative relation between hen and chick as though both usages were as thoroughly sedimented in the English language. Only with hindsight do we note that it was relation as applied to the kin connection that was the relative novelty. So, for all that the conceptual notion of relation can be borrowed back so effectively from the domain of kin relations, the historical question remains: How did it come to be applied to kinspersons in the first place? Recent technological developments have perhaps added to the reasons that make the question worth asking. While kinship and knowledge provide figurative resources for one another, borrowed back and forth, the historical direction in which the concept of relation expanded—from knowledge production to kinship connection—would seem to have left traces in a certain persistent asymmetry. Not only for this reason, but perhaps including it, knowledge holds the privileged position.

III

Kinship and Knowledge

Analogies are not relations of cause and effect; concepts do not—pace the Calverts—procreate. People carry them across domains, often because there is some argument to pursue. Analogies are relations of resemblance. Now that does not mean their fancifulness is idle—on the contrary, much of culture is a fabrication of resemblances, a making sense through indicative continuities. It follows that appreciating the power of a parallel between conceptual and familial relations does not depend on demonstrating the direct derivation of one from the other. It is conceivable, for example, that the terms *relation* and *relative* migrated into kinship from their more general usage at the time for *associates*, persons connected through mutual acknowledgment. Like the circle of scientists, the circle of persons who publicly recognized one another perhaps anticipated some of the class overtones of kinship so evident by Jane Austen's time. The point is that once tropes and images are lodged in a particular context or domain, they are capable of summoning other contexts regardless of derivation. It would seem that since early modern times English speakers have thus kept knowledge and kinship in tandem. Each still seems to offer people the power of drawing the other into itself. Asymmetrically, however, they do not work on each other to quite the same effect. Here we have to consider how terms come to be naturalized in their new domains and analogy is submerged.

Familial and procreative language in philosophy and science have long been naturalized to refer not only to classificatory schema but to nonhuman processes of reproduction. Some of these terms were widely used in natural historical and anatomical writings *before* they became applicable to human relations, one such being *reproduction* itself (cf. Jordanova 1995, 372). But then no one blinks an eye at referring to mother and daughter cells. Such terms have a technical job to do, and any figurative recall will seem for the most part irrelevant. Consider this 1833 description of the planets:

> When we contemplate the constituents of the planetary system from the point of view which this relation affords us, it is no longer mere analogy which strikes us, no longer a general resemblance among [the planets]. . . . The resemblance is now perceived as a true *family likeness*; they are bound up in one chain—interwoven in one web of mutual relation. (Chambers 1969 [1844], 10–11, from John Herschel's *Treatise on Astronomy in Lardener's Cyclopaedia*, quoted in Beer 1983, 169, original emphasis)

Herschel wanted to displace a weak sense of analogy between planetary bodies (they look alike) by a strong sense of the affinity between them (their orbits are related to one another). The first "relation" in this passage is a mathematical deduction between distances from the sun and revolutions around it, while the second sounds as though it could have acquired resonances of kinship.[32] But equally well, he could have simply been reinforcing habitual usage. A family is an assemblage of objects, and all he was insisting on was their necessary or systemic connection. It does not have to be expressly as kinship that such ideas are embedded in knowledge practices.

Ideas about knowledge embedded in kinship practices is another matter altogether; they are there *as* knowledge. Certainly for English speakers, a peculiarity of knowing in kinship terms is that information about origins is already grasped as knowledge. The information constitutes what they know about themselves. Facts about birth imply parentage, and people who find things out about their ancestry, and thus about their relations with others, acquire identity by that very discovery. This means that information about kin is not something that can be selected or rejected *as information* (cf. Strathern 1999). Since kinship identity is realized within a field of relationships, knowing about one's kin is also knowing about oneself. One has no option over the relationships. So information can only be screened out at the cost of choosing whom to recognize. This leads to a sense in which we may say that relationships come into being when the knowledge does. As a proposition about kin, it can be taken quite literally.

The potential for analogy depends on the domains (kinship, knowledge) being kept separate, and as long as they are separate, each endows the other with its own distinct properties. I return to the asymmetry. To say that knowledge is a part of contemporary kinship thinking in a way that kinship is not a part of knowledge—my general point here—reminds us of the relation and its direction of expansion. My interest in the early modern material has been, all along, its pointers to practices of creating knowledge. What I do not know is how we might or might not, historically speaking, align this with creating kinship. Let me conclude with a situation where anthropologists do know something about kinship practices. It returns us to present-day arguments, to practices stimulated by the new reproductive technologies and to the arena of litigation. We might read the situation either as a move toward greater abstraction (a new form of relatedness without relatives) or as a move toward greater concreteness (where value is recovered for kinship substance, indeed where one might say that kinship is being turned back from knowledge into information). We also find moments when the domains of kinship practice and knowledge practice cannot be kept separate, and analogy becomes impossible again.

The Informational Family

In 1992, American Donna Safer sued [the estate of] her father's physician for not having informed her of his condition. She had been diagnosed with the same cancer from which her father died; had she known, she might have been able to take precautionary measures. The New Jersey trial court concluded that a doctor had no legal duty to warn the child of a patient of a genetic risk. The appeal court disagreed: there was an obligation to inform in instances of genetic disorders "where the individual or group at risk is easily identified." It went on:

> [T]he duty [is appropriately] seen as owed not only to the patient himself ... [but] extends beyond the interests of a patient to members of the immediate family of the patient who may be adversely affected by a breach of that duty. (quoted by Dolgin 2000, 557)

Of diverse cases, Dolgin considers this the most radical. It compels her to identify an emergent phenomenon, the genetic family.

A kinship system, which has a propensity to base relatedness on what can be known about people's connections, was bound to be intensely interested in the new certainties afforded by genetic testing. The *genetic family*, persons proved or presumed to be genetically related, is at once held together by the substance people ascribe to genes and by the information these supposedly contain. What is newly important about the genetic tie is that it gives family members information about one another. Whereas warning parents about children's genetic conditions reflects general understandings of the parent-child relationship, the reverse case not only removes the doctrine of patient confidentiality among adults but imposes an obligation on third parties to warn family members about the medical condition of others. In this undermining of individual privacy, family members are treated as an undifferentiated group.[33] Knowledge here is knowledge of genetic makeup. There is no option as to the ensuing facts of relationship. But while information about origins automatically becomes knowledge for the person, under circumstances such as these it can revert to information again. It becomes similar to other kinds of information acquired from outside sources. Indeed, Dolgin stresses, nothing else need be known about the relationship between parent and child than the fact that the body of one holds information useful to the other.

Like finding direct evidence of inspiration from within a literary work, genes offer direct knowledge of heredity unmediated by parentage. Yet, in practice, personal knowledge of a family's genetic history is the route by which people may start inquiring into their own susceptibilities or find out more about afflictions already on them (Finkler 2000). So why is Dolgin so struck by the novel properties of what she calls the genetic family, at least as it is legally constructed? Relatives have become like their genes; value lies in the information they carry, and here persons appear substitutable for one another. What is lost is the concreteness of specific relationships. "Genes suggests nothing about social relationships. They are simply data" (2000, 544). The genetic family, she goes on, challenges the presumption that the law can safeguard families of choice—the ideal of solidarity and lasting commitment—for the construct of the genetic family precludes choice and is indifferent to the character of family life. Indeed, the genetic family is neither America's traditional family with its hierarchy and community nor its modern family consciously holding autonomous individuals together. Instead, information about any one member is merged with information about them all.

> The links connecting Donna to her father—or any member of a genetic family to any other—are a-moral links that neither define nor depend upon the scope and meaning of social relationships among family members. (2000, 561)

As repositories of information, persons are replicas of one another: relatedness without relatives one might say.[34]

The genetic family is also being lived outside the American courtroom. And genetic information that appears to extract relatedness from relationships can equally encourage people to seek out far-flung connections—which may or may not be turned back into active relationships.[35] The point is that they do not have to be. "In contemporary society people have tended to become separated from kin, if not from their immediate family, and family and kinship have taken on an amorphous cast, for multiple reasons, the most obvious being geographic dispersal." Kaja Finkler's (2000, 206) general observations on the American family follow with the specific comment that notions of genetic inheritance may move it together again. Women diagnosed as having a hereditary disease search for information from relatives with whom they might have long been out of close contact.[36] But the recorporealization of the family comes with the proviso that

> interaction with family and kin may no longer be required in order for people to recognize relatedness and connection. . . . To the sense that one forms part of a family chiefly because one shares the same genes, requiring no social participation nor sense of responsibility to those who are related except to provide blood samples for testing purposes, removes the moral context of family relations. (2000, 206)

More than this, her expectation that people would blame their ancestors for passing on faulty genes was upturned: the women she interviewed said their families were not accountable for their affliction. Genes are amoral entities.[37] For there is a sense in which they are equally a-relational: "They are another kind of thing, a thing in itself where no trope can be admitted" (Haraway 1997, 134).

The routines of family life have usually meant that relationships without responsibilities tend to fade away. A truism about knowledge can keep them in view: the genes that carry the data informing you what you are at the very same time comprise the mechanisms that bring about what you are. This looks like a reworking of an old theme, the constitutive nature of kinship knowledge. But to find kinship knowledge in the gene is, so to speak, to find it in itself. Knowledge and kinship become momentarily inseparable. They are not analogues of each other—even more so than Herschel's planets, resemblance dissolves into an identity. Only an extraneous factor could prize them apart again. And Justice Kennard's winkling out of the analogy between conceivers of ideas and conceivers of children introduces just such a factor. It was property ownership that showed them up to be different: it would actually have to be argued that knowledge may be regarded as belonging to persons in the same way as they might imagine their genes belonging to them.

In the background to Donna Safer's suit for the wrong done to her by the withholding of genetic information lies increasing nervousness about setting precedents for ownership. Two issues, among many others, concern commentators in the United States. On the one hand, legal instruments (such as statutes) that define genetic information as the property of those to whom it pertains do so with concerns about individual privacy in mind. On the other hand, the very idea that people should claim property in genetic information is vigorously opposed by sections of the biotechnology industry—the imposition of ownership rules on genetic information would require a record-keeping regime that could inhibit research, provide a context for litigation, and interfere with profits. It has been proposed in the United States that ownership should be replaced by the doctrine of informed consent. Informed consent rules grant people the right to know about uses to which others will put information about their genes, as Dolgin notes. Another analogy: like the division proposed in the United Kingdom between license to publish and copyright, this could divide the owners of rights to exploit the information (who would enjoy the economic benefit of, say, developing technology) from the persons giving informed consent (who would enjoy a kind of moral right, an identification with their genes and a potential safeguard to their genetic privacy). For the latter, and it is a cultural commonplace, what seems supremely at issue in gene information is that this core bit of kinship should be accessed as knowledge for, belonging to, and about themselves.

Conclusion

Kinship practices and knowledge practices comprise fields that, since early modern times, have provided figurative ammunition for one another. The complex possibilities of terms such as *conceive* had long been in place, while others—of which I have singled out *relation*—appear to have been formed at this time. Conceptual relations have enjoyed some historical priority over kinship relations. *Relation* already denoted intellectual practice—referring back to something, making a comparison—before it became applied to ties of blood and marriage. This was the period when *relation* in its conceptual sense was to be given a long chain of effects in new practices of knowledge-making. Over time, analogies between domains may be submerged, revived, and submerged again. I ended with a recent social phenomenon, the genetic family, where kinship identity can be imagined as literally embodied in an informational code and information can be imagined as a kinship substance. It is as though the analogies between knowledge and kinship were compacted into one another. But that elision is brought into being by circumstances that hardly exhaust everything one might want to say about either knowledge or kinship. I noted that *property* started up fresh analogies. Let me briefly go back to the beginning, and to a different fate for ideas about genetic substance.

The American women who hoped to sell their eggs for $50,000 were prepared to turn one kind of substance (genetic material) into another (money). In the United Kingdom, where, by law, that conversion is not possible, egg donors do different kinds of conversions. One potential conversion is into connections, but connections created outside a premise of kinship. Here new separations emerge as well. If one starts not with kin, people whom one knows, but with people whom one does *not* know, fresh scope for relational reasoning also emerges.

In meeting various egg donors in Britain, Monica Konrad was struck by the vagueness with which they talked about the connection between donors and recipients; she suggests that it is out of the very condition of anonymous diffuseness that people conceive relations of a kind (Konrad 1998, 652). "As ova substance is disseminated in multiple directions to multiple numbers of recipients . . . donors and recipients are partaking collectively in an exchange order of non-genealogical relatedness" (1998, 655). In this process, substance may be leached of biological significance (the eggs are "not like a physical thing that have come from my body" [quoted 1998, 651]). What signifies is being the origin of a process that another carries forward. Women aim to help others whom they do not and largely do not want to know. The wish to assist "a someone" contains the essence of their own agency, an extension of themselves that takes effect across a dispersed universe of unidentified others. In short, Konrad describes persons forming themselves through an extensional relatedness via multiple persons who are separated from them by being neither locatable nor nameable. Ova donors need effect no specific transaction in order to value their action. "What appears as the agency of these donors does so as the value of multiple and untraceable circulations of persons and body parts anonymized as (an)other's action, as a generalized, diffuse relatedness" (1998, 661). This relatedness may not have relatives, but it does have signifying others.

women as would-be mothers: the donors see the situations of both donors and recipients as parallel.

Reaching out to an audience of multiple recipients sounds not so far removed from the aspirations of authors. But unlike authorial identity, at least of the scientific kind, the basis for these particular donors' relations with the women they saw themselves as helping was that their accountability would have no forward effect: their gesture contained its own definition of responsibility (to help a someone). The relations did not translate into interaction, and the eggs did not need a name.[38] Hence, it seemed possible to leave quite undefined whether or not what they were giving away was something they felt they owned. In other words, the parallelism rests on what is also an unbridgeable gulf between them: in this sense, donors and recipients are in a relation of analogy.

Many of the British women's feelings have no doubt been echoed on the American side (cf. Ragoné 1994). All I do is underline the obvious, that there are always new domains with which to make connections and thus new material for analogies. In the prevailing (Euro-American) view, technology and its scientific basis has had a tremendously inventive impact in creating new material. Intriguing, then, is the way in which some analogies endure. The expansion of the term *relation* is a case in point. So I come back to wanting to ask about kin connections between English speakers in early modern times. To what kinship practices did the new concept of relation speak; what emergent interactions might its properties have addressed? From the perspective of kinship, anthropologically speaking, the sciences of the time come to look rather interesting.

Acknowledgments

This chapter is based on the 2000 Rothschild Lecture to the Department of History of Science, Harvard University, under the title "Emergent Properties: New Technologies, New Persons, New Claims." My warm thanks to both the chair and the department for the inspiration. Its argument continues an essay on "The Relation" (1995); it owes much to colleagues who have worked on kinship and related issues: notably to Debbora Battaglia, Barbara Bodenhorn, Janet

Carsten, Jeanette Edwards, Sarah Franklin, Frances Price, Heléna Ragoné, and to the Wenner Gren Conference organized by Sarah Franklin and Susan McKinnon on *New Directions in Kinship*, to which parts were presented. Paul Connerton gave the original adroit scrutiny, as did Susan Drucker-Brown, Joyce Evans, Eric Hirsch, and Annelise Riles. My direct debts to Janet Dolgin and to Mario Biagioli should be evident. The Cambridge and Brunel joint research team working on *Property, Transactions, and Creations: New Economic Relations in the Pacific*, funded by the U.K. Economic and Social Research Council, has been an indirect but emphatic intellectual support.

Notes

1. The appropriation of legal opinion for cultural understanding conceals the extent to which judges' written opinions (the bulk of the material referred to here) are produced in the very awareness "that what they write gets picked up as the stuff of cultural criticism" (Annelise Riles, pers. Comm.).
2. *In re Marriage of John A. and Luanne H. Buccanza*, see Dolgin (2000); I appreciate her sending me the record from the California Court of Appeal, March 10, 1998.
3. This was argued in accordance with the existing ruling that a husband's consent to his wife's artificial insemination makes him the lawful father. By consenting to the medical procedure, the couple had put themselves into a position similar to an IVF husband. Note the reference also to their "initiating" the procedure: "Even though neither [of the couple] are biologically related to [the child], they are still her lawful parents given their initiating role as the intended parents in her conception and birth" (72 Cal. Rptr. 2d at 291).
4. I am grateful for permission to cite so liberally from the as yet unpublished paper.
5. In Biagioli's view, the basic problem of how to divide attributable claims from acknowledging the support that made them possible is not solved by the corporate model. The issue is an epistemological one about the relationship between the specificity (of a particular piece of work) and the general conditions of its possibility. Cf. Haraway (1997, 7): "Only some of the necessary 'writers' have the semiotic status of 'authors' for any 'text.'"
6. The U.K. Association of Learned and Professional Society Publishers' draft (1999) "license to publish" suggests publishers relinquish copyright to authors. Authors could publicly self-archive their work, and would be free to give it away, while all rights to sell (on paper or online) would be held by the publisher.
7. Moral rights point clearly to the originator but, unlike property rights,

cannot be sold or otherwise assigned (they may be waived). Long established in much of the rest of Europe, although foreign to U.S. copyright law, moral rights came into English law through the U.K. 1988 Copyright, Design and Patent Act.

8. Lone literary authors can of course take a collective view, although they may have principally in mind a community that comprises audience as well as fellow writers. A Draft Declaration from the U.K. National Consultation of Academic Authors opens: "Academic authors communicate and share ideas, information, knowledge and results of study and research by all available means of expression and in all forms. They recognise that participants in this scholarly communication process include academic editors, publishers and presentation experts."
9. I have drawn from this before (e.g., Strathern, 1995, 1999). Derek Morgan sent me a report of *Anna Johnson v. Mark Calvert et al.* (Cal. 1993) 851 P. 2d 776–800 (May 20 1993) and I have since received the printed version from Janet Dolgin. Thanks to them both.
10. "We conclude that although the Act [Uniform Parentage Act, California, 1975] recognizes both genetic consanguinity and giving birth as means of establishing a mother-child relationship, when the two means do not coincide in one woman, she who intended to procreate the child—that is, she who intended to bring about the birth of a child that she intended to raise as her own—is the natural mother under California Law" (851 P. 2d 776).
11. The commentator continued: "The mental concept must be recognized as independently valuable; it creates expectations in the initiating parents of the child, and it creates expectations in society for adequate performance on the part of the initiators as parents of the child" (851 P.2 d 782). Another had argued that reproductive technology extends "affirmative intentionality" and that intentions voluntarily chosen should be determining of legal parenthood. Dolgin (2000) points out that the doctrine of intent can thus support either a traditional view (the likelihood of enduring relationships) or a modern view of the family (suggesting choice and negotiation).
12. I take culture as a field constituted through domains of experience, practice, and knowledge, which can be at once differentiated from one another and transversible without cognitive dissonance. How much dissonance qualifies as a [cultural] shift is variously a matter of scale or perspective.
13. Opposers to this view argued that there could be no property without the "thing" (Rose 1993, 70; Woodmansee 1994, 49–50). Whether or not copyright can be property is still questioned. This is partly because of its unusual legal status (it exists not in fact but only in law; it can be infringed but not "stolen," and rather than being a thing protected for as long as it exists, it ceases to exist at the end of its term), but partly because "a sizeable body of otherwise intelligent persons . . . argue from the mistaken premise that something cannot be truly 'property' unless it is solid and has the attributes of a physical presence" (Phillips and Firth 1990, 107).

14. Rose says that the analogy could never have got very far when the issue of authors' rights turned to the pursuit of profit. However, another historian notes (Jordanova 1995, 378): "Many eighteenth century commentators did indeed see production as a form of reproduction; they could therefore conceptualise children as commodities," although she qualifies this by referring to capital, that is, something into which parents invest.
15. As Rose (1993, 89) quotes Blackstone, here defending the argument that duplicates of an author's work make it no less the author's original work *in conception*: "Now the identity of a literary composition consists intirely in the sentiment and the language: the same conceptions, cloathed in the same words, must necessarily be the same composition" (from Blackstone's *Commentaries*, 1765–69, emphasis removed).
16. Along with a span of other connotations: e.g., the claim that the author's right was based on the fact that he "created" rather than just discovered or planted his "land" (Rose 1993, 56–57, 116).
17. Possessiveness lay generally in identity or likeness, a sense of "ownness" between parent and child and, in the ideas of the time, parental authority and power over the child. However there were also specialized debates over the consequences of identity. Some of the seventeenth-century philosophers discussed by James (1997, 248–52) distinguished the spiritual unification of oneself with one's object of knowledge, likened to the benevolent love of the father, from physical union, as in the mother's effects on the unborn child, which expose the mind to "inescapable afflictions of sense" and the person to too much influence from others to be able to form a clear knowledge of a world.
18. It is to be understood that these are cultural/social categories, not psychological ones. (Outram [1987, 21] has pointed out the evolution of salon circles in eighteenth-century France, where scholars found a kind of second family, often involving removal from the biological father, for the fledgling savant "the freedom to pursue innocent knowledge . . . could only occur as a result of . . . rejection of parental authority.") Woodmansee's argument ends with a comment on the concomitant emergence of the notion that work could be read in order to uncover the author's personality. Coombe (1998, 219) can thus generalize—like the commentator cited by Justice Kennard—that copyright laws came to protect works "understood to embody the unique personality of their individual authors."
19. Relocation, displacement, making the once present absent, withholding what others are expecting—these can all capacitate the contexts in which people act (Battaglia 1995, 1999).
20. Focus is not on authorial vision but on the quality of information, verifiable by comparison with other pieces of information. The procedure is not, of course, restricted to science; in discussing the nature of evidence, Hume ("On the Association of Ideas," 1748) succinctly remarks that a reason for a fact will be another fact. Haraway's (1997) critique of the modest witness lies precisely in observing that the juncture at which facts become visible is the juncture at which the witness becomes invisible.

21. My term: a rhetoric of equality had displaced old canons of authorization. "The Royal Society's 'modern' rejection of authority in scientific matters quite specifically mobilized codes of presumed equality operative in early modern gentle society. Just as each knowledge-claim was to make its way in the world without help or favoritism, so all participants played on a level field" (Shapin 1994, 123). From another time and place, the young savants mentioned above (n. 20) were specifically fledgling scientists and, in escaping their birth origins, were escaping "the tainted world of career-making, patronage, and advantage" (Outram 1987, 21).
22. "It is evident that there is a principle of connection between the different thoughts and ideas of the mind, and that . . . they introduce each other with a certain degree of method and regularity" (Hume, "On the Association of Ideas," 1748). All objects of human inquiry may, he avers, be divided into two kinds—relations of ideas and matters of fact. As far as "connections among ideas" are concerned, we find three principles: resemblance, contiguity, and cause and effect. When it comes to reasoning over matters of fact, this is largely founded on the last: "by means of that relation [of cause and effect] alone we can go beyond the evidence of our memory and senses."
23. Outram (1995, 53) quotes de Condillac (*Treatise on Sensations,* Paris, 1754): "Ideas in no way allow us to know beings as they actually are; they merely depict them in terms of their relationship with us." There are parallel properties in social relations, specifically relations of kinship, but I do not expand on this here.
24. Connections "within" may be seen as another example of connections "between." Ollman quotes Leibniz: "There is no term so absolute or so detached that it doesn't enclose relations and the perfect analysis of which doesn't lead to other things and even to everything else, so that one could say that relative terms mark expressly the configuration which they contain" (1971, 31).
25. The phrasing is from Ollman (1971, 27) on Marx's attempt to distinguish two types of relations.
26. Thanks to Eric Hirsch and Paul Connerton for their observations here.
27. From the perspective of certain seventeenth-century philosophers, for instance, it has been argued that it would be a mistake to treat knowledge as an intellectual matter divorced from emotion. Rather, "[t]he view that emotions are intimately connected to volitions enabled the philosophers . . . to make space for a conception of knowledge as feeling" (James 1997, 240).
28. At least if we can go by the citations in the *OED* [1971 edition]. These do not work simply as figures of speech, although through explicit analogy they may become so. Note that *kinship* is a thoroughly modern term. (*Kin* and *kinsfolk* are ancient, but *kinship* as both a relationship by descent or consanguinity and a relationship in respect of quality or character was a nineteenth-century coinage.)
29. In the dual senses of *receiving seed* (becoming pregnant) and *taking something into the mind* (grasping an idea); only later, and it is recorded thus

from the seventeenth century, is *conceive* used more loosely to cover both conception (of a woman) and *begetting* (of a man).

30. A usage that seems to have become prevalent in Jane Austen's circles. Handler and Segal (1990, 33) suggest that *connection* stressed the socially constructed and mutable (their phrasing) dimension of the kinship tie as opposed to its natural basis in blood. *Family* seems to have referred to the household and to those related through common descent before it became a term for an assemblage of items in the seventeenth century.
31. The original reads: "Having the notion that one laid the egg out of which the other was hatched, I have a clear idea of the relation of dam and chick between the cassiowaries in St. James's Park; though, perhaps, I have but a very obscure and imperfect idea of those birds themselves" (1690, 237). The cassowaries were one in a long line of unusual creatures kept in public view, many of which set puzzles for the "classifying imagination" (Ritvo 1997).
32. Scientists who were dealing with living, reproductive organisms had the advantage of being able to close whatever gap "between metaphor and actuality" existed. Beer cites *The Origin of Species*, where the idea of family is given a genetic actuality when descent becomes "the hidden bond of connexion which naturalists have sought under the term of the Natural System" (1983, 170). Kinship was no figure of speech, but conveyed "true affinities" between living things. Darwin argued that all living forms could be grouped together, several members of each class being connected "by the most complex and radiating lines of affinities" (quoted from *The Origin of Species*, Beer 1983, 167). Consider present-day artificial life workers: "Kinship terms from the Euro-American lexicon have been read onto biogenetic connections and *then used to structure knowledge about biogenetic categories themselves*. One genetic algorithmist . . . did not stop with *parents* and *children* in describing relationships between bit strings but added terms like *grandparent*, *aunt*, *cousin*" (Helmreich 1998, 152, my emphasis).
33. She has already pointed out the possibility that hereditary traits that appear to apply to overall ethnic or racial groups could be taken as evidence applying to individual members of them; see Rabinow 1996, chap. 6.
34. This resonates with what is happening in the way people have been setting up new procreative units: one can have reproductive relatedness (quasi kin, friends as family) without relatives: the new kin detach relationships from kinship (Weston 1991).
35. The positive aspect of having breast cancer was for one woman her "relationship with the extended family. I'm stuck with this. It's nice to know that I'm back in the family" (Finkler 2000, 98).
36. Sometimes to embrace all those connected as kin, at other times to detach relatedness from kinship, as in the case of a woman who was urged by her genetic counselor to contact various people she did not count as her relatives (although she referred to them as her "cousin," "uncle," "aunt"; Finkler 2000, 67). Otherwise loosely connected kin are relinked through the emphasis given to shared body and blood bonds.

This may overlay existing ties: "People are compelled to recognize consanguinity even when in the lived world . . . [the] family . . . may be grounded in friendship or sharing of affect and interest" (2000, 206).

37. The breast cancer patients uniformly absolved their ancestors from responsibility for transmitting genetic disease. At the same time, DNA encourages neither the reinvention of the self nor the embellishment of past ancestry (Finkler 2000, 208). It may, however, allow one to claim as an ancestor someone with whom one has no traceable connection but through the DNA, that is, through a history of disease. These data refer to negotiation in family relations; in other circumstances, the revelation of genetic connection may lead to expressions of solidarity—even injunctions of the order that the demonstration of common kinship should lead us to all assuming responsibility for one another. (I am gratedful to Adam Reed [pers. comm.] for this observation.)
38. By interesting contrast with the emphatic kinship perspective recorded by Edwards (2000). As one egg donor put it, "I've just provided the means for the pregnancy, and as far as I am concerned once my eggs have gone, that's fine by me" (quoted in Konrad 1998, 652).

Bibliography

Battaglia, Debbora. "On Practical Nostalgia: Self-Prospecting among Urban Trobrianders." In *Rhetorics of Self-Making*, edited by D. Battaglia. Berkeley and Los Angeles: University of California Press, 1995.

———. "Towards an Ethics of the Open Subject: Writing Culture in Good Conscience." In *Anthropological Theory Today*, edited by H. L. Moore. Cambridge: Polity Press, 1999.

Beer, Gillian. *Darwin's Plots: Evolutionary Narrative in Darwin, George Eliot and Nineteenth Century Fiction.* London: Routledge & Kegan Paul, 1983.

Biagioli, Mario. "The Instability of Authorship: Credit and Responsibility in Contemporary Biomedicine," Life Sciences Forum. *The FASEB Journal* 12 (1998): 3–16.

———. "Scientists' Names as Documents." In *Documents: Artefacts of Modern Knowledge*, edited by A. Riles. Durham, N.C.: Duke University Press, in press.

Chambers, Robert. *Vestiges of the Natural History of Creation.* 1844 reprint. Victorian Library edition, Leicester University Press, 1969.

Coombe, Rosemary. *The Cultural Life of Intellectual Properties: Authorship, Appropriation, and the Law.* Durham, N.C.: Duke University Press, 1998.

Dolgin, Janet. "Choice, Tradition, and the New Genetics: The Fragmentation of the Ideology of Family." *Connecticut Law Review* 32 (2000): 523–66.

———. *Defining the Family: Law, Technology, and Reproduction in an Uneasy Age.* New York: New York University Press, 1997.

———. "Just a Gene: Judicial Assumptions about Parenthood." *UCLA Law Review* 40 (1990): 637–94.

Edwards, Jeanette. *Born and Bred: Idioms of Kinship and New Reproductive Technologies in England.* Oxford: Oxford University Press, 2000.

Finkler, Kaja. *Experiencing the New Genetics: Family and Kinship on the Medical Frontier*. Philadelphia: University of Pennsylvania Press, 2000.

Franklin, Sarah. "Making Transparencies: Seeing through the Science Wars." *Social Text* 46–47 (1996): 141–56.

Handler, Richard, and Daniel Segal. *Jane Austen and the Fiction of Culture: An Essay on the Narration of Social Realities*. Tucson: University of Arizona Press, 1990.

Haraway, Donna. *Modest_witness@second millennium.femaleman(c)_meets oncomouse [tm]: Feminism and technoscience*. New York: Routledge, 1997.

Helmreich, Stefan. *Silicon Second Nature: Culturing Artificial Life in a Digital World*. Berkeley and Los Angeles: University of California Press, 1998.

Hume, David. "An Inquiry Concerning Human Understanding." 1748. Reprinted in *Essays, Literary, Moral, and Political*. London: Ward, Locke & Co, n.d.

James, Susan. *Passion and Action: The Emotions in Seventeenth Century Philosophy*. Oxford: Clarendon Press, 1997.

Jordanova, Ludmilla. "Interrogating the Concept of Reproduction in the Eighteenth Century." In *Conceiving the New World Order: The Global Politics of Reproduction*, edited by F. D. Gisburg and R. Rapp. Berkeley and Los Angeles: California University Press, 1995.

Konrad, Monica. "Ova Donation and Symbols of Substance: Some Variations in the Theme of Sex, Gender, and the Partible Person." *Journ. Royal Anthrop. Institute* 4 (1998): 643–67.

Latour, Bruno. "Visualization and Cognition: Thinking with Eyes and Hands." *Knowledge and Society: Studies in the Sociology of Culture Past and Present* 6 (1998): 1–40.

Locke, John. *An Essay concerning Human Understanding*. 1690. New ed., London: Ward, Lock & Co., n.d.

Morgan, Derek. "A Surrogacy Issue: Who Is the Other Mother?" *International Journal of Law and the Family* 8 (1994): 386–412.

Ollman, Bertell. *Alienation: Marx's Conception of Man in Capitalist Society*. Cambridge: Cambridge University Press, 1971.

Outram, Dorinda. "Before Objectivity: Wives, Patronage, and Cultural Reproduction in Early Nineteenth Century French Science." In *Uneasy Careers and Intimate Lives: Women in Science 1789–1979*, edited by P. Abir-Am and D. Outram. New Brunswick, N.J.: Rutgers University Press, 1987.

———. *The Enlightenment*. Cambridge: Cambridge University Press, 1995.

Phillips, Jeremy, and Alison Firth. *Introduction to Intellectual Property Law*. London: Butterworth, 1990.

Rabinow, Paul. *Essays on the Anthropology of Reason*. Princeton, N.J.: Princeton University Press, 1996.

Ragoné, Helena. *Surrogate Motherhood: Conception in the Heart*. Boulder, Colo.: Westview Press, 1994.

Ritvo, Harriet. *The Platypus and the Mermaid, and Other Figments of the Classifying Imagination*. Cambridge, Mass.: Harvard University Press, 1997.

Rose, Mark. *Authors and Owners: the Invention of Copyright*. Cambridge, Mass.: Harvard University Press, 1993.
———. "Mothers and Authors: Johnson versus Calvert and the New Children of our Imagination." *Critical Inquiry* 22 (1996): 613–33.
Shapin, Steven. *A Social History of Truth: Civility and Science in Seventeenth-Century England*. Chicago: University of Chicago Press, 1994.
Strathern, Marilyn. *The Relation: Issues in Complexity and Scale.* Cambridge: Prickly Pear Pamphlets no. 6, 1995.
———. "Refusing Information." In *Property, Substance, and Effect: Anthropological Essays on Persons and Things*, edited by M. Strathern. London: Athlone Press, 1999.
Weston, Kathleen. *Families We Choose: Lesbians, Gays, Kinship*. New York: Columbia University Press, 1991.
Woodmansee, Martha. *The Author, Art, and the Market: Re-reading the History of Aesthetics*. New York: Columbia University Press, 1984.

8.

Beyond Authorship

Refiguring Rights in Traditional Culture and Bioknowledge

PETER JASZI AND MARTHA WOODMANSEE

An author in the modern sense is the creator of unique literary or artistic works, the originality of which warrants their protection under laws of intellectual property—Anglo-American *copyright* and European *authors' rights*. This notion is so firmly established that it persists and flourishes even in the face of contrary experience. Experience tells us that our creative practices are largely derivative, generally collective, and increasingly corporate and collaborative. Yet we nevertheless tend to think of *genuine* authorship as solitary and originary.

This individualistic construction of authorship is a relatively recent invention, the result of a radical reconceptualization of the creative process that culminated less than two centuries ago in the heroic self-presentation of Romantic poets. In the view of poets from Herder and Goethe to Wordsworth and Coleridge, genuine authorship is *originary* in the sense that it results not in a variation, an imitation, or an adaptation, and certainly not in a mere reproduction, but in a new, unique—in a word, "original"—work, which, accordingly, may be said to be the property of its creator and to merit the law's protection as such.[1]

With its emphasis on originality and self-declaring creative genius, this notion of authorship has functioned to marginalize or deny the work of many creative people: women, non-Europeans, artists working in traditional forms and genres, and individuals engaged in group or collaborative projects, to name but a few. Exposure of these exclusions—the recovery of marginalized creators and underappreciated forms of creative production—has been a central occupation of cultural studies for several decades. But the same cannot be said for the law. Our intellectual property law evolved alongside of and to a surprising degree in conversation with Romantic literary theory. At the center—indeed, the linchpin—of Anglo-American copyright as well as of European authors' rights is a thoroughly Romantic conception of authorship.[2] Romantic ideology has also been absorbed by other branches of intellectual property law, such as the law of patent and trademark; and it informs the international intellectual property regime. In patent it survives today both in figurations of the inventor and in the emphasis, which this body of law shares with copyright, on the "transformative" moment in the creative process.

We suggested above that cultural production necessarily draws upon previous creative accomplishments. For the better part of human history this derivative aspect of a new work was thought to contribute to, if not virtually to constitute, its value. Writers, like other artisans, considered their task to lie in the reworking of traditional materials according to principles and techniques preserved and handed down to them in rhetoric and poetics—the collective wisdom of their craft. In the event that they chanced to go beyond the state of the art, their innovation was ascribed to God, or later to Providence. Similarly, in the sphere of science, invention and discovery were viewed as essentially incremental—the inevitable outcome of a (collective) effort on the part of many individuals applying inherited methods and principles to the solution of shared problems.[3]

It was not until the eighteenth century, and then chiefly in western Europe, that an alternative vision of creative activity focusing on the endowments and accomplishments of the individual genius began to take shape. In a sharp departure from the self-understanding of writers of previous generations, authors in the new Romantic mode viewed their task as one of *transforming* the materials of personal sense experience through the operation of their unique, individual genius. This change of emphasis mystified the writing process, obscuring the reliance of these writers on the work of others. The notion that a technological or scientific breakthrough owes its existence to the genius—the unique creative abilities—of an individual inventor seems to be even more recent. It appears to date only to the third quarter of the nineteenth century.[4] Borrowed from literary discourse, this notion similarly obscures the collective or collaborative element in scientific invention and discovery. Both misrepresentations of creative activity appear to have fostered and been fostered by modern intellectual property law. Like copyright, modern patent emphasizes individual achievement—chiefly by rewarding the identification of a single *genuinely* transformative moment in a process that, instead, has been usually viewed as collaborative, incremental, and continuous.

As a consequence, this body of law tends to reward certain producers and their creative productions while devaluing others. Especially hard hit in this regard is the creative production characteristic of developing areas of the world. This North-South inequity in the distribution of intellectual property is the subject of the present essay. We aim, first, to bring attention to its scope and to the central role of the author/inventor construct in sustaining it. We then turn in Part II to review some of the most visible recent initiatives to redress this inequity. Arguing that such initiatives tend to get dispersed in the "force field" of Romantic proprietorship, we explore in Part III some other ways of thinking and talking about creative production that could prove useful in the coming discussion of an alternative legal order.

I

Consider, first, the way in which our laws of intellectual property dispose of the *cultural heritage*—including stories, sounds, and images of all kinds—of peoples of the so-called developing world as well as of indigenous groups within North American and western European societies. In 1992, the firm of Ferolito, Vultaggio & Sons, known for its AriZona brand iced teas, introduced a new high-alcohol beverage under the label "Original Crazy Horse Malt Liquor." In addition to the name and purported likeness of the revered Tasunke Witko, or Crazy Horse, the label features a generic Indian in a headdress, a beadwork design, the sacred Lakota "medicine wheel" symbol, and (on the verso) the text: "The Black Hills of Dakota, steeped in the history of the American West, home of Proud Indian Nations. A land where imagination conjures up images of blue clad Pony Soldiers and magnificent

Native American Warriors. . . . A land where wailful winds whisper of Sitting Bull, Crazy Horse, and Custer." When it appeared in stores, the new niche beverage, packaged in a large, whiskey-style bottle bearing this label, met resistance from various Native-American communities with which Tasunke Witko had been associated. Throughout his life their revered leader had opposed the introduction of alcohol into Indian communities, they protested, and he had also forbidden the representation or reproduction of his image.[5]

Our stores are full of merchandise created by drawing on traditional cultural materials in this way. So accustomed have we become to seeing it that we may fail to notice the problem it poses: the traditional communities in which valued images, patterns, designs, and symbols of this kind originated rarely share in the profit from, and often, as in this example, may not even condone their exploitation by entrepreneurs in the creation of new products of "value."

Under our reigning national and international laws of intellectual property, traditional communities like the Lakota Sioux do not have rights in their cultural heritage. Were copyright to be recognized in the artwork that constitutes this heritage, doctrines of "economic right" would enable these communities to forbid its commercial exploitation, or to dictate the terms and conditions under which exploitation could occur. In most countries they would also enjoy a measure of additional protection under parallel and independent doctrines of moral right, giving them (and their successors) legal authority to prevent the misattribution or derogatory distortion of their works—even by those who have been authorized to exploit the works economically. But in the absence of a work of authorship none of these legal doctrines can apply.

Traditional patterns and symbols like those reproduced by Ferolito, Vultaggio & Sons are not works of authorship because to qualify, a text must have been created by an identifiable individual or individuals—or a corporation acting as an individual—and must exhibit "originality," as copyright doctrine terms the traces of new creativity that are entailed by such a provenance. The source of the medicine wheel and other symbols at issue in the collective culture of the Sioux community precludes the identification of individual authors and prevents them from qualifying as original—indeed, their cultural value resides in their fidelity to, rather than any divergence from, the age-old symbols that have been transmitted over generations within this community.[6]

From the point of view of intellectual property law, these symbols reside in the public domain, so in appropriating them to market its new beverage Ferolito, Vultaggio & Sons is legally within its rights. Even as the law offers little or no aid to the indigenous community from which the symbols have been extracted, it rewards such entrepreneurs who "add value" by revising or recontextualizing traditional imagery. Such marginal added value constitutes original authorship, justifying a copyright in the resulting design as a so-called derivative work. In consequence, if another distributor of beverages were to copy that design, Ferolito, Vultaggio & Sons could bring suit for infringement of its copyright (to say nothing of the additional trademark rights it enjoys above and beyond copyright as a result of its commercial use of the symbols represented on that label). By virtue of the emphasis it places on innovation (however insignificant in quantity or quality), intellectual property law thus not only fails to discourage the appropriation of traditional culture, but actually rewards and promotes it.

Let us turn to the way in which intellectual property law disposes of the scientific heritage of traditional communities.[7] We refer to the appropriation of their bioknowledge by northern pharmaceutical, biotech, agricultural, and personal care industries in search of newer and better pesticides, cosmetics, and cures for the world's illnesses. The huge number of plant species—which is estimated at between 250,000 and 750,000 worldwide[8]—makes random "prospecting" for those with commercial potential unfeasible, so these industries depend on the bearers of traditional knowledge to identify those plants likely to prove useful. According to one estimate, three-quarters of the plants that provide the active ingredients in our prescription drugs first came to the attention of researchers because of their use in traditional medicine.[9] Yet here again, those who led them to these plants—the communities in which knowledge of the plants' curative potential originated and has been handed down—do not share in the huge profits that these prescription drugs produce when they are brought to market. To date, such ethnobotanical prospecting has led primarily to the development of "new" compounds, including pharmaceuticals and pesticides, that employ chemicals harvested from plants as their active ingredients. In the future, however, we can expect more and more of these new compounds to employ synthetic versions of the chemicals originally isolated from wild plants, rather than actual derivatives.

Consider the much publicized case of the rosy periwinkle. This plant species was first harvested in Madagascar for pharmaceutical use, and the two complex alkaloids isolated from it (vinblastine and vincristine) now form the basis of compounds used in anticancer chemotherapy. Formulations of these active ingredients have proved particularly effective against childhood leukemia and Hodgkin's disease and now earn the Ely Lilly pharmaceutical company an estimated $100 million a year.[10] But while Lilly still harvests the periwinkle to produce these medicines, it has left Madagascar behind.[11] Lilly no longer relies on the island as the primary source of this "raw material." The plant, which grows readily in warm climates throughout the world, is now widely cultivated in the Philippines and Texas. Carrying this process of alienation one step further, in a trend that almost certainly represents the future of drug development, France's Pierre Fabre Laboratories has created an entirely synthetic version of one of the periwinkle-derived alkaloids for the treatment of bronchial and breast cancers.[12]

However the drug is formulated, what has been appropriated in the process of its development and commercialization is not so much the botanical materials as something more abstract and intangible: indigenous peoples' knowledge of the beneficial properties of those materials. Such bioknowledge is exactly the sort of commodity of the mind that intellectual property law values and protects. As useful scientific information it falls squarely within the domain of patent law. Yet under patent doctrine it is not eligible for protection. Why?

Much as copyright requires the agency of an individual creative "author," so patent demands the agency of a personalized "inventor" whose genius produces innovations that surpass the prior art by virtue of their novelty. Through his or her efforts, the inventor transforms known preexistent raw materials—as traditional bioknowledge would be figured in patent discourse—into something useful and new. Thus, the people of Madagascar, the custodians of the crucial knowledge of the periwinkle's curative properties, do not count as inventors under patent doctrine any more than do the Lakota Sioux as authors, and they are not eligible for patent protection. Protection goes rather to the entrepreneurial pharmaceutical, Ely Lilly, which, having relied on their knowlege to identify the promise of the periwinkle, has gone on to engineer its active chemical ingredients so as to improve it for commercial application. Such improvements, although marginal, qualify Lilly as an inventor, justifying the award of a patent. The availability of such patent protection is what makes it possible for the company to reap profits on such a large scale.

The people of Madagascar, meanwhile, have received nothing of significance in exchange for their knowledge—not even an assured income from the sale of the plants themselves. These desperately poor islanders are thus rapidly deforesting their country to gain arable land on which to grow subsistence and market crops. Today, less than twenty percent of Madagascar's original forest cover remains. And although ethnobotanical teams of African scientists and students are hurrying to record popular knowledge about the curative properties of other plants, it seems inevitable that much of this lore will be lost with the island's biodiversity.

Herein lies a further disadvantage of the present intellectual property regime. The developing areas of the world in which most of the as yet untapped plant species are most prevalent, the great tropical forests, are typically also the poorest. With few available sources of income, not even from their valuable bioknowledge—profits from which go to the northern drug companies—the peoples in these areas of the world have no choice but to consume their heritage in an effort to survive. When this occurs, we all lose—peoples of the developed and developing world alike. For with the disappearance of the great forests, popular knowledge of the curative properties of their diverse flora—their crucial biolore—will rapidly disappear as well, leaving the drug companies to prospect randomly in what remains of nature—a scenario that is not financially feasible.

Such nonoptimal outcomes are the product of our intellectual property regime, and more particularly, of the conception of creative production that lies at its center. This body of law figures creative production as essentially individual and originary. Accordingly, it views the critical creative moment in both of these examples to lie in the transformative activity of the two entrepreneurs—Ferolito, Vultaggio & Sons, and Ely Lilly. Having been handed down by tradition, the designs, images, and lore on which these companies operate lack an identifiable author or inventor. Intellectual property law thus regards them as naturally occurring raw materials, which lie available to all for the taking. Not in themselves the locus of value, they acquire value through the creative activity of the entrepreneurs who transform them into beverage brands and internationally marketable drugs.

II

There have been a number of efforts to address this problem over the past three decades. Here we will review only a few of the most visible and suggest why they have foundered. Until recently the primary focus of such efforts has been traditional *cultural* heritage. Thinking about recognizing legal rights in the *scientific* heritage of indigenous peoples is, by contrast, still in a very early stage.

An early, tentative effort to address the dilemma of indigenous intellectual property may be seen in the Act of the Berne Convention for the Protection of Literary and Artistic Works (1971). While protection of so-called folkloric works—that is, "traditional creations of a community such as the so-called folk tales, folk songs, folk music, folk dances, [and] folk designs or patterns"[13]—is not mandated by the treaty, Article 15(4)(a) does give countries bound by the Berne Convention the option of adopting local legislation to afford protection "in the case of unpublished works where the identity of the author is unknown, but where there is every ground to presume that he is a national of a country of the Union. . . ." Where it might be extended, therefore, such protection would be available only on the basis of the legal fiction that the work is in fact the creation of one or more "unknown" (but otherwise qualifying) individual authors.

At the most practical level, the difficulty with this invitation to shoehorn traditional culture into national law lies in the potential for resistance in the core copyright concepts that are not addressed in the provision. Though the idea of authorship may bend a little, it will not bend much, with the result that most of the content of traditional culture would fail to qualify (by virtue of its lack of "originality") even under the fictionalized standard of Article 15(4)(a). It is difficult to imagine, for example, how the fiction could accommodate the Lakota medicine wheel. In fact, Article 15(4)(a) does not appear to have inspired any domestic legislation. Nevertheless, its general approach to the problem of inserting traditional culture into the scheme of copyright, and the shortcomings of that approach, are reflected in subsequent proposals to extend intellectual property protection to traditional cultural heritage.

The misfit between copyright and the forms of creative production that are most characteristic of peoples of the developing world found explicit international acknowledgment in 1982, which saw adoption by the World Intellectual Property Organization (WIPO) and UNESCO of a set of recommended Model Provisions for National Laws on the Protection of Expressions of Folklore against Illicit Exploitation and Other Prejudicial Actions. The "expressions of folklore" to which the Model Provisions were designed to apply include "productions consisting of characteristic elements of traditional artistic heritage developed and maintained by a community . . . or by individuals reflecting the traditional artistic expectations of such a community." The terms of the Model Provisions would penalize unauthorized economic exploitation of such materials outside the traditional or customary context, and against what might be thought of as "moral rights" offenses—for example, false attribution, or the kind of derogatory distortion of materials drawn from folkloric tradition. The right to enforce these prohibitions might be allocated differently in different national implementations of the provisions—in some to the communities that are the custodians of a tradition, and in others to a state agency or state-designated "competent authority."[14]

Unfortunately, however, there has been relatively little significant implementation of the WIPO-UNESCO Model Provisions.[15] The reason, we suspect, is that despite the drafters' recognition that copyright cannot easily be applied to protect traditional cultural materials, the sui generis approach of the Model Provisions does not really go far—or at least not far enough—to escape the "force field" of copyright. Although the Model Provisions do not employ the terminology of copyright discourse—terms such as *author, work,* and *originality*—they preserve the general structure of copyright doctrine, with its conventional subdivisions of economic and moral rights. The Model Provisions focus exclusively on the thing itself—the expression of folklore—and by necessary implication on protection of the creative investments that went into its production, rather than on preservation of the cultural processes that gave rise to it and the values it expresses. Although the author-function of conventional copyright discourse is displaced onto representatives of the community, or a designated competent authority, it is still recognizable as such. So while the Model Provisions incorporate more sophisticated insights into the nature of the problem of providing appropriate legal protection for traditional cultural materials than does Article 15(4)(a) of the Berne Convention, they ultimately stumble on the same obstacle.

To turn to efforts to craft protection for the scientific heritage—the bioknowledge—of indigenous peoples, there has been a recent, if tentative, initiative of significance at the international level: the United Nations Convention on Biological Diversity, concluded at the Earth Summit in Rio de Janeiro in 1992. Article 8(j) of the treaty mandates signatories to take measures to "respect, preserve and maintain knowledge, innovations, and practices of indigenous and local communities embodying traditional lifestyles relevant for the conservation and sustainable use of biological diversity," and requires governments to assure that such knowledges are used with the approval of the communities in question, and consistent with the principle of "equitable sharing of the benefits" resulting from their use. Whether and how these principles will be implemented, and what role intellectual property rights may play in that implementation, remains to be seen. Clearly, however, they need not be implemented through the adaptation of existing intellectual property rights or by the articulation of new ones. Thus, in the very tentativeness of its approach, which opens a space for the development of new nonintellectual property-based legal mechanisms, the Biodiversity Convention arguably represents an advance over earlier efforts to protect traditional cultural and scientific heritage by incorporating it into Eurocentric models of rights in intangibles—especially when we compare it to the provisions of the Agreement on Trade-Related Aspects of Intellectual Property Rights (TRIPS Agreement) which constituted Annex IC of the Marrakesh

Agreement Establishing the World Trade Organization, concluded in April 1994. TRIPS binds signatory nations to provide enhanced protection for pharmaceutical and chemical innovations of companies that exploit traditional bioknowledge, but it contains no imperative for the protection of that bioknowledge itself.

Specifically, Article 27(1) of TRIPS requires protection for inventions "without discrimination as to the . . . field of technology," a reference designed to assure (among other things) that nations that did not protect pharmaceuticals by patent would be required to do so.[16] But because traditional bioknowledge is not new and does not involve an inventive step, it falls outside the category of mandatory patent subject matter that TRIPS Article 27(1) defines.

In addition, TRIPS Article 27(3) requires the protection of "plant varieties either by patents or by an effective *sui generis* system or by any combination thereof." In other words, national laws must provide for Western-style intellectual property protection, premised on innovation, in new versions or adaptations of naturally occurring plant species—a mandate that leaves no space for the protection of traditional bioknowledge.[17] Herein lies the key to the difference between the approach of TRIPS and that of the Biodiversity Convention. As one South African commentator put it, the TRIPS agreement "sees knowledge as belonging to the public domain [and] views Indigenous Knowledge in terms of Intellectual Property which should be protected within the Intellectual Property Rights regime, based on Western notions of individual ownership. The [Biodiversity Convention] on the other hand, focuses on communal ownership. Accordingly, knowledge is viewed as being owned by the local community in whose customs, practices and traditions it is embedded."[18]

Increasingly, activists in the cause of promoting biodiversity through the protection of traditional knowledge have come to view TRIPS as not merely irrelevant to their objectives but potentially inimical. One recent commentary, for example, asserts that by requiring life patents and plant variety protection, TRIPS overrides two basic assumptions of the Biodiversity Convention: "that intellectual property is a matter of national sovereignty and policy, and that life forms are part of the public domain," because "biodiversity represents a cultural and ecological heritage developed over generations and upon which our collective survival depends. Subjecting this heritage to a legal regime of commercial monopoly rights under TRIPS will destroy the conditions for its conservation and sustainable use, especially by the communities, and thereby destroy society's access to diverse food and medicine."[19]

Due in part to the impetus of the Biodiversity Convention, a reconsideration of approaches to the legal protection of indigenous knowledges and traditional cultural materials is underway. Academic literature on the topic is proliferating,[20] and WIPO has created a new Global Intellectual Property Issues Division, whose charge includes promoting intellectual property rights for new beneficiaries, and whose jurisdiction cuts across the traditional categories of expressions of folklore and bioknowledge. Notably, representatives of the peoples and communities who are the custodians of such bodies of cultural heritage are directly involved in the discussion—both at the invitation of international organizations and as the result of their own initiatives. An outstanding example is the Mataatua Declaration of the First International Conference on Cultural and Intellectual Property Rights of Indigenous Peoples, drawn up by an assembly of over 150 delegates from fourteen countries meeting in New Zealand in June 1993. The declaration includes the statement that "indigenous peoples are the exclusive guardians of their knowledge," and as such must be the ones to define it, must be first the beneficiaries of it, must be respected for their right to create new knowledge or discover new aspects of traditional knowledge, and must be the ones to decide whether to protect, promote, or develop their knowledge.

Yet another factor contributing to the present sense of urgency surrounding issues of indigenous knowledge and cultural heritage is the coincidence of the fiftieth anniversary of the Universal Declaration of Human Rights. Article 27.2 of the Declaration affirms the right of every person to "protection of the moral and material interests resulting from any scientific, literary or artistic production of which he is the author." The inadequacy of this formulation—which may also be found in Article 15.1 of the International Covenant of Economic, Social and Cultural Rights—will be immediately apparent: it constructs creative activity individualistically, placing the creative production most typical of indigenous peoples squarely outside the scope of the Declaration.[21]

Finally, we would call attention to a fortuitous geopolitical coincidence that has probably done more than any other recent development to put the issue of legal protection for traditional knowledges and cultural materials on the world agenda and create a real possibility that, sometime within the next five years, a new international treaty addressing rights in cultural heritage may be concluded. This was the procedural linkage of this issue with a substantively unrelated issue—that of protection for databases—that occurred at the December 1996 WIPO Diplomatic Conference in Geneva, Switzerland.

The agenda of the Diplomatic Conference called for the delegates of the 127 nations represented in the WIPO to consider three draft treaties. Two of these, dealing primarily with issues of copyright and neighboring rights in the digital environment, were concluded and signed: the WIPO Copyright Treaty and Treaty on Protection of the Rights of Performers and Producers of Phonograms. The third, a proposed agreement on Rights in Collections of Information, which had been injected into the agenda at the last moment by the United States and the European Union on behalf of their domestic database industries, was not. This initiative met the resistance of delegates of developing nations who perceived that it would mandate new international and domestic sui generis protection for data compilations which, consisting of unoriginal facts, have always fallen, by definition, outside the scope of conventional copyright law. In denouncing the initiative, they pointed out that the problem of securing effective protection for traditional cultural materials and knowledges had been under international discussion, without significant progress, for a generation, notwithstanding its importance to developing peoples and nations. Why, they asked, should the conceptually equivalent problem of data rights, in which the developed nations have the chief stake, receive priority?

The Diplomatic Conference concluded without reaching agreement on the merits of the proposed database treaty, but a procedure and general timetable were established for study and resolution of the issues it raised, and an equivalent procedure was mandated for advancing progress on issues related to the protection of indigenous knowledges and traditional cultural materials. This has already led to the convening of the UNESCO-WIPO World Forum on the Protection of Folklore at Phuket, Thailand, in April 1997, and the WIPO Roundtable on

Intellectual Property and Indigenous Peoples, held in Geneva in July 1998. Having gotten linked to progress toward an international agreement on something as important to the information industries of the developed world as database protection, some kind of treaty protecting the characteristic creative productions of traditional communities, including those in developing countries, now seems likely.

What exactly must such a treaty achieve? There seems to be substantial consensus: Most participants in the discussion agree that what is needed is balance, or, in the memorable phrase of Hong Yongping, a Chinese presenter at the 1997 WIPO Forum, rules assuring "effective protection with reasonable use"[22]—a scheme of protection that simultaneously reflects the special cultural concerns of indigenous peoples and other custodians of traditional knowledge and at the same time permits continued utilization of their works on reasonable terms as the basis of new cultural productions, pharmaceuticals, crop varieties, and the like. There is also widespread acceptance that any scheme of protection should respond to the principle of "fair sharing of benefits" articulated in the Biodiversity Treaty. The question is how to accomplish such balance between control and access, while assuring equitable distribution of the fruits of exploitation. The terms of the coming discussion—the dominant metaphors and tropes around which it will be organized—are crucial.

III

In the past, public discussion about control over and access to productions of the mind had been personalized around such metaphorical figures as the author and the inventor. But the figure of the individual creative genius cannot be used to structure discussion about legal rights in traditional knowledges. Still, metaphorization of the discussion seems inevitable, so the choice of an organizing trope matters. Already, a battle for discursive dominance is under way between two diametrically opposed alternatives drawn from the realm of economic discourse: the notion that "information wants to be free" and the opposing notion of the "tragedy of the commons." The two tropes have a common starting point in their characterization of traditional knowledge prior to legal intervention as a *public good*—a commodity that is not fenced off by any barriers to impede public access and use. For the purpose of both tropes, the original state of this information is figured as a version of the commons. The tropes part company in the conclusions they draw from this characterization.

The notion that information wants to be free, familiar to those who read into the history of copyright, has been given a new lease on life by the spread of electronic communication. John Perry Barlow appeals to it when he urges that the Internet be left alone—unregulated.[23] Barlow, and other commentators who deploy this essentializing trope, make the further claim that, especially in the electronic environment, attempts to regulate information are not only unavailing but threaten the good information order. "[T]he increasing difficulty of enforcing existing . . . laws," he writes, "is already placing in peril the ultimate source of intellectual property—the free exchange of ideas."[24]

This way of figuring the nature of information is generally associated with progressive positions on issues relating to the legal status of traditional knowledge. In her recent book, *Biopiracy: The Plunder of Nature and Knowledge*, Vandana Shiva invokes the "free" character of genetic information to denounce Western efforts to reduce traditional knowledge to ownership through the patenting of new derivative pharmaceuticals and plant varieties: "Biotechnology, as the handmaiden of capital in the post-industrial era," she writes, "makes it possible to colonize and control that which is autonomous, free, and self-regenerative."[25] The same position was expressed in the much publicized controversy surrounding W. R. Grace's patenting of a pesticide made of ground Neem seeds that critics claim has been used in India for centuries. "The real battle," Jeremy Rifkin, who spearheaded a challenge of the patent, is quoted in the *New York Times* as saying, "is whether the genetic resources of the planet will be maintained as a shared commons or whether this common inheritance will be commercially enclosed and become the intellectual property of a few big corporations."[26]

However, transnational corporations and governments acting on their behalf also mobilize the trope of "free" information to some effect. A particularly notorious example was the April 14, 1992, memorandum from vice-presidential staff members John Cohrssen and David McIntosh to Dan Quayle's Chief of Staff Bill Kristol, written "to alert [Kristol] to serious problems with the draft international convention on biological diversity." The pertinent part of the memo—which played a significant role in delaying the U.S. signing of the treaty—claimed that under the treaty:

- Special legislation would need to be passed for the benefit of indigenous populations, i.e., American Indians, since the draft convention has special provisions for them [and]
- It would greatly increase litigation because of new compensation legislation [that] would need to be passed[,] as the draft treaty contains a vaguely worded provision to establish liability and a right to compensation for damage to biodiversity.
- The draft convention proposes to regulate biotechnology, in a manner totally unacceptable to the US: to restrict domestic and international commerce in biotechnology related products.[27]

This text is a powerful invocation of the trope of inherently, essentially, naturally "free" information. But it mobilizes that trope to purposes dramatically different from those of Shiva and Rifkin—to argue that any interference with the ability of U.S. companies to exploit indigenous bioknowledge represents an unacceptable departure from the status quo.

This bivalent trope clearly has limitations, then, as an organizing structure around which to build discussion about future legal regulation of access to traditional knowledge. Not the least of the trope's limitations is that discussion organized around it will not escape the "force field" of the author-inventor figure that has long exerted such a powerful influence over discussion of rights in information. To figure information—including traditional knowledge—as not having been created by anyone at all and thus not susceptible to ownership is simply to invert the trope of authorship.

The notion that information is free, a public good like air and water that one ought to be able to draw upon at will, also gives rise to a powerful countermetaphor, the so-called "tragedy of the commons"—invoked to justify reducing commonly owned (or unowned) things to the status of property. The trope became popular in environmental literature during the 1960s, where it was argued that since one only takes care of things one owns, resources held in common—unowned and unprotected by anyone—are (inexorably) doomed to be over-exploited.[28] Although the utility of the "tragedy of the commons" metaphor has been extensively questioned in scientific and economic literature,[29] it appears to be achieving new currency in the law—including intellectual property law—where it functions as an easy-to-grasp and poignant shorthand for the larger neoclassical economic principle that, to quote Neil Netanel, "private entitlements can best promote allocative efficiency when would-be users must pay the price agreed upon by the entitlement holder in a voluntary exchange."[30]

Like its mirror image, the trope of free information, this trope too is bivalent. Just recently it was successfully invoked by large corporate copyright owners to argue for a twenty year extension of the term of copyright—the Sonny Bono Copyright Term Extension Act of 1998. In the congressional testimony of Disney, Time-Warner, and others, the *public domain*—a commons resulting from the expiration of limited terms of protection in copyrighted works—was consistently figured as a kind of informational dumping ground, littered with abandoned movies, songs, and the like that, because no owner had an economic motivation to bring them to market, were in practice unavailable for public use.[31] Yet the trope is also being mobilized in defense of what might be viewed as progressive objectives. Thus, one writer has recently invoked it to argue for new legal norms to promote the preservation of cultural heritage by discouraging the black market in stolen artifacts.[32] More emphatically—and more controversially—Joseph Henry Vogel has argued from the "tragedy of the commons" that the best hope for the preservation of biodiversity lies in the creation of a comprehensive scheme of intellectual property rights, modeled on existing patent and copyright regimes, in genetic information. However, his advocacy of this market model gives cause for suspicion. Among his "Ten Principles for Conserving Genetic Information" is this one: "Endorse legislation giving equal protection to artificial and natural information [and] at the same time attenuate the ability to alienate the new property rights"[33]—that is, endow indigenous communities with rights in their bioknowledge, but restrict their freedom to commercialize their new property. This extraordinary qualification reflects Vogel's doubt that indigenous peoples will be able to enact their part as rational profit maximizers in his scheme of conservation-by-privatization, and indeed his doubt may be well placed. The relationship of the bearers of cultural traditions to their traditions is surely more complex.

Conceptually, this bivalent trope of the "tragedy of the commons" does not escape the gravitational pull of possessive individualism any more than the competing argument to the effect that information wants to be free. In the mode of analysis associated with the "tragedy of the commons," effective social ordering is closely linked to property ownership. In this discourse one of the primary characteristics of the property owner is that his or her relationship to the thing owned is rooted in self-interest. The person in whom rights are vested in an effort to avoid over-exploitation of a resource is presumed to be motivated to put that resource to its best and highest use—in order to maximize benefits and minimize his or her costs. Similarly, by virtue of his or her creative investment, the author of copyright law—the exemplary "possessive individual"—is literally responsible for a work, both reaping the benefits of its exploitation and bearing the associated costs (such as the risk of censure or prosecution).[34]

Arguments for protection of the environment through the privatization of genetic information ignore the possibility that factors other than immediate self-interest may shape the relationship of indigenous peoples to their intangible heritage. While emphasizing how indigenous groups may promote the conservation of nature as rights holders bargaining with prospective users in a transactional marketplace, such arguments fail to recognize the importance of these groups' role as the custodians, for the time being, of living traditions. By denying these custodial interests, which escape the market, reliance on the pro-enclosure "tragedy of the commons" metaphor as an organizing trope would seriously distort the coming discussion of new rights regimes for the protection of traditional culture and bioknowledge. Just as, inevitably, it would focus attention on the cultural or informational objects to be protected, and away from the processes that produce or sustain them.

We conclude by sketching the outlines of an alternative metaphor for organizing discussion of future law governing access to traditional knowledge and cultural heritage—that of "sustainable development," familiar from the environmental literature of the last three decades.

Simply put, this notion addresses the observation that the environment cannot sustain the current pace and manner of economic expansion, that this pattern of development is, in a word, unsustainable. But it does not address this problem of environmental degradation by prioritizing environmental protection pure and simply; rather, acknowledging the continuing need for development—for industrialization of impoverished parts of the world especially—it urges instead the balanced approach captured in the notion of sustainable development: a "process of change," to quote from *Our Common Future*, the 1987 report of the World Commission on Environment and Development that first brought global attention to the idea, "in which the exploitation of resources, the direction of investments, the orientation of technological development, and institutional change are all in harmony and enhance both current and future potential to meet human needs and aspirations."[35]

Since the appearance of this report there has emerged a substantial body of interdisciplinary literature devoted to defining and developing political, economic, and legal instruments to achieve this goal of continued, but sustainable development. The notion vaulted into prominence, however, at the Rio Summit that led to the United Nations Convention on Biological Diversity—because it seemed a useful vehicle for harmonizing North-South political differences.

How might interdisciplinary conversation about the development of norms and practices for the protection of traditional knowledge and cultural heritage be advanced by adopting cultural sustainability as the organizing metaphor? What advantages does this trope have over authorship (and other cognate concepts) around which intellectual property law historically has been organized? And why might it be a more fruitful basis for discussion than either free information or the "tragedy of the commons," the new economic tropes that we have identified as false alternatives to authorship as controlling metaphors, ineffective precisely because they fail to escape the gravitational pull of the authorship concept itself?

However tentatively, we would suggest that a discussion refracted through the lens of cultural sustainability might succeed—where one organized by means of other metaphors ultimately would fail—in transcending the near-exclusive emphasis on the nexus between the maker and the specific products of his or her creative efforts that dominates conventional intellectual property discourse. As we have argued, that discourse is marked by a strong individualistic emphasis, which makes it difficult to think and talk clearly about instances in which cultural work is carried forward by or within groups. Just as, characteristically, intellectual property law thinking tends to approach issues of cultural policy by defining issues and solutions in terms of things to be (or not to be) protected; in doing so it risks missing what is both most valuable to, and most valuable about, the cultural work of indigenous communities: the means by which their custodianship over various cultural objects and bodies of information is carried forward. Thus, for example, the larger question raised by the case of the rosy periwinkle is not how the bearers of a specific item of knowledge about the properties of a specific plant might have been afforded some economic return in connection with its exploitation, but what measures would have been necessary to maintain the systems within which that item of knowledge and others like it were preserved, to assure their continued availability to the human community at large.

Adoption of the metaphor of cultural sustainability would represent an acknowledgment that maintenance of traditional knowledge systems within living communities should be the first-order goal of any new legal initiatives to safeguard traditional culture, and that, compelling as are equity arguments for compensation to indigenous peoples whose knowledge is commercialized, such compensation is only a means—and only one means—by which to accomplish that goal; in some situations, it will be better served by affording greater rights to traditional communities, endowing them (for example) with the absolute authority to withhold sacred knowledge from the marketplace. By the same token, however, because (like other invocations of the sustainability concept) cultural sustainability is premised on balancing the need to use resources with the need to assure their continued availability, the metaphor also inherently recognizes the potential risk of overprotection: a potential rights regime that gave traditional communities the ability to bar dissemination of the proverbial botanical AIDS cure would be subject to criticism within the discursive framework established by adopting the proposed metaphor. Indeed, as we have suggested, the central tension in the policy discussion concerning indigenous cultural rights—as with that relating to any system of knowledge regulation—is between the impulse toward control and the impulse toward access. Unlike other available metaphors, sustainability has the important advantage of containing a built-in recognition of both of these conflicting impulses.

Specifically, a sustainability-based approach might help everyone engaged in the discussion of initiatives to safeguard traditional culture to:

- Recognize more fully the critical custodial role that indigenous peoples play in maintaining valuable traditions and bodies of knowledge, and acknowledge the ways in which that complex role differs from one of conventional ownership or proprietorship;
- Ask and answer questions about how a wide range of possible social or legal policies (including, but not limited to, new rights regimes) might encourage desirable forms of collective social behavior in relation to traditional knowledge and cultural heritage;
- Refocus attention in connection with legal measures and initiatives away from the consideration of individual entitlements, and toward an accounting of the cultural requirements of particular traditional communities;
- Avoid the unfruitful binary of "ownership–no ownership" in considering whether (and if so, what) regulation of the use of traditional knowledge and cultural heritage may be appropriate;
- Take into account the collective interests of reusers and consumers of information, both outside traditional communities and within them.

A discussion conducted in terms of cultural sustainability would be appropriate to the consideration of new rights regimes, such as the proposals for "Community Intellectual Rights" (CIR), which were first proposed by the Third World Network in 1994 and are now gaining currency in Latin America and parts of Africa.[36] Likewise, it could guide further discussions of initiatives to mobilize the content of existing legal regimes (including, but not limited to those of intellectual property) into bundles of rights that could be deployed by traditional communities to protect their knowledge—the so-called Traditional Resource Rights (TRR) approach.[37]

Moreover, the lens of cultural sustainability could be profitably employed to examine proposals and projects to address the gaps in national and international legal safeguards for traditional culture through private legal ordering, such as the well-publicized 1992 agreement between Costa Rica's National Institute of Biodiversity (INBio) and the Merck pharmaceutical company.[38] Likewise, it could be applied in assessing the benefit-sharing approach adopted in the mid-1990s by the Shaman pharmaceuticals firm,[39] or the more recently announced contract between an Indian government research institute and a local traditional community to share the benefits of a medicine based on the active ingredient of a plant to which its members directed research scientists.[40]

In addition, and perhaps most importantly, any discussion of the future of legal measures to safeguard traditional heritage conducted in terms of the metaphor of cultural sustainability would, by its nature, be one in which traditional communities and their representatives would be full participants. Only through the fullest possible consultation will it be possible for policymakers to determine what legal measures actually will function to help maintain the processes by which culture is conserved, transmitted, and elaborated within those communities—as any inquiry based on cultural sustainability requires. Perhaps because conventional intellectual property rights constitute part of the conventional framework of Western law, legal experts in developed countries have long been ready to prescribe intellectual property rights–based approaches to traditional culture and traditional science. Not surprisingly, as we have detailed above, these proposals have been largely ineffective. Increasingly, however, traditional communities are finding their own voices. The deep logic of cultural sustainability would help to reinforce their demands to be heard.

Notes

The authors would like to acknowledge gratitude to the Max Planck Institute for the History of Science for providing the impetus and a lively forum for this work at its International Conference on Science, Technology, and the Law, Berlin, August 23–27, 1999.

1. See Martha Woodmansee, "The Genius and the Copyright: Economic and Legal Conditions of the Emergence of the 'Author'"; reprinted in Woodmansee, *The Author, Art, and the Market*, 35–55.

2. See Peter Jaszi, "Toward a Theory of Copyright: The Metamorphoses of 'Authorship.'"
3. See Christine MacLeod, "Concepts of Invention and the Patent Controversy in Victorian Britain." MacLeod quotes Isambard Brunel's succinct expression of this view in his 1851 memoirs: "I believe that the most useful and novel inventions and improvements of the present day are mere progressive steps in a highly wrought and highly advanced system, suggested by, and dependent on, other previous steps, their whole value and the means of their application probably dependent on the success of some or many other inventions, some old, some new. I think also that really good improvements are not the result of inspiration; they are not, strictly speaking, inventions, but more or less the results of an observing mind, brought to bear upon circumstances as they arise, with an intimate knowledge of what already has been done, or what might now be done, by means of the present improved state of things, and that in most cases they result from a demand which circumstances happen to create" (147). See also Edith Tilton Penrose, *The Economics of the International Patent System*, esp. 19–41.
4. MacLeod, "Concepts of Invention," esp. 150–53. See also MacLeod, *Inventing the Industrial Revolution*, esp. chaps. 10 and 11.
5. On this episode, see Nell Jessup Newton, "Memory and Misrepresentation"; Peter Jaszi and Martha Woodmansee, "The Ethical Reaches of Authorship," esp. 961–63; and Rosemary Coombe, *The Cultural Life of Intellectual Properties*, 199–207.
6. The venerable age of the medicine wheel, which contributes so much to its cultural value, also weighs against its eligibility for meaningful protection because many of the rights awarded to creators under intellectual property law are limited in duration—economic rights under copyright, for example, endure for the lifetime of an author plus 70 years.
7. On the differences between science and lore, see Arun Agrawal, "Dismantling the Divide between Indigenous and Scientific Knowledge."
8. These are the figures of Manuel F. Balandrin et al., "Natural Plant Chemicals," 1157.
9. See Steven R. King, "The Source of Our Cures," 19.
10. See Edward O. Wilson, "Threats to Biodiversity," 116.
11. Ironically, the information about the properties of the rosy periwinkle that first drew Ely Lilly's researchers to Madagascar did not even come from the indigenous knowledge base of that society. As it turns out, the investigation of the periwinkle began because Filipino and Jamaican folklore suggested that a tea brewed from its leaves could be a remedy for diabetes (see Karen Ann Goldman, "Compensation for Use of Biological Resources," 717 n. 131). If anyone deserves compensation for appropriated bioknowledge in the case of the periwinkle, then it is perhaps the Filipino and Jamaican communities in which this folklore was preserved rather than the people of Madagascar.
12. See Anne Jeanblanc, "Fighting Cancer on Many Fronts," 42–43.
13. WIPO, "1967, 1982, 1984," 5.

14. On the Model Provisions and their limitations, see Christine Farley, "Protecting Folklore of Indigenous Peoples," 44–46. The history of the Model Provisions is addressed, among other topics, by contributors to Peter Seitel, ed., *Safeguarding Traditional Cultures: A Global Assessment*, the proceedings of a conference convened in connection with the ten-year review of the 1989 UNESCO "Recommendation on the Safeguarding of Traditional Culture and Folklore." Another useful recent status review is the WIPO's *Intellectual Property Needs and Expectations of Traditional Knowledge Holders*, based on a series of regional conferences and fact-finding missions.
15. Darel A. Posey and Grant Dutfield report that "a number of African countries, such as Nigeria, have enacted legislation based, at least in part, on the Model Provisions" but do not give specifics (*Beyond Intellectual Property*, 100).
16. Article 27 (2), which permits states to make limited exclusions from patentability, applies only to inventions that would, if commercialized, threaten public order or morality. The only concessions to less-developed countries on this issue are those found in Articles 65 and 66, which allow such countries four to ten years (depending on their stage of development) to phase in TRIPS-compliant domestic legislation.
17. Although originally intended to apply to new plant varieties resulting from human manipulation of biological materials, this provision could be viewed as an invitation to enact national laws of a more comprehensive character, applicable to naturally occurring species as well. The problem is that under TRIPS the object of such protection would be the plant varieties themselves, and not human knowledge concerning their properties.
18. Mongane Wally Serote, "One Fundamental Threshold," 3.
19. Genetic Resources Action International.
20. See esp. Graham Dutfield's 1,400-entry *Annotated Bibliography*. See also Tom Greaves, ed., *Intellectual Property Rights for Indigenous Peoples*; Miges Baumann et al., eds., *The Life Industry*; and Darel Posey and Graham Dutfield, *Beyond Intellectual Property*.
21. Erica-Irene Daes, "Discrimination against Indigenous Peoples," 29.
22. Hong Yongping, "The Experience of Asia and the Pacific Region."
23. Barlow enlists Thomas Jefferson in defense of his cause, quoting Jefferson's characterization of information as by nature a public good: "If nature has made any one thing less susceptible than all others of exclusive property, it is the action of the thinking power called an idea, which an individual may exclusively possess as long as he keeps it to himself; but the moment it is divulged, it forces itself into the possession of everyone, and the receiver cannot dispossess himself of it. Its peculiar character, too, is that no one possesses the less, because every other possesses the whole of it. He who receives an idea from me, receives instruction himself without lessening mine; as he who lights his taper at mine, receives light without darkening me" ("The Economy of Ideas," 85).
24. Ibid., 86.
25. Vandana Shiva, *Biopiracy*, 45.

26. As quoted by John F. Burns, "Tradition in India vs. a Patent in the U.S." See also Michael D. Lemonick, "Seeds of Conflict"; and Richard H. Kjeldgaard and David R. Marsh, "Claims upon Nature."
27. John Cohrssen and David McIntosh, "Major Problems with the Draft Convention on Biological Diversity."
28. The resurrection of this old idea is generally credited to Garrett Hardin's 1968 article on population ecology, "The Tragedy of the Commons."
29. See esp. E. P. Thompson, *Customs in Common*, 107.
30. Neil Weinstock Netanel, "Copyright and a Democratic Civil Society," 319.
31. See Peter Jaszi's discussion in "Goodbye to All That," 611.
32. Claudia Caruthers, "International Cultural Property," esp. 167–69.
33. Joseph Henry Vogel, *Privatisation as a Conservation Policy*, 123.
34. The disciplinary roots of copyright are explored in Carla Hesse, "Enlightenment Epistemology." The same impulse to discipline by assigning responsibility for texts is explored by Mario Biagioli with regard to the conventions of attribution in contemporary science ("The Instability of Authorship").
35. World Commission on Environment and Development, *Our Common Future*, 46. See David Hunter et al., *International Environmental Law*, esp. chap. 3.
36. Under a CIR regime, local communities that were the custodians of particular bodies of knowledge would be required to share that knowledge with other like communities so long as it is not sought for commercial purposes; commercial users would be required to pay the local community (if registered) or the state (as trustee, in lieu of such registration) a stipulated royalty on sales, or a nonmonetary equivalent to be determined by local custom, practice, and usage; where more than one community is the custodian of a particular body of knowledge, payments in connection with its commercialization would be shared among them; and firms commercializing local knowledge would be barred from seeking to control it through the exercise of Western intellectual property rights such as patents. See Gurdial Singh Nijar, *In Defense of Local Community Knowledge and Biodiversity*; the draft Community Intellectual Rights Act; and Manuela Carneiro da Cunha, "The Role of UNESCO in the Defense of Traditional Knowledge." The effect of such legislation would be to create for traditional culture a version of a mechanism much discussed but little implemented in connection with conventional intellectual property law: the so-called domaine publique payant. See Christine Farley, "Protecting Folklore of Indigenous Peoples," 49–50.
37. The TRR approach draws on "basic human rights; the right to self-determination; collective rights; land and territorial rights; religious freedom; the right to development; the right to privacy and prior informed consent; environmental integrity; intellectual property rights; neighboring rights; the right to enter into legal agreements; rights to protection of cultural property, folklore and cultural heritage; the recognition of cultural landscapes; recognition of customary law and practice;

and farmers' rights" (Programme for Traditional Resource Rights, "What Are Traditional Resource Rights?"). See also, generally, Darrell A. Posey and Graham Dutfield, *Beyond Intellectual Property*.

38. According to one recent account, "In return for plant and insect extracts and other samples, Merck gave INBio a $1.14 million research and sampling budget, undisclosed royalties on any new drugs that emerge, and technical assistance and training for Costa Rican scientists. In turn, INBio agreed to donate 10 percent of the upfront payment and half of any royalties they receive to conservation efforts in Costa Rica" (Mary Parlange, "Eco-nomics"). The Merck-INBio agreement has been criticized for failing to take into account the interests of indigenous peoples within Costa Rica as well as the collective national interest in ecological conservation, and for promoting secrecy and exclusivity rather than information-sharing.

39. See Peter Jaszi and Martha Woodmansee, "The Ethical Reaches of Authorship," 967–68. For a somewhat jaundiced view of Shaman's later fortunes, see "Ethnobotany: Shaman Loses Its Magic," 77.

40. This plant, Trichopus zeylanicus, found in the tropical forests of southwestern India, is collected by the Kani tribal people. Writing for *Science* (March 12, 1999) Pallava Bagala reports that scientists at the Tropical Botanic Garden and Research Institute (TBGRI) in Trivandrum, Kerala, having isolated and tested the ingredient, incorporated it into a compound, which they named "Jeevani," giver of life. The tonic is now being manufactured by the Aryavaidya Pharmacy Coimbtore Ltd., a major Ayurvedic drug company. Quoting Graham Dutfield, Bagala writes that "the process marks perhaps the first time that cash benefits have gone directly to the source of the knowledge of traditional medicines. 'It is a replicable model because of its simplicity,' Dutfield says about a chain of events that began well before the international biodiversity treaty was signed. TBGRI scientists learned of the tonic, which is claimed to bolster the immune system and provide additional energy, while on a jungle expedition with the Kani in 1987. A few years later, they returned to collect samples of the plant, known locally as arogyapacha, and began laboratory studies of its potency. In November 1995, an agreement was struck for the institute and the tribal community to share a license fee and 2 percent of net profits. Another agent from the same plant is undergoing clinical tests for possible use as a stamina-building supplement for athletes" (Bagala, "Indian Deal Generates Payments").

Bibliography

Agrawal, Arun. "Dismantling the Divide between Indigenous and Scientific Knowledge." *Development and Change* 26 (1995): 413–39.

Bagala, Pallava. "Indian Deal Generates Payments." *Science* 283 (March 12, 1999): 1614.

Balandrin, Manuel F., James A. Klocke, Eve Syrkin Wurtele, and William Hugh Bollinger. "Natural Plant Chemicals: Sources of Industrial and Medicinal Materials." *Science* 228 (June 7, 1985): 1154–60.

Barlow, John Perry. "The Economy of Ideas: A Framework for Rethinking Patents and Copyrights in the Digital Age (Everything You Knew about Intellectual Property Is Wrong)." *Wired* 2.03 (March 1994): 84–90, 126–29.

Baumann, Miges, Janet Bell, Florianne Koechlin, and Michel Pimbert, eds. *The Life Industry: Biodiversity, People, and Profits*. London: Intermediate Technology Publications, 1996.

Biagioli, Mario. "The Instability of Authorship: Credit and Responsibility in Contemporary Biomedicine." *FASEB Journal* 12 (1998): 3–16.

Boyle, James. *Shamans, Software, & Spleens: Law and the Construction of the Information Society*. Cambridge, Mass.: Harvard University Press, 1996.

Brush, Stephen B., and Doreen Stabinsky, eds. *Valuing Local Knowledge: Indigenous People and Intellectual Property Rights*. Washington, D.C.: Island Press, 1996.

Burns, John F. "Tradition in India vs. a Patent in the U.S." *New York Times*, 15 September 1995, sec. D, p. 4, col. 1.

Caruthers, Claudia. "International Cultural Property: Another Tragedy of the Commons." *Pacific Rim Law and Policy Journal* 7 (1998): 143–69.

Cohrssen, John, and David McIntosh. Memorandum to Bill Kristol on "Major Problems with the Draft Convention on Biological Diversity." 14 April 1992.

Community Intellectual Rights Act. <http://users.ox.ac.uk/~wgtrr/cira.htm>

Coombe, Rosemary. *The Cultural Life of Intellectual Properties: Authorship, Appropriation, and the Law*. Durham, N.C.: Duke University Press, 1998.

The Crucible Group. *People, Plants, and Patents: The Impact of Intellectual Property on Biodiversity, Conservation, Trade, and Rural Society*. Ottawa: International Development Research Centre, 1994.

da Cunha, Manuela Carneiro. "The Role of UNESCO in the Defense of Traditional Knowledge." In *Safeguarding Traditional Cultures*, edited by Seitel, 143–48.

Daes, Erica-Irene. "Discrimination against Indigenous Peoples: Study on the Protection of the Cultural and Intellectual Property of Indigenous Peoples." UNESCO Commission on Human Rights. Sub-Commission on Prevention of Discrimination and Protection of Minorities. 28 July 1993. E/CN.4/Sub.2/1993/28.

Day, Kathleen. "Rain Forest Remedies: More Drug Companies Turning to Tribal Healers for Medicines." *Washington Post*, 19 September 1995, E-4.

Dutfield, Graham. *Annotated Bibliography on Traditional Resource Rights, Intellectual Property Rights, and Conservation of Biodiversity*. Unpublished ms. 1999. Abbreviated version available at <http://users.ox.ac.uk/~wgtrr/bib1.htm>

"Ethnobotany: Shaman Loses Its Magic." *The Economist*. U.S. ed., 20 February 1999, 77.

Farley, Christine Haight. "Protecting Folklore of Indigenous Peoples: Is Intellectual Property the Answer?" *Connecticut Law Review* 30 (fall 1997): 1–57.

Genetic Resources Action International. *Global Trade and Biodiversity in*

Conflict 1 (April 1998). <http://www.grain.org/publications/gtbc/issue1.htm>

Goldman, Karen Ann. "Compensation for Use of Biological Resources under the Convention on Biological Diversity." *Law and Policy in International Business* (January 1994): 695–726.

Greaves, Tom, ed. *Intellectual Property Rights for Indigenous Peoples: A Sourcebook.* Oklahoma City: Society for Applied Anthropology, 1994.

Hardin, Garrett. "The Tragedy of the Commons." *Science* 162 (1968): 1243–48.

Hesse, Carla. "Enlightenment Epistemology and the Laws of Authorship in Revolutionary France, 1777–1793." *Representations* 30 (1990): 109–37.

Hunter, David, Jim Salzman, and Durwood Zaelke. *International Environmental Law and Policy.* Westbury, N.Y.: Foundation Press, 1998.

Jaszi, Peter. "Toward a Theory of Copyright: The Metamorphoses of 'Authorship.'" *Duke University Law Journal* (April 1991): 455–502.

———. "Goodbye to All That." *Vanderbilt Journal of Transnational Law* 29 (1996): 595–611.

Jaszi, Peter, and Martha Woodmansee. "The Ethical Reaches of Authorship." *South Atlantic Quarterly* 95 (1996): 947–77.

Jeanblanc, Anne. "Fighting Cancer on Many Fronts." *World Press Review* (May 1993): 42–43.

King, Steven R. "The Source of Our Cures." *Cultural Survival Quarterly* 15 (1991): 19–22.

Kjeldgaard, Richard H., and David R. Marsh. "Claims upon Nature." *Intellectual Property Magazine* (winter 1996): 36–39.

Lemonick, Michael D. "Seeds of Conflict." *Time Magazine* 146, 25 September 1995, 50. <http://cgi.pathfinder.com/time/magazine/archive/1995/ 950925/950925.science.html>

MacLeod, Christine. "Concepts of Invention and the Patent Controversy in Victorian Britain." In *Technological Change: Methods and Themes in the History of Technology*, edited by Robert Fox. Amsterdam: Harwood Academic Publishers, 1996.

———. *Inventing the Industrial Revolution: The English Patent System, 1660–1800.* Cambridge: Cambridge University Press, 1988.

Netanel, Neil Weinstock. "Copyright and a Democratic Civil Society." *The Yale Law Journal* 106 (1996): 283–387.

Newton, Nell Jessup. "Memory and Misrepresentation: Representing Crazy Horse." *Connecticut Law Review* 27 (1995): 1003–54.

Nijar, Gurdial Singh. *In Defense of Local Community Knowledge and Biodiversity.* Penang, Malaysia: Third World Network, 1996.

Parlange, Mary. "Eco-nomics." *New Scientist* (6 February 1999): 42-?.

Penrose, Edith Tilton. *The Economics of the International Patent System.* Baltimore: Johns Hopkins Press, 1951.

Pinel, Sandra Lee, and Michael J. Evans. "Tribal Sovereignty and the Control of Knowledge." In *Intellectual Property Rights for Indigenous Peoples*, edited by Greaves, 43–55.

Posey, Darrell A., and Graham Dutfield. *Beyond Intellectual Property: Toward*

Traditional Resource Rights for Indigenous Peoples and Local Communities. Ottawa: International Development Research Centre, 1996.

Programme for Traditional Resource Rights. "What Are Traditional Resource Rights?" <http://users.ox.ac.uk/~wgtrr/trr.htm>

Seitel, Peter, ed. *Safeguarding Traditional Cultures: A Global Assessment.* Washington, D.C.: Center for Folklife and Cultural Heritage, Smithsonian Institution, 2001.

Serote, Mongane Wally. "One Fundamental Threshold." Document prepared for the WIPO Roundtable on Intellectual Property and Indigenous Peoples. 28 August 1998. WIPO/INDIP/RT/98/4C

Shiva, Vandana. *Biopiracy: The Plunder of Nature and Knowledge.* Boston: South End Press, 1997.

Thompson, E. P. *Customs in Common: Studies in Traditional Popular Culture.* New York: The New Press, 1993.

Vogel, Joseph Henry. *Privatisation as a Conservation Policy: A Market Solution to the Mass Extinction Crisis.* South Melbourne: CIRCIT, 1992.

Wilson, Edward O. "Threats to Biodiversity." *Scientific American* (September 1989): 108–16.

WIPO (World Intellectual Property Organization). *Intellectual Property Needs and Expectations of Traditional Knowledge Holders: WIPO Report on Fact-Finding Missions on Intellectual Property and Traditional Knowledge (1998–1999).* Geneva: WIPO, April 2001.

———. "Model Provisions for National Laws on the Protection of Expressions of Folklore against Illicit Exploitation and Other Prejudicial Actions."

WIPO (International Bureau of WIPO). "1967, 1982, 1984: Attempts to Provide International Protection for Folklore by Intellectual Property Rights." Document prepared for the UNESCO-WIPO World Forum on the Protection of Folklore. 17 March 1997. (UNESCO/FOLK/PKT/97/19).

Woodmansee, Martha. *The Author, Art, and the Market: Rereading the History of Aesthetics.* New York: Columbia University Press, 1994.

———. "The Genius and the Copyright: Economic and Legal Conditions of the Emergence of the 'Author.'" *Eighteenth-Century Studies* 17 (1984): 425–48.

World Commission on Environment and Development. *Our Common Future.* Oxford: Oxford University Press, 1988.

Yongping, Hong. "The Experience of Asia and the Pacific Region." Paper prepared for the UNESCO-WIPO World Forum on the Protection of Folklore. 17 March 1997. (UNESCO/FOLK/PKT/97/15).

9.

Uncommon Controversies

Legal Mediations of Gift and Market Models of Authorship

CORYNNE McSHERRY

I. Introduction

Is a work of scientific authorship a gift or a commodity?[1] Perhaps the most vexing feature of authorship in academic science is its ability to instantiate and traverse two visions of scholarly exchange. According to one vision, scientific authors participate in a gift economy, a system of exchange premised on reciprocity, reputation, and responsibility in which the commodification of scholarly work is immoral (Hyde 1983; Hagstrom 1965). Pierre Bourdieu (1988), however, argues persuasively that the academic knowledge economy can be better understood not as a web of moral obligations, but as a system of capital accumulation and investment. In Bourdieu's view, the value of that capital depends on the continuing ability of the academy to define and guarantee a market. Taking the laboratory rather than the university as the unit of analysis, Bruno Latour and Steve Woolgar (1979) portray authorship as the linchpin of this market, or, more directly, of a "cycle of credit" wherein "knowledge" is made available in exchange for credit (recognition), which can be reinvested in the means of production of more knowledge.

In what follows, I consider how copyright doctrine, the body of law most directly concerned with scientific authorship, enables and addresses particular tensions between gift and market economies. To ground this inquiry, I offer a copyright dispute between a junior professor and her mentor, the resolution of which would involve the mobilization and concealment of a set of assumptions about the nature of scientific authorship and, most remarkably, the reconstruction of scholars as celebrities. Authorship narratives shared by students and professors on the campus of a major U.S. university supplement the legal tale.[2] Using these stories as navigational tools, I map the production of scientific authorship at the gift/market border. I show how, as law reconstructs some of its foundational categories to apply to the university—categories rooted in the market economy, yet imbued as well with values and rhetorics associated with the gift economy—law helps reconfigure knowledge work and knowledge ownership.

II. An Uncommon Controversy

Heidi Weissmann began working with Leonard Freeman in 1977 when she was a fourth-year radiology resident and he was chief of nuclear medicine at Montefiore Medical Center. Their collaboration resulted in several published and unpublished works, among them a syllabus created for a course they co-taught at Harvard Medical School in 1980.[3] This syllabus was revised several times by both parties for review courses at several institutions.

In 1985, Weissmann published a version of the syllabus under her name alone. This syllabus reorganized the original work and added new illustrations, captions, references, and text. Unbeknownst to Weissmann, Freeman reproduced this version of the syllabus for his 1987 review course, listing himself as its sole author. When Weissmann learned of the reproduction, she demanded that the syllabus be withdrawn from the course materials. Freeman complied, but not before the work had been circulated among a few people. Weissmann filed suit for copyright infringement, arguing that her changes were significant enough to grant her individual copyright in the piece as a derivative work. Freeman counterargued that the piece was jointly authored, a product of their research partnership. He further argued that, even if the syllabus were not the product of that collaboration, his use of it was a fair use rather than an infringement. Who was right?

According to the *Chicago Tribune*, the answer to this question mattered less than the fact that the case, along with several others like it, had been brought to the courts at all (Grossman 1997). In a lengthy article, the newspaper suggested that the "human architecture of higher education" was being dismantled (C1). The agents of this change were resentful graduate students and junior researchers who had turned to the legal system to resist the appropriation of their research by senior professors anxious to boost their own publication rates. In a world of shrinking research budgets, the *Tribune* contended, professors and advanced graduate students had become competitors rather than collaborators. It was perfectly logical, therefore, for students to try to defend their position in this competition by asserting property rights. If the trend continued, the report concluded, the "medieval" guild structure of academe might at last be dragged into modernity.

If the university is no longer a guild, does it look more like a factory or a limited partnership? Or perhaps a temp agency? These are questions worth keeping in mind, for they point to the stakes of the *Weissmann* case. The gift model of academic exchange grounds the university's enviable and strategic position as producer and guarantor of valuable knowledge. Put simply, modern universities are crucial "knowledge resources" precisely because of their "reputations for neutrality," a reputation based on their location "outside" of the realm of economic interest (Walshok 1995, 191).[4] At the same time, the university and its inhabitants must ultimately operate in the market and assert themselves as property owners if they are to reap the benefits of this position. Scholars may "reach beyond the walls," as former Stanford president Donald Kennedy (1997, 241) puts it, but their feet had better stay firmly planted within those walls if the academy is to retain its position outside the messiness of "society" and its products are to retain the value generated from that location (241). By bringing the case forward, Heidi Weissmann positioned herself as an autonomous individual property owner rather than an aspiring member of a gift community. She pulled the market right into the walls of academe, thereby exposing the unstable foundations of the walls themselves.

Without Foundations

The task of the courts, then, was to shore up those foundations, or so it would seem from the reasoning advanced by the two judges who heard the case. The District Court for the Southern District of New

York viewed the dispute as a simple matter of misguided ego, a standpoint that was given sense by the court's equally misguided effort to ignore basic copyright doctrine.[5] The Court of Appeals for the Second Circuit would sharply rebuke the lower court on this point but, as we shall see, the appellate court offered a strange and contradictory rhetorical strategy of its own.[6]

To parse these interpretations, we need to know something about joint authorship doctrine, a body of law that is itself fraught with contradiction. In general, copyright law assumes and invokes a highly individualized model of creative production (Edelman 1979; Rose 1993; Jaszi 1994; Woodmansee 1994).[7] Yet copyright law does acknowledge that some works, such as a song that emerges from a partnership between a lyricist and a musician, are collaborative. All identifiable contributors are considered authors of such joint works, and each has an undivided right of ownership thereto. In principle, then, joint authorship seems to carve out room for sociality in copyright law. To be identified as a joint author, however, each individual author must have contributed an independently copyrightable element.[8] In addition, each author is granted property rights in the work as if he or she were the sole author, and need not consult with other authors regarding subsequent use so long as he or she "shares" the profits of that use. Authorship, then, is only "shared" in a financial sense (Jaszi 1994).

Under this doctrine, two questions should have been asked in *Weissmann*: First, did both authors intend for the piece to be a joint work? Second, were Weissmann's changes substantive and original enough to transform the syllabus into a derivative work?[9] Strangely enough, these questions were not prominent in the first phase of the case. Instead, the opinion issued by Judge Milton Pollack focused on Weissmann and Freeman's relationship and their individual credibility.

Judge Pollack pointed to the pair's history of research and publication, arguing that the work was an "evolutionary stock piece" that had "evolved" from that collaboration. In other words, the syllabus was not an individual effort but a product of a set of reciprocal obligations that emerged in and through community activity (1261). Yet, legally speaking, the appearance of an evolutionary process should have been less important that the question of whether the work was original enough to count as a new subspecies. Fortunately, Judge Pollack declared, the court was provided with the "best qualified expert opinion [on the matter], that of the defendant, the acknowledged outstanding expert in the field" (1257). Such an eminent scientist would, of course, be capable of forming an objective opinion on the originality of Weissmann's changes. Freeman had found her changes to be trivial. Enough said.

Well, almost. Made uneasy, perhaps, by his own quick acceptance of the objective viewpoint of the defendant, Judge Pollack was careful to emphasize Weissmann's lack of credibility as a witness and as a scientist. Weissmann had testified that Freeman had not contributed at all to the piece. "This is my words, my work, my expression," she said, "Dr. Freeman had no participation in it" (1258). This claim, combined with her hostile demeanor, fatally damaged Weissmann's credibility in the court's eyes. First, physical evidence of Freeman's contribution of visual material existed. Second, Freeman was listed as a coauthor on previous versions of the syllabus. Besides, Judge Pollack stressed, Freeman *had* demonstrably contributed in another way: it was his name as principal investigator that made the research possible, and, as such, he was "the person with whom 'the buck stops'"(1259). Obviously, Weissmann had lied on the stand when she said Freeman had not participated in the creation of the work. Freeman's claim to joint authorship of the piece was affirmed.

Empirical studies of scientific authorship suggest that Weissmann's claim was not so incredible, and that Freeman's position as coauthor might have been based on minimal or no written contribution to the work (Shapiro, Wenger, and Shapiro 1994; Tarnow 1999). I want to defer the question of contribution, however, in order to address a different question. Namely, what could justify Judge Pollack's preoccupation with the relationship between the authors and their individual credibility rather than with the object of the dispute—the work itself? The answer is: a set of assumptions about the exceptional nature of academic authorship and the economy of knowledge within which it is situated.

The Gift

> This essay of mine, though it will be added to the inventory of my own intellectual capital, my curriculum vitae, and hopefully will count towards enhancing my academic status and income—is still a gift, to be consumed and circulated in the gift culture of research and scholarship; no one will pay me for writing it and I will not sell it.
>
> —Jim Swan

Judge Pollack characterized the case as an "uncommon controversy." The only thing really uncommon about it was the profession of the parties involved, a profession in which it is inappropriate to identify one's creations as private property. Copyright law in general assumes that authors need and deserve monetary profit and fosters a market economy in intellectual commodities. Few bonds of trust exist in this market, and one of the principal objects of copyright law is to make up for that lack by defining the respective economic rights of market actors.

Academic authors, by contrast, are supposed to write for honor, and the academic system of exchange is supposed to be based on the reciprocal and personalized exchange of gifts, not the impersonal selling of private property (Hyde 1983; Mauss 1967). Through the quality and generosity of one's giving, receiving, and repaying, one demonstrates authority, spiritual favor, and especially honor, for reputation is the heart of this system of obligation. As Marcel Mauss observes, "men could pledge their honour long before they could sign their names" (36). Reputation is invested in and guaranteed by things, and that investment stands as guarantee, in turn, of future prosperity. As a marker of obligation, moreover, gifts remain bound up with the donor, such that the donor's identity works to animate the gift. This close relationship between donor and gift reflects a prior duty on the donor's part, for individuals owe themselves, as well as their possessions, to the community. Impelled by the same duty, other community members must in turn accept the gift and thereby create a channel to relieve themselves of their own obligation to give to others.

Searching for remnants of a gift economy in liberal societies, Mauss points to the "liberal professions," within which, he implies, "honor, disinterestedness, and corporate solidarity are not vain words" (1967, 67). Applied to the academic professions, "the gift" is most often invoked in the context of a boundary, as in Jim Swan's suggestion that university administrators must negotiate "the boundary between two cultures, between the 'feminine' economy of the gift culture and the market economy of risk and exploitation" (Swan 1994, 77). A recent study by the Pew Higher Education Roundtable describes the "gift" side of the border as occupied by "communit[ies] of devotees bound by a common interest . . . [each hoping] to win the regard of other members" (1998, 3).

As the above suggests, the gift-donor identity relation is crucial to academic exchange. Academic efforts are sent out into the world to bring an equivalent gift back to the donor. "The thing given is alive and . . . strives to bring to its original clan and homeland some equivalent to take its place" (Mauss 1967, 10). Once published, an article can garner recognition and status for the donor and the more recognition the gift (and therefore the donor) receives, the greater the value of the original and subsequent gifts (Hyde 1983; Hagstrom 1965).[10] The community, in other words, determines value, a fact to which we will return below. At the same time, as Warren Hagstrom (1965) notes, a scientist cannot publicly admit to any expectation of reciprocity, lest she be suspected of a less than perfect devotion to the production of truth.

By instantiating mutual obligations (to truth, to persons), the academic work also recreates a receptive community even as it marks the donor's status in that community. For Lewis Hyde (1983), this operation is made visible in the breach. Hyde argues that the production of knowledge as a commodity situates the producer as "less a part of the community" (81). "Community appears," he insists, only "when part of the self [in the form of research] is given away" (92). Community ties are further affirmed through repayment in the form of reciprocal papers, citations to the work, and financial support for the creation of new gifts.

The Creator-Work Relation

Ironically, perhaps, the gift-donor relation marks a point of shared meaning between legal and scientific discourses. A brief comparison between plagiarism and copyright infringement may help clarify this connection. Copyright infringement cases tend to focus on the work, and particularly on the existence of substantial similarity between the original and the infringing work. Guilt depends on fairly extensive borrowing of fixed expression and is not excused by attribution: the work, rather than the author, is the primary focus of analytical attention. Plagiarism, however, involves ideas as well as expression; explicit borrowing may be slight or nonexistent, and attribution will generally resolve the issue. This last aspect is crucial, for it bespeaks an older and very different system of valuation. Plagiarism was condemned in ancient Rome and Greece, where "literary theft" was characterized as an appropriation of another's honor and "immortal fame" (Long 1991, 856). The term derives from *plagiarius*, to kidnap, and signifies breaking a connection between the author's name and the work (Stearns 1992; St. Onge 1988). To sever this connection is to destroy the basic requirement of the gift: that it be imbued with the spirit of the giver and remain connected to that person. This connection is one reason gifts matter: it is what makes gifts risky to give and receive and helps give them value.

Yet honor is one of the elements of authorship that allows the concept to traverse gift and market models. A tight linkage between authorial identity and the work was one of the pillars of modern copyright. Seventeenth- and eighteenth-century advocates of copyright constructed the work as the embodiment of the author, "the objectification of the writer's self" (Rose 1993, 121). This object could be copied, of course, and the ideas within it circulated, but the author's expression remained her own. Thus, copyright discourse treated the author and the work as simultaneously linked and autonomous. It is not surprising, therefore, that Mauss identified a remnant of gift culture in intellectual property rights. Both regimes acknowledge, indeed depend on, an intimate relationship between the law of persons and the law of things.

In both models, this creator-work relation is the basis of value. Intellectual property law creates and maintains the exchange value of texts by policing reproduction. This activity, and the market economy of knowledge it engenders and secures, rests on the fiction of the singular creator. The promulgation of codes forbidding plagiarism, coupled with unspoken pressure not to accuse suspected perpetrators, similarly work to maintain the value of gifts by constructing them as the true products of the giver.

In a gift economy, however, the author's name guarantees both a product and a truth-seeking process (Stearns 1992). Copyright law holds an infringer responsible whether or not there was any deliberate effort to deceive, because the copyright holder's economic interest has been damaged. Because it undermines the code of conduct in and through which rational discourse is produced, plagiarism is a crime against reason and academic science itself rather than an infringement on individual economic rights. Money is not at issue in plagiarism cases, at least not overtly—what is at issue is truth. Thus, if the copying is demonstrably accidental, it may be excused, for while it may still harm the originator, it is no longer a harm to the process of creation.

This emphasis on the violation of process was present in the comments of researchers with whom I spoke in 1998 in the course of a larger study of academic intellectual property formation. Most researchers reported at least one experience with suspected plagiarism. Of those, only three pursued the issue beyond expressing irritation to close colleagues. One protested to an editor after seeing whole pages of a manuscript the researcher had sent to a colleague reproduced in that colleague's next book. This researcher described the violation as an act of "intellectual rape," a description that resonates with the comments of another senior researcher who had also been plagiarized:

> I had submitted a grant proposal . . . and on one of those submissions I got a review back that didn't match the quality of the science, it was much more negative than it should have been, the science was good and I rebutted it, the thing was ultimately funded. But then I got to review myself a grant from another investigator submitted to [a foundation] a very prominent investigator, someone I knew, whom I compete with but I knew not to be a very honest broker, and in his proposal verbatim were sections from my proposal . . . and basically all I did, although I was pretty animated about it at the time, was decide that you only end up smelling when you get in a fight with a skunk. (interview with senior researcher)

He didn't want to be known for that complaint, he said, but for his work. Several other researchers who felt they had been plagiarized expressed similar concerns. "My experience is that people who make a big stink about these things are considered to be hotheads," said one (interview with senior researcher). Indeed, there is something unseemly about accusing another of plagiarizing your work, for it implies a degree of desperation. A really creative scientist does not need to control one idea—she has others. She can also, in theory, rely on an informal enforcement network: three researchers expressed the expectation (and, in some cases, the experience) that plagiarizers eventually acquire a bad reputation. In the meantime, the victim must take care to protect her own good name.[11]

Sociality

Against this background, we can begin to understand Judge Pollack's argument. In a gift economy, objects are meaningful in the context of relationships. The judge seemed to suggest that because Weissmann and Freeman had coauthored numerous "scholarly scientific works" they had established a system of mutual exchange as well as a position in the wider gift culture of research and scholarship (1315). Weissmann's attempt to claim the revised syllabus as her own signaled a denial of that exchange relationship.

Gift logic also explains the court's focus on reputation. The logic of the gift inextricably binds persons to things. The court began from the standpoint that Freeman had invested his reputation in the syllabus, and that that investment was as important as the actual writing. Having invested reputation—imbued the syllabus with his spirit—Freeman remained bound to it, and "trivial" modifications could not sever that relationship.

A gift culture model also explains the sense of moral outrage that infused Judge Pollack's opinion. Gifts must be permitted to circulate; the gift cannot be withdrawn from circulation (i.e., transformed into capital) without losing its status as a gift. Weissmann's effort to treat the syllabus as property by claiming copyright ownership was, in a gift context, immoral. Implicit in the court's reasoning was a claim that Freeman had given her a gift—the use of his reputation. "It was the defendant who opened the doors for Dr. Weissmann," Judge Pollack observed, "making all of her research and writing possible and professionally recognized" (1258–59). By claiming a sole property right in what the court saw as the product of a set of mutual obligations, Weissmann had transformed the gift into capital.

In the final paragraph of the opinion, Judge Pollack at last referred directly to the unpardonable sin of academic life. In claiming that she wrote the entire piece, Weissmann was, he argued, accusing Freeman of plagiarism: "Plainly the overbroad position she took resulted in a grave insult to her mentor and professional colleague. Dr. Freeman had neither motive nor need to plagiarize, considering his preeminent grasp of the subject" (1263). This, Judge Pollack suggested, was an attack on Freeman's reputation, and the judge's concern for it marks a final invocation of the gift rhetoric to resolve a contest for private property.

In short, faced with an outrageous transgression of the gift/market boundary, Judge Pollack deployed the language of the gift to resolve the dispute and thereby restabilize that border. Judge Pollack's importation of gift rhetoric into a body of law organized around the production and protection of commodities depended on and exposed the identity-work relation upon which both gift and market economies of knowledge are partially founded. Reproduced as well was a vision of scholarly creation as a space of sociality rather than of private property. The work remained a gift in the eyes of the court, imbued with the spirits of both Dr. Freeman and Dr. Weissmann.

There was a deep irony to Judge Pollack's reasoning, however, for at the end of the day, gift rhetoric was ultimately used to secure Freeman's *property* right in the work—his right to use, enjoy, and commodify the work. In this regard, it would seem that this process of translation was also, as perhaps it had to be, a process of betrayal, and what was betrayed was the very gift culture Judge Pollack seemed to want to defend. The gift did not remain a gift, but rather was transformed into an object of private property rights.

Betrayal signifies revelation as well as treachery, and we might want to ask what is revealed about gift and market knowledge economies in this border dispute. Joint authorship doctrine exemplifies copyright law's refusal to take account of collaborative cultural production (Jaszi

1994). To the extent that Judge Pollack set claims of sociality above those of the individual author-subject, his opinion might be celebrated as a resistant approach to collaborative authorship. If so, the court wrote a cautionary tale.

Credibility and Status

The case, for Judge Pollack, began and ended with status. The dispute, he declared, had to be understood in light of "the parties' relationship, the stature of the defendant . . . defendant's supervision, guidance and control [of Weissmann's career]" and Freeman's role as principal investigator on the joint projects in the context of which the syllabus developed. Weissmann was the "developing junior member of the association." Freeman was its supervisor, "lending credibility . . . by his standing, reputation, knowledge, perception and experience"(1251). Authority, in this discursive formation, is the crux of authorship.

Judge Pollack's reasoning acknowledged that while gift cultures may be communitarian in some respects, they are not egalitarian. In fact, gift giving is one of the principal strategies for recreating hierarchy—one's gifts mark one's rank, as does the ability to accept a gift. The worth of both gift and donor is constructed in and through repayment—an uncited work, for example, has less value and accrues little honor for its maker. The risk, however, is that the repayment may impoverish the donor. If another researcher takes the work and refutes or transcends it, her gift may trump the original gift, and her status will improve, while that of the donor will drop. "Political and individual status . . . and rank of every kind, are determined by the war of property" (Mauss 1967, 35). Chiefs exchange with other chiefs, family leaders with other family leaders, and with every exchange risk losing status through the inability to repay.

As Weissmann's "chief," Judge Pollack argued, Freeman was ultimately responsible for the gift. The junior associate's effort to transform it into her property disrupted the hierarchy that responsibility implies. No wonder the court was shocked at Heidi Weissmann's lawsuit. Indeed, in his evaluation of damages the judge reemphasized that the suit could only have been motivated by some kind of malice. The court took the extraordinary step of stating, for the record, its desire to know why Weissmann had brought the suit. "Judging by the hostility evident in Weissmann's demeanor and testimony," the court observed, "the answer has to be that this action was brought for personal reasons." The case, the court concluded, was an "unfortunate lapse of judgment" on Weissmann's part (1258). But the case can also be read as a tactic of subversion, an effort to resist dependence. A decision in Weissmann's favor would have suggested that mentor and mentee were equal and autonomous individuals (property owners), thereby necessarily undermining a trust relationship based on dependence and obligation. In the face of the contest for the meaning of scholarly production this assertion engendered, Judge Pollack reconstructed the university as a site of collaboration—but also of hierarchy.

III. Out of the Guild and into Modernity?

In its report on the first phase of the Weissmann case, the *Chicago Tribune* referred to Judge Pollack's opinion as a prime example of the "chilly reception" students, research assistants, and other junior scholars suing for copyright infringement could expect from the courts. The same article pointed to the subsequent appellate opinion, written by Judge Cardamone for the Second Circuit Court of Appeals, as the first step in a long delayed transition in the academy from feudalism to modernity. Summarizing this and several other cases involving junior and senior academic authors, the *Tribune* declared that the "serfs" might be poised to "topple the lords."

Judge Cardamone rejected Judge Pollack's evolutionary theory of authorship, arguing that coauthorship of preexisting works did not automatically confer authorship in subsequent works. "If such were the law," argued Judge Cardamone, "it would eviscerate the independent copyright protection that attaches to a derivative work" (1317). The issue, then, was intention and content of the work. With regard to the former, the court invoked Judge Learned Hand's famous rule that all contributors must "plan an undivided whole [in which] their separate interests will be as inextricably involved, as are the threads out of which they have woven the seamless fabric of the work."[12] Weissmann clearly did not intend for this particular work to be jointly authored; she neither submitted the syllabus to Freeman for comments nor published the syllabus under both names. The authors could not have intended for the work to remain "forever indivisible," Judge Cardamone continued, because scientific research, by definition, "is a quest for new discoveries"(1319). Weissmann, as a scientist, had embarked on such a quest and left her mentor behind. In so doing, she had made the new product "her own."

Having dispensed with intent, Judge Cardamone turned to originality to determine whether Weissmann's revisions were sufficient to make the piece a derivative work. Citing Fyodor Dostoyevsky and John Stuart Mill as authorities, he asserted "originality is and always has been rightly prized" (1321). Credibility does not determine originality, Judge Cardamone noted; changes must merely be nontrivial. Weissmann's changes must have been nontrivial, else Freeman would not have copied the syllabus but used an earlier version of the work.

In sum, the Second Circuit treated scientific authors like any other joint authors. More directly, it treated the lawsuit as it would treat a commercial dispute, ignoring the gift economy model of the university and the status of the defendant. One would expect, therefore, that the decision marked a fundamental disconnect between legal and academic representations of authorship. A closer examination of the context of authorship suggests, however, that scientific authors do indeed look as much like property owners as donors, and the knowledge economy looks as much like a market as a community of devotees.

Individuation and Investment

Copyright law is often criticized for its refusal to acknowledge or accommodate collaborative practice and, as noted, joint authorship doctrine is considered a case in point (Jaszi 1994; Lunsford and Ede 1994; Boyle 1996). Yet scientific discourse itself produces authors as autonomous individuals competing in a market, as even proponents of the gift model of exchange concede. Hagstrom, for example, notes that academic recognition "is awarded to the individual . . . who freely selects problems and methods and who evaluates the results" (1965, 69). It is awarded, in other words, to the individual who looks the most like a fully autonomous rational liberal subject. In academic science as in law, then, collaborative ownership of an object of knowledge depends on the individuation of subjects of knowledge.

As market theorists of academic exchange observe, these rational liberal subjects operate very much like commercial entities. Scientists, argue Latour and Woolgar (1979), are like corporations, and their curriculum vitae are like annual budget reports. Authorship credit, they suggest, is defined as credibility—recognition of an "ability to do science" rather than simply a "job well done." This credibility, or scientific capital can be accumulated and then invested in support of someone else's work, in research proposals, or in getting subsequent work accepted. If it is invested wisely, it will garner a return in the form of, for example, research funding. Wise investments are those that respond most effectively to the laws of supply and demand. Scientists are figured as both employers and employees: their funding sources remain the ultimate power in this market over which they have limited control.

Successfully invested, scientific capital creates more capital in the form of authorship credit on publications to which one has, or, as Judge Pollack put it, "lent authority." Students and professors often describe this investment in nonfinancial terms—as a matter of time spent discussing research directions, editorial feedback, problem formulation, even writing. Nevertheless, intellectual and financial investments are closely linked. One professor stated that she would not let a student put out a paper without her name on it "because I had to bring in money for that student," adding that her approach was "an unspoken rule of the culture" (interview with senior researcher). Another professor (I'll call him Professor Richards) described his irritation when another faculty member (I'll call him Professor Colling) was listed as a fellow coauthor on a student's paper. Professor Richards said he had "fed" the student "intellectually" and given the student financial support as well. The student had admitted to Richards that Colling had had little input on the paper but the student "wasn't in a position to assert." Richards's discomfort was mitigated, however, by the fact that Colling had provided significant financial support early in the student's career, before Richards "took it over" (interview with senior researcher). Clearly, financial investment counts as a contribution.

One student put the issue rather more starkly. His advisor, he said, had given him little or no intellectual support. Why was the advisor still listed as an author?

> He is providing money for me so it's a little difficult to say "hey, you're not getting on this paper." . . . He needs his name on papers to get more research grants coming in . . . Tenure is completely based on whether they're bringing in money. In order to bring in money you have to have a huge list of papers you've published. (interview with doctoral student)

Several other students echoed these comments, though most felt that funding was not the only reason for listing their advisors as coauthors. "[Publications are] the real mechanisms for demonstrating to [sponsors] that you've spent their money and actually done something with it" said one student. "Everybody has a vested interest in it, it's just sort of part of the game, of the system" (interview with doctoral student).

The *Chicago Tribune* characterized the academic architecture of human relations as feudal: in exchange for minimal financial support, the "serfs" till the soil while the "lords" reap the profits. Scientific authorship more closely parallels corporate authorship, whereby corporations claim copyright in works produced under their auspices. The presumption is that the work was produced under the direction of the corporate body: the corporation, then, is the point of origination. Ill-paid students, as "employees," are permitted to claim authorial status, but must also be sure to acknowledge their professors as points of origination, providing their professors and their professors' sponsors with a profit.

Although he did not refer to it, this practice gives sense to Judge Cardamone's refusal to treat scientific authorship as special. Perhaps the *Tribune*'s hyperbolic celebration of the academy's entrance into "the modern" was right on target, at least to the extent that modernity can be taken to denote the ascendance of a rhetoric of creativity that assumes and affirms private property rights. The appellate court's opinion dislodged the gift economy model of academic scholarship in favor of a market-oriented model. In so doing, it replicated the academy's own production of scientific authors as individualized, morally autonomous property owners who are legally if not structurally equal—a creation that exists simultaneously with the gift model of authors as communitarian, hierarchical, morally constrained gift givers and receivers. A more modern academy was thereby produced and normalized. Or was it?

IV. The Specter of the Gift

In a final twist, the specter of the gift continued to haunt the case, providing the means for a partial reconciliation of competing visions of scientific authorship. Having resolved the question of ownership, the Second Circuit was faced with another question: Even if Weissmann could legitimately claim copyright in the work, did Freeman's use of it for educational purposes constitute a fair use, that is, a reasonable exception to copyright rules? Judge Pollack had paid comparatively little attention to this issue in the first phase of the case, but Judge Cardamone did not. Four factors determine fair use: (1) the purpose and character of the use, including whether such use is of a commercial nature or is for nonprofit educational purposes; (2) the nature of the copyrighted work; (3) the portion used in relation to the copyrighted work as a whole; and (4) the effect of the use upon the potential market for or value of the copyrighted work.[13] Judge Pollack had argued that the syllabus was intended for nonprofit "educational use." Further, because the piece was "factual and scientific," the law was predisposed to facilitate its free dissemination over, for example, "works of fiction or fantasy." Because it was not in fact used in the class, moreover, the syllabus had "no market value" and its use could have no market effect (1262).

Again, the Second Circuit Court of Appeals came to a rather different conclusion, and the logic of that conclusion rested upon a conceptualization of the community of scholarship in both gift and market terms. "Monetary gain is not the sole criterion" of profit, Judge Cardamone argued:

> Dr. Freeman stood to gain recognition among his peers in the profession . . . he did so without paying the usual price that accompanies scientific research and writing, that is, by the sweat of his brow. Particularly in an academic setting, profit is ill-measured in dollars. Instead what is valuable is recognition. (1324)

Judge Cardamone's assertion that monetary gain was irrelevant to academic work was contradicted, somewhat, by his insistence a few sentences later on the importance of the economic incentives of copyright protection to scientific production. Truly curious reasoning, however, emerged in Judge Cardamone's assessment of the "market value" of the syllabus. "The particular market at issue here—namely, the world of scientific research and publication" he said, operates to encourage the circulation of scientific work through incentives of promotion and advancement. Recognition, in this formulation, was the "fruit of one's labor"(1326). The syllabus was a way of producing recognition for Weissmann; thus it had a market value aside from any direct monetary remuneration. In short, the Second Circuit located the academic setting outside of the money economy and simultaneously reconstructed that setting as a market.

The Academic Market and the Professor as Celebrity

The ramifications of this double movement come into relief if we look closely at existing legal parallels for the Second Circuit decision. The economic rights the court identified are not, of course, entirely new in the history of the professions (Larson 1977). What is startling, however, is the construction of "recognition," or reputation, as "the fruit of one's labor" and, therefore, a form of intellectual property. More directly, what is startling is the application of such reasoning to the professoriate. In the United States, it has been more often found to apply to celebrities within the regime of publicity rights.

Put simply, the right of publicity is the right of a person to control the commercial use of his or her identity. The courts have reasoned that celebrities, like private persons, have a right of protection from unauthorized commercial intrusion. In addition, since celebrities invest time, labor, and money in the construction of their public selves, they have a moral right to reap "the fruits of their labor" while others have a moral obligation not to "reap where they have not sown" (Nimmer 1954). In the past several decades, the right has become a "real" private property right: "fully assignable and descendible, as well as potentially perpetual" (Gordon 1993, 153 n. 14). In addition, the scope of publicity rights has been dramatically extended. Initially, publicity rights were narrowly construed, covering only name, likeness, photograph, voice, and signature. In California at least, publicity rights now cover evocations of identity, including advertising slogans, objects associated with a celebrity (e.g., a car), singing styles, and so forth.[14]

Publicity rights take the creator-work relation a step further: the work is the star persona, the objectification of the author's self. In essence, publicity rights inhere in elements of identity so distinctively personal that they can be alienated. In *Weissmann*, it is a scientist's reputation that is considered sufficiently personal to be produced as a kind of commodity.[15]

As it happens, scientific authors, particularly those actively involved in education as well as research, might learn lessons from celebrities. The ability to claim ownership in reputation has enabled movie stars to gain firmer control over the circulation of performances. A central struggle to define the academy in an information economy turns on control of academic labor. Educators in the sciences and the humanities argue that they must assert and defend their copyright in their lectures and other written works to prevent the appropriation of that work by university administrators, distance-learning corporations, and other private entities bent on "commercializing" that work (Noble 1997; Leatherman 1998). Increasingly conscious of a lack of control over the conditions of their labor, professors, like stars, seek mastery through the assertions of intellectual property rights. Judge Cardamone's reasoning indicates one possible mode of assertion.

Yet there is a price to be paid for this strategy, one that reminds us of the broader implications of Cardamone's decision. The moral force of publicity rights rests upon the "unquestionable truth" of individual labor. In other words, publicity rights work to reinforce the same individuated creator-work relation upon which gift and market economies of knowledge are founded. And the question is, or should be, what kinds of labor, performed by whom, are construed as valuable, by whom, and for what purposes? Like Judge Cardamone, jurists deciding publicity rights cases consistently treat fame as a product of individual achievement; celebrities "painstakingly build" their public personalities through years of effort, "assiduously cultivating" their reputations.[16] One court declared in 1970 that "a celebrity must be considered to have invested his years of practice and competition in a public personality."[17] Yet it is not clear why, exactly, a celebrity or an academic must be considered to have done so (Madow 1993).

One effect of this focus upon individual labor is that it obscures the work of technicians, assistants, and audiences (or a community of scholars) to produce valuable reputations. Fame is conferred by others—one does not become famous on one's own (Madow 1993). Audiences themselves labor, making selections among potential and alternate meanings, and their selections influence the reproduction of authorized identities. In his study of the construction of the star persona, Richard Dyer (1986) notes that fan clubs, fan magazines, and audience research techniques channel practices of reception into practices of production. The work of audiences is even more evident in academic science, where the value of the work (hence the value of the scientist's reputation) depends on reading practices (e.g., noting order of authorship), experimental testing, and citation. Further, this focus on individual labor obscures the activities of corporate entities from the University of California to the Disney Corporation to construct and profit from fame.

Thus, Judge Cardamone's support of Weissmann's right to profit from her reputation both exposed and reinforced a fundamental shared premise of gift and market models of creative exchange: the assumption that creative work can and should be individuated. That individuation engenders a sharply impoverished view of knowledge production and legitimates a broader discourse whereby, as critical legal theorist Bernard Edelman (1979) puts it, "the claim to describe [the author] becomes the practice of the owner" (25).

V. The Gift, the Market, and the Information Economy

> In what ways and with what effects can the university, both inside and outside the market economy, useful and useless, function as a surplus that the economy cannot comprehend?
>
> —Robert Young

My intention in this essay has been to partially answer Young's query so as to reframe it. The *Weissmann* case indicates a need to ask: In what ways and with what effects can the market economy comprehend the university precisely because of its uselessness, that is, its location outside of the realm of economic interest? How is the boundary between gift and market models of academic exchange disrupted and maintained?

Set against the background of academic practice, authorship as a concept appears to be deeply embedded in competing visions of scholarly work, yet flexible enough to circulate between these notions. These competing visions are never reconciled. Rather, they maintain an uneasy coexistence. Heidi Weissmann's lawsuit challenged this uneasy accord because its resolution seemed to demand the displacement of the gift in favor of the market. Faced with this visible transgression, the district court sought to reinscribe a gift model of scientific authorship. The appellate court recognized the impossibility of this strategy and repositioned Weissmann and Freeman as proper liberal subjects, autonomous property owners. Yet Judge Cardamone, too, was finally unable to leave the gift behind, choosing instead to awkwardly knit the two models together. In a final turn that is symptomatic as well as productive of scientific authorship's multivalence, he constructed gift and market as both opposed and hopelessly intermingled in the space of a few sentences.

This confusion of meaning adds a touch of poignancy to the *Chicago Tribune*'s search for signs of modernity in the academy, if we keep in mind Latour's (1993) observation that "the modern world has never happened" in the sense that its central tenets have never been fulfilled (39). That is, modernity is not built on dichotomies, such as that between gift and market. Rather, modernity is built on hybrids "made possible by absolute investment in dichotomies," its social arrangements stabilized by the concepts that intermediate between them.

To make all of this activity work, however, these hybrids have to be concealed. One of the things that helps authorship mediate gift and market economies is the shared assumption that the two models do not share important assumptions. This is what was challenged in *Weissmann*. The case highlighted the productive tensions between gift and market economies and perhaps even the dangerous incursion of a market mentality into the public domain of the gift. Read against the grain, however, another threat can be discerned: that the hybridity of scientific authorship might be exposed.

Heidi Weissmann's property claim, as an extraordinary instance of boundary crossing, demanded an equally extraordinary feat of reconstruction. The district court's argument was the first effort in this direction, but its importation of the gift model into a property context threatened the foundational opposition between gift and market. The Second Circuit's final argument was a more significant rescue operation, one that simultaneously recognized academic authors as private property owners and reconstructed the academic knowledge system as a gift economy. This rescue was made possible by the final betrayal it encapsulated: the very complex of shared assumptions it attempted to conceal.

Notes

1. The original version of this essay appeared in Corynne McSherry, *Who Owns Academic Work? Battling for Intellectual Property* (Cambridge, Mass.: Harvard University Press, 2001). Reprinted by permission of the author and publisher.
2. A few words on the methodology of the empirical study: I used snowball sampling to generate a list of staff, students, and researchers located at two research centers in a major university on the West Coast. In interviews lasting one to two hours, I asked my respondents to tell me about the preparation of research articles and conference papers, how they decided who should be named as an author in a given publication, in what order, and who was and was not included in that decision-making process. In most of the interviews, we also discussed experiences with secrecy and plagiarism. Two-thirds of the interviews discussed here were conducted at a small research unit, only a few years old and populated primarily by electrical engineers. I chose to focus on the field of electrical engineering because, as an applied science with a tradition of relatively close ties to private industry and concomitant anxiety about its status as "science," academic engineering's investment in boundary maintenance is particularly visible. Several of the professors interviewed had worked in private industry prior to being recruited to the university, all held Ph.D.'s, all had numerous publications, and most were senior researchers, meaning they held tenure. The students were all doctoral candidates in electrical engineering or computer science with one or more publications on their curricula.

   The second site was a large computing research facility populated by electrical engineers but also biologists, neuroscientists, computer scientists, and physical scientists who, for various reasons, use advanced computing technologies in their research. I chose the latter site in part because my conversations there gave me a chance to compare practices in electrical engineering with those in other fields. Authorship practices vary across fields and subdisciplines, of course, and the comments included here should not be taken as perfectly representative of "technoscience" as a whole. Nor should they be taken as representative of academic authorship in toto—writers in the humanities, for example, are more likely to publish as sole authors and engage in different dynamics of attribution. Please see McSherry, *Who Owns Academic Work?*, for further methodological details.

3. *Syllabus*, in this context, means a paper reviewing the recent literature in a given field, rather than simply a listing of topics to be covered in a course. This kind of syllabus accompanies lectures given in a course and is used by students to study for medical board exams.
4. Clinical trials of new drugs, for example, are validated in part by the scientific integrity of the laboratories conducting the trials. Investment decisions are based in large part upon faith in the results of those trials, and the university recognizes this in its careful policing of the use of its name for advertising purposes. More broadly, basic research in academic science is expected to generate unexpected inventions and commercially useful data precisely because it is not always concerned with the bottom line. Or, as an executive of a major oil company put the matter to a researcher looking to move to the private sector: "You're worth a lot more to us working in the university coming up with good ideas," the executive said, "than you'd be on our research staff being forced to work on projects that have already been decided by management" (interview with senior researcher).
5. 684 F. Supp. 1248 (1988).
6. 868 F. 2d 1313 (1989).
7. The Anglo-American tradition of literary property was founded upon the idea that authorship involved "imprinting . . . an author's personality" on a thing; a process verified by the thing's "originality" (Rose 1993, 114). This process of imprinting involved individual mental labor, carried on "separated . . . from the rest of mankind" (Daniel Defoe in Rose, 39). Labor, argued William Enfield in 1774, "gives a man a natural right of property in that which he produces: literary compositions are the effect of labor; authors have therefore a natural right of property in their works" (in Rose, 85). Creative individuals, through investment of labor, created something that had not previously existed.
8. *Childress v. Taylor*, 945 F. 2d 500, 1991. See P. Jaszi, *On the Author Effect II: Contemporary Copyright and Collective Creativity*, in *The Construction of Authorship: Textual Appropriation in Law and Literature*, eds. M. Woodmansee and P. Jaszi (Durham and London: Duke University Press, 1994).
9. A derivative work is one that transforms, adapts, revises, or otherwise modifies a preexisting work such that the new product represents an original work of authorship.
10. Recognition carries with it the burden of responsibility. Mario Biagioli has argued that the gift economy of scientific authorship depends upon the credit/responsibility dualism. Credit for a new discovery attaches to a scientist's name, but that named scientist is also, theoretically, responsible for the truth of that claim. The practice of granting "courtesy" author-

ship to individuals only loosely associated with the project undermines this form of responsibility, as a rash of cases of scientific fraud in biomedicine have illustrated (Biagioli 1998; LaFollette 1992).

11. These norms, and their informal enforcement, are not confined to technoscience. Consider, for example, a scandal that took place at Texas Tech, in which a history professor was exposed as a dedicated plagiarist. Several of his writings, including a book manuscript, were revealed to have been plagiarized. The professor was denied tenure, but little further action was taken. The book manuscript was published, with few revisions. The professor was later hired to evaluate other people's research as a grant monitor for the National Endowment for the Humanities. His plagiarism did at last become "public" when reviews of the published book called attention to it (one of the people asked to review the book was the same person who had exposed the suspicious passages in the first place—the historian who wrote the work that had been copied). Despite the scandal and a subsequent accusation of plagiarizing yet another work, the professor kept his job at the NEH for several years (Mallon 1989).
12. *Edward B. Marks Music Corp. v. Jerry Vogel Music Co.*, 140 F. 2d 266 (2d Cir. 1944), at 267.
13. 17 U.S.C. § 107.
14. In 1974, the Ninth Circuit declared that an ad showing a race car with distinctive markings was sufficient to evoke the identity of race car driver Lothar Motschenbacher. *Motschenbacher v. R. J. Reynolds*, 498 F. 2d 821 (9th Cir. 1974). The decision in *John W. Carson v. Here's Johnny Portable Toilets,* 698 F. 2d 831 (6th Cir. 1983), affirmed Johnny Carson's right of publicity claim in the phrase "Here's Johnny" (used nightly by Ed McMahon to introduce Carson). A few years later, courts accepted arguments from Bette Midler and Tom Waits that commercials that used "sound-alikes" had infringed on their right of publicity. *Midler v. Ford Motor Co.*, 849 F. 2d 460 (9th Cir. 1988) cert. denied; *Waits v. Frito-Lay, Inc., and Tracy Locke Inc.*, 978 F. 2d 1093 (9th Cir. 1992).
15. As yet, the academic persona does not seem to be fully alienable (as is the star persona), and this is a crucial distinction. It took only two decades to make this shift in Hollywood, however, and it is not improbable that as the academic economy of knowledge is reconstructed on market terms, professors will find it lucrative to license their names.
16. *Lombardo v. Doyle,* 396 N.Y.S. 2d 661, 664 (1977); and *Hirsch v. SC Johnson and Co.*, 280 N.W. 2d 129, 134–35 (1979).
17. *Uhlaender v. Henriksen*, 316 F. Supp. 1277, 1282 (1970), cited in Madow 1993, 183.

Bibliography

Biagioli, M. "The Instability of Authorship: Credit and Responsibility in Contemporary Biomedicine." In *FASEB Journal* 12 (1998): 3.

Bourdieu, P. *Homo academicus.* Translated by P. Collier. London: Polity, 1988.

Boyle, J. *Shamans, Software, and Spleens: Law and the Construction of the Information Society*. Cambridge, Mass.: Harvard University Press, 1996.
Coombe, R. "Embodied Trademarks: Mimesis and Alterity on American Commercial Frontiers." In *Cultural Anthropology* 11 (1996): 202.
Dyer, R. *Heavenly Bodies: Film Stars and Society*. New York: St. Martin's, 1986.
Edelman, B. *Ownership of the Image: Elements for a Marxist Theory of Law*. London: Routledge, 1979.
Gordon, W. "A Property Right in Self-Expression: Equality and Individualism in the Natural Law of Intellectual Property." *The Yale Law Journal* 102 (1993): 1533.
Grossman, R. "In Academe, the Serfs Are Toppling the Lords." *Chicago Tribune*, 24 August 1997, C1.
Hagstrom, W. O. *The Scientific Community*. New York: Basic Books, 1965.
Hyde, L. *The Gift: Imagination and the Erotic Life of Property*. New York: Vintage, 1983.
Jaszi, P. "On the Author Effect: Contemporary Copyright and Collective Creativity. In *The Construction of Authorship: Textual Appropriation in Law and Literature*, edited by M. Woodmansee and P Jaszi. Durham and London: Duke University Press, 1994.
Kennedy, D. *Academic Duty*. Cambridge, Mass.: Harvard University Press, 1997.
LaFollette, M. *Stealing into Print: Fraud, Plagiarism, and Misconduct in Scientific Publishing*. Berkeley: University of California Press, 1992.
Larson, M. *The Rise of Professionalism: A Sociological Analysis*. Berkeley: University of California Press, 1977.
Latour, B. *We Have Never Been Modern*. Translated by Catherine Porter. Cambridge, Mass.: Harvard University Press, 1993.
Latour, B., and S. Woolgar. *Laboratory Life: The Social Construction of Scientific Facts*. Beverly Hills, Calif.: Sage, 1979.
Leatherman, C. "Shared Governance under Siege: Is It Time to Revive It or Get Rid of It?" *Chronicle of Higher Education*, 30 January 1998, A8.
Long, P. "Invention, Authorship, 'Intellectual property,' and the Origin of Patents: Notes Towards a Conceptual History." *Technology and Culture* 32 (1991): 847.
Lunsford, A., and L. Ede. "Collaborative Authorship and the Teaching of Writing." In *The Construction of Authorship: Textual Appropriation in Law and Literature*, edited by M. Woodmansee and P. Jaszi. Durham and London: Duke University Press, 1994.
Madow, M. "Private Ownership of Public Image: Popular Culture and Publicity Rights." *California Law Review* 81 (1993): 125.
Mallon, T. *Stolen Words: Forays into the Origins and Ravages of Plagiarism*. New York: Ticknor and Fields, 1989.
Mauss, M. *The Gift: Forms and Functions of Exchange in Archaic Societies*. Translated by I. Cunnison. New York: Norton and Co., 1967.
Nimmer, M. "The Right of Publicity." *Law and Contemporary Problems* 19 (1954): 203.
Noble, D. "Digital Diploma Mills: The Automation of Higher Education."

First Monday. 1997. <http: /firstmonday.org/issues/issue3_1/noble/index.html> (2 June 1999).

Pew Higher Education Roundtable. *Policy Perspectives*, special issue (March 1998).

Rose, M. *Authors and Owners: The Invention of Copyright*. Cambridge, Mass.: Harvard University Press, 1993.

St. Onge, K. R. *The Melancholy Anatomy of Plagiarism*. Boston and London: University Press of America, 1988.

Shapiro, D. W., N. S. Wenger, N. F. Shapiro, "The Contribution of Authors to the Multiauthored Biomedical Research Papers," *Journal of the American Medical Association*, 1994; 1 271: 438–42.

Stearns, L. "Copy Wrongs: Plagiarism, Process, Property and the Law." *California Law Review* 80 (1999): 513.

Swan, J. "Touching Words: Helen Keller, Plagiarism, Authorship." In *The Construction of Authorship: Textual Appropriation in Law and Literature*, edited by M. Woodmansee and P. Jaszi. Durham and London: Duke University Press, 1994.

Tarnow, E. "The Authorship List in Science: Junior Physicists' Perceptions of Who Appears and Why." *Science and Engineering Ethics* 5 (1999): 73.

Walshok, M. *Knowledge without Boundaries: What America's Research Universities Can Do for the Economy, the Workplace, and the Community*. San Francisco: Jossey-Bass, 1995.

Young, R. "The Idea of a Chrestomathic University." In *Logomachia: The Conflict of the Faculties*, edited by R. Rand. Lincoln: University of Nebraska Press, 1992.

PART III

The Fragmentation of Authorship

10.

Rights or Rewards?

Changing Frameworks of Scientific Authorship

MARIO BIAGIOLI

This essay is about the attribution of authorship in academic science, with special emphasis on the extensive collaborative projects typical of "Big Science." These environments are characterized by large-scale multiauthorship, and may produce articles with hundreds of names stretching the author's byline over a few pages.[1] These cases are particularly interesting because they foreground with great clarity the problems of attribution typical of scientific authorship in general. After a discussion of the general problems of scientific authorship, I analyze two new definitions (one from particle physics and one from biomedicine) that may be pointing to a radical transformation in what it means to be a scientific author today.

The Problem

Authorship is a particularly thorny issue in science because of the specific logic of its reward system—a logic that is quite distinct from (and usually complementary to) that of intellectual property law. Definitions of scientific authorship are not codified in a corpus of doctrine like intellectual property law (IP) but change across disciplines and institutions.[2] However, while the many disciplinary expressions of scientific authorship are indeed varied and apparently contradictory, the logic underneath those positions is fairly consistent and therefore analyzable.

Like copyright, scientific authorship concerns something fixed in a medium (an article, a book, an abstract). But the analogy between scientific claims and the objects of copyright ends very soon. Most of the differences between the two can be traced to the fact that scientific authorship is not about property rights but about true claims about nature. This fundamental distinction is played out at many levels, some theoretical, some mundane. To begin with a mundane example, a nonscientific work is protected by copyright just by virtue of its being fixed in a tangible medium (without the further requirement of publication), but a scientific claim does not count as such unless it is made public and subjected to peer evaluation. In the case of copyright, an author obtains rights in the material inscription of his or her originality precisely because it is produced by something—personal expression—that is his or hers to begin with. Whether or not other people see or appreciate it as a result of its publication is not relevant to the author's rights in it.[3] Instead, a scientific claim is not rewarded as the material inscription of the scientist's personal expression, but a nonsubjective statement about nature. Consequently, it cannot be the scientist's property. This means that he or she does not have inherent rights in a scientific claim in the way a "normal" author has rights in the product of his or her personal expression simply by virtue of being the creative producer of that inscription. From this, it follows that unless it is published and evaluated by peers, a scientific claim does not count as such and does not bring rewards to the scientist who produced it. In sum, scientific authorship is not a right but a reward. And such a reward is not bestowed by one specific nation (according to its law), but by an international community of peers (according to often tacit customs).

That academic scientific authorship is about rewards, not property rights, is reflected in the fact that scientific credit is usually said to be "symbolic."[4] Probably this is not the right adjective, but it tries to capture the fact that scientific credit is about professional recognition that can be transformed into money (in the form of jobs, fellowships, and grants) but is not money-like in and of itself. Some have argued that science works like a gift economy in which a scientist give publications to his or her peers (as a gift) and receives credit from them (as a counter-gift).[5] But whether or not the notion of the gift can capture the peculiar logic of scientific rewards, what is clear is that credit is attached to qualitative notions such as truth, novelty, and scientific relevance, which have been proven very hard to quantify precisely because they operate (and need to operate) in an economy that is distinct from capitalistic economy. Accordingly, truth is priceless not only in the sense of being such an expensive commodity that no amount of money can buy, but in the sense that it has to be priceless because it cannot belong to the logic of interest and its ubiquitous unit of measure—money. The dichotomy between truth and interest is one of the standard topoi of the logic of scientific authorship.

Once we rule out the possibility of quantification through something like money (and especially when we exclude the logic of exchange value from science), the attribution of scientific credit and authorship becomes a very tricky matter of qualitative judgment. As a deputy editor of JAMA puts it, "the coin of publication has two sides: credit and accountability. On the credit side no one has the least idea of what the coin is worth, or who should be awarded the coins, or how the coins should be lined up for inspection. . . ."[6] Traditionally, peer review has been cast as the process through which scientific credit is reliably assessed, but recent studies have opened up this venerable blackbox, showing its many limitations, especially when a publication has been produced by many people with different expertises and disciplinary affiliations.[7] The frequent complaints that the quantity rather than the quality of a candidate's publications seems to be the major factor in promotion cases stem from these difficulties.[8]

The in-depth evaluation of a candidate's work is a time-consuming process, but time constraints cannot fully explain the widespread (if much criticized) tendency to rely on quantitative assessments of a candidate's publications. Especially in large-scale multiauthorship contexts, the qualitative evaluation of a candidate's work turns out to be a conceptual nightmare, not just a very onerous task. Evaluation is a complex and inherently contestable process even in the case of a single-authored publication. But when a vitae includes dozens of articles coauthored with dozens of other scientists, the complexity and ambiguity of evaluation grows exponentially, thus stretching (or breaking) the credibility of the entire process.

Evaluators have to contend with two thorny and potentially

intractable questions: What is the overall value of the article I'm reading? And what "share" of this value should I attribute to the candidate? It seems that precisely because of the difficulties produced by defining scientific credit as something that cannot be quantified, scientific credit often ends up being quantified by default and in the most crude manner: by adding up the articles bearing the candidate's name. Scientists, editors, and administrators realize very clearly that this situation is irreconcilable with their views about how science ought to operate. And yet it is far from clear how these problems could be solved within the very logic of the scientific economy they wish to uphold.

Another peculiarity of the problem of attribution of credit and authorship in science is that it is deemed inseparable from the attribution of responsibility. A scientist gets credit, but has to take epistemological (and perhaps legal) responsibility for the truth of the claims he or she publishes. These issues have become particularly urgent in the wake of numerous cases of scientific fraud and misconduct. The development of large-scale collaborations and the publication of articles with hundreds of authors has only escalated the problem by making it harder to figure out which names listed on the byline should carry the burden of responsibility. Some proclaim that each coauthor should be responsible for the entire publication. Others, instead, contend that responsibility should be limited to the extent of one's contribution. As with the definition of credit, these discussions are still waiting for closure and it is not clear how (or whether) that closure will come about.[9] What is clear, however, is that the pressure is building toward the reform of (or revolution in) the definition of scientific authorship.[10]

The Peculiar Economy of Scientific Authorship

In a liberal economy, the objects of IP are artifacts, not nature. One becomes an author by creating something original, something that is not to be found in the public domain. Copyright is about "original expression," not content or truth. Scientists, therefore, cannot copyright the content of their claims, as nature is a fact, and facts are in the public domain. The only thing researchers (or journals) can copyright about scientific publications is the form they use to express their claims. Also, saying that scientists are authors because their papers reflect personal creativity and original expression (the kind of claim that justifies copyright) would actually disqualify them as scientists because it would place their work in the domain of artifacts and fictions, not truth. A creative scientist (in the sense that IP gives to creativity and originality) is a fraudulent one.

Like copyright, patents too reward novelty as they cover novel and nonobvious claims. But, unlike copyrights, such claims need to be potentially useful to be patentable. Scientists, then, can become authors as patent holders, but cannot patent theories or discoveries per se either because they are "useless" by virtue of being "pure science," or because they are about something that belongs to the public domain.[11] While it is increasingly common for scientists (mostly geneticists) to patent what might appear to be natural objects, they do so by arguing that these objects have been extracted from their original state of nature and packaged within processes (usually diagnostic tests) that are deemed useful.[12] Scientists can patent useful processes stemming from their research, and yet academic scientific authorship is defined (at least for the time being) in terms of the truth of scientific claims, not of their possible usefulness in the market. In sum, according to the categories and tools of IP, a scientist as academic scientist is, literally, a nonauthor.

Intellectual property rights are justified by saying that the author takes as little as possible from the public domain (or "previous art") and that, by adding to and tranforming what he or she has taken from the public domain, he or she produces an original work or nonobvious useful device or process.[13] But a scientist is not represented as someone who transforms reality or produces "original expressions." And contrary to patent applicants who try to minimize their overlap with "previous art," scientists buttress their new claims by connecting them as much as possible to the body of previous scientific literature.[14] Fencing off a work from the commons of the public domain or "previous art" is a smart move if what you want to achieve is private property. But it is a plainly self-defeating tactic if the claim you are putting forward is not about property, and if it can bring you credit only by being endorsed, used, and cited (but not bought as property) by your peers.

Perhaps the business practice that comes closest to science may be the "free software" movement.[15] Another partial analogy between science and IP may be found in the legal notion of "compulsory licensing," as the author, in exchange for a certain reward, relinquishes the right to control who may use his or her work (though in science one does not get monetary rewards but only citations from such licensing).[16]

Author as Cause or Authorship as Reward?

The definition of scientific authorship is further complicated by the fact that notions of credit and attribution of authorship are not only fuzzy, but their fuzzinesses are codependent. In IP, the definition of the author in terms of his or her creative contribution and personal expression provides the legal axiom for construing his or her products as objects in which the author ought to have rights. For instance, the 1976 Copyright Act does not define *author*, but uses it as a primitive notion.[17] Ownership issues begin with the axiom that "an 'author' is one to whom anything owes its origin."[18] The author is the prime mover who "causes" the product, thereby constituting it as his or her intellectual property. But, as I have argued, such a causal framework is inapplicable to science, as it would undermine its epistemological authority by casting its claims in the category of artifacts. This creates a no-win situation—though a conceptually intriguing one.

The inapplicability of the traditional figure of the author as creator sets the definition of scientific authorship adrift because it is not clear what notions of authorial agency could be put in its place to draw the line and articulate the connection between the author and the credit he or she is due while simultaneously upholding the epistemological status of scientific claims as nonfictional. One of the consequences of this conundrum is that what becomes conceptually destabilized is not just the definition of authorship, but also that of authorial credit. This problem is evidenced in the current debates among scientists, editors, and science administrators. While in IP the articulation of authorial rights follows from the assumption about who an author is and what he or she does, in science we see that that relationship is not one of one-way causality, but oscillates back and forth between the definition of author and that of his or her credit.

For instance, it is not uncommon to see the author defined in terms of what kind of credit is deemed to be authorial.[19] This would be like having IP start with rights and then move back to picture what kind of subject those rights could be attached to. For instance, if you say that data collection constitutes authorial credit, then the data collector is entitled to have his or her name in the byline. If not, he or she ceases to be an author and ends up listed in the acknowledgment section. Depending on the discipline, one may encounter either scenario. In sum, the scientific author oscillates between being the producer and the product of the products he or she produces. (This dovetails with my previous suggestion that scientific authorship is not about rights, but about rewards.)

The Coupling of Credit and Responsibility

A reader familiar with the discourse of IP—a discourse that focuses on rights rather than responsibilities—might be surprised to see how frequently the inseparability of authorial credit and responsibility is invoked in discussions of scientific authorship.[20] If a claim about nature were like a product its author could sell in the market, then responsibility for its "faults" could be negotiated legally and monetarily in terms of liability. But this cannot apply to claims about nature because they are not owned by anyone, cannot be sold, and therefore appear to be alien to the logic of monetary liability. While it sounds quite natural to say that a scientist should be responsible for what he or she publishes, it is much more difficult to figure out exactly what that means. Scientific responsibility sounds good, but what kind of object is it?

Technically, scientific fraud amounts to lying about nature. But what crime or misdemeanor is that? As a thought experiment, one could say that fraud is like libeling nature, but then nature is not exactly a legal subject entitled to the legal protection of its reputation. One could also look at other scientists—not nature—as the damaged party and argue that a fraudulent paper misleads other scientists into wasting time and resources doing work that relied on those fraudulent claims. But those scientists did not purchase that fraudulent paper the way a consumer may have purchased a flawed product. The fraudulent paper was in the public domain, and it was those scientists' choice to pick it up and use it. Of course things are much more complicated than this, especially because the economy of science is inherently based on trust and it is not clear whether it could operate outside of that framework. The point of my little casuistic exercise here is that, like credit, responsibility is simultaneously essential to the operation of science and yet impossible to reduce to one clear definition. I find it interesting that despite the sense of moral outrage stirred by cases of scientific fraud, there are few tools to punish its authors besides firing them, denying them access to future funding, or, in certain cases, asking them to pay back the funds they have misused.[21] Most of these actions are, in effect, forms of exile or ostracism from the scientific community, but carry few or no tangible legal consequences.

Both in the case of credit and responsibility the problem is that a scientific claim is neither simply natural nor simply artifactual (in the sense that *natural* and *artifactual* assume within a logic that opposes public domain and private property). A scientific claim is not nature itself nor an artifact in the traditional (and legal) sense of the word. As such, it operates in a legal no man's land. As in the case of credit, the default solution to the dilemma posed by the attribution of responsibility has been to attach it permanently (whatever "it" means) to the scientist's name. Intellectual property rights (and responsibilities) can be tranferred contractually, but scientific credit and responsibility are seen as inalienable, that is, inseparable from the name of the original author. But while the coupling of credit and responsibility to the scientist's name is, I believe, a default move, it is not an arbitrary one.

Because it is not clear what axioms one could use to define credit and responsibility in science and to determine how they should be related, it appears that those categories can be defined only in the negative, as categories that are complementary to their counterparts in IP: scientific authorship is not like IP authorship, scientific credit is not like IP rights, scientific responsibility is not like financial liability, scientific credit cannot be transferred like IP rights, and so on. In sum, the coupling of credit and responsibility and their inalienable link to the scientist's name may be seen as a desperate one—one that is overdetermined by the lack of other possibilities.

If you can't treat scientific authorship as IP authorship nor can you say that the author of science is nature itself, then you need to redefine the authorial function of the scientist in a way that does not turn him or her into an IP-style author and yet acknowledges the human cause of that claim about nature. This, I believe, has been achieved by treating the scientist not as a legal subject (who operates in an IP context), but only as a body with a name. Of course I am not saying that the people who practice science are not legal subjects, but simply that, in so far as they work as scientists, they operate in a peculiar economy in which what matters is their name (and the fact that there is a real person behind that name), not the rest of the "bundle of rights" that, as legal subjects or citizens of specific nations, they may have attached to their names.[22] To put it differently, scientists qua scientists are humans, but not quite legal subjects.

Too Many Names, Too Few Names

Until the emergence of large-scale multiauthorship, science administrators and editors were able to treat scientific authorship as a non-problem, as something similar to its literary cousin. It seemed plausible to think of the scientist as the person who had the idea, did the work, wrote the paper, and took credit and responsibility for it. Despite all the differences between credit and responsibility in science and literature, the individuality of the scientific author seemed to provide a containment vessel for its hard-to-define functions.

Multiauthorship has unhinged this unstable but plausible-looking conceptualization, and has produced divergent reactions among science administrators and practicing scientists. Science administrators have tried to hold on to traditional notions of individual authorship and to treat multiauthorship as an aggregate of individual authors. For instance, the ICMJE (International Committee of Medical Journal Editors), an influential body representing hundreds of anglophone biomedical journals, has required that each name listed in an article's byline (no matter how long that byline might be) refer to a person who is fully responsible for the entire article (not just for the task he or she may have performed).[23]

This stance emerged also as a response to the finger-pointing that tends to develop among coauthors accused of having published fraudulent claims. In some of these cases, senior authors listed in the byline have argued that they were either unaware that their names had been added to the author list (a sort of "inverse plagiarism" aimed at increasing the publication chances of the article), or that, although they did participate in the research, they had nothing to do with the fraudulent aspects of the publication.[24] While these claims were found ad hoc and self-serving in some instances, they did match the investigators' findings in others.[25]

Additionally, the ICMJE has been concerned with what it saw as the inflation of authorship credit due to multiauthorship. For instance, how can one be sure that all these names refer to people whose diverse skills were actually necessary for and contributed to such a large project? The ICMJE's overall response has been to put forward stringent definitions of authorship in an attempt to control the scale of multiauthorship, rein in inflation, and facilitate the enforcement of authorial responsibility. Rather than developing a radical redefinition of authorship in the light of the new conditions of production brought about by large-scale collaboration, the ICMJE has gone back to and reinforced the figure of the individual author—the only figure it saw fit to sustain the credit-responsibility nexus.

Accordingly, what qualifies a person for authorship are his or her intellectual contributions, not other forms of labor that are deemed non-intellectual:

> Authorship credit should be based only on substantial contributions to (1) conception and design, or analysis and interpretation of data; (2) drafting the article or revising it critically for important intellectual content; and on (3) final approval of the version to be published. Conditions 1, 2, and 3 must be all met. Participation solely in the acquisition of funding or the collection of data does not justify authorship. General supervision of the research group is also not sufficient for authorship.[26]

That is, the scientific author is separated from and placed above those "workers" who contributed to the production of that text but did not contribute to its "uniqueness," to the specificity of its claims and its epistemological status.[27]

Several practitioners have objected to this definition, while others never noticed it.[28] The critics' position has been that they cannot be responsible for those aspects of a project that fall outside of their work and expertise.[29] They have also argued that a narrow definition of authorship is unfair to many scientific workers who, while not engaged in the conceptualization and writing of a certain publication, still made such work possible.[30] If these contributors do not receive authorship credit, they would receive no credit at all. In sum, researchers in large-scale biomedicine projects tend to think of authorship in corporate terms, that is, as stocks in a company that carry credit and responsibility in proportion to their share of the total value of the enterprise. To them, their names are, literally, their stocks.

But while one can empathize with the critics, their position is fraught with as many tensions as that of ICMJE. Their "corporate" perspective would require a means to demarcate and quantify their contributions and responsibilities that flies in the face of the current logic of the economy of science (especially that of responsibility). In some ways, they are trying to apply the categories of liberal economy to something that, instead, is complementary to it. At the same time, the ICMJE's attempt to control the problems of authorship simply by controlling the number of authors smacks as well-intentioned magical thinking and is at odds with the changing realities and intricacies of large-scale collaborative research.

A coauthored scientific publication makes for a very unusual pie whose features resist, in different ways, what both the ICMJE and its critics would like to do to it. Surprising as it may sound, cutting it in thin slices does not necessarily reduce the value of each slice, but it also leaves that value undetermined. As a result, multiauthorship does not produce credit inflation (as the ICMJE fears), nor does it allow for a quantitative division of the "shares" (as the critics would like). Mutatis mutandis, this is not unlike what we find in copyright law, where all "authors of a joint work are co-owners of copyright in the work," which means that "each joint owner of a work may exercise all the rights of a copyrights owner with respect to that work."[31] Of course, an author of a joint work cannot simply sell it and take off with the bundle. She is legally accountable to the other joint authors. For instance, she has to share the profits with them and may not sell or license the work in a way that would curtail the rights of the other joint authors (as by giving out an exclusive license to a third party).[32] What is interesting here is that even copyright law, despite the range of legal categories it can draw upon, is unable to divide up the pie of authorial rights among the coauthors. All it can do is make each joint author responsible for splitting the income deriving from the uses of those rights (though even then the modalities of that split remain a matter of negotiation).

While scientific authorship is not about rights (and therefore the IP doctrine of the undivisibility of copyright among coauthors cannot be applied to it) I still think we have a family resemblance here in the sense that, like the rights in a coauthored work, scientific multiauthorship is not a zero-sum game. The main difference in these two cases is that while with a coauthored work one can draw the line between the indivisible rights in the work and the monetarily divisible income from those rights, in the case of scientific multiauthorship such a line is nowhere to be found because a scientific claim is not about property rights.[33] So adding a name to the byline of a scientific article does not reduce the value of the other authors' contributions by any tangible amount because it's not clear what the overall value of that text (or of its parts) might be.[34] In the end, scientific authorship seems to work like a hologram in which each fragment "contains" the whole.[35] However, it is not that each name contains full authorship in a determinable, positive sense. It works that way, but only as a negative, default effect. In science, a coauthor becomes a full author because it is not clear how one could deny him or her that status given the chain of indeterminacies surrounding the function of the scientist's name and the value of a scientific work.

From Authorship to Contributorship and Guarantorship

Recently, two new frameworks for scientific authorship have been put forward and implemented, if only within limited communities. While it is unlikely that they will settle all debates about authorship, at least they are expanding both the practical options and the conceptual vocabulary for dealing with these issues.

The first one comes out of debates within the biomedical community. In a recent article published in *The Journal of the American Medical Association* (JAMA), Drummond Rennie (one of JAMA's deputy editors) and his collaborators argued that:

> Because the current system of authorship is idiosyncratic, ambiguous, and predisposed to misuse, we propose in its place a radical change: a new system that is accurate and discloses accountability. We propose the substitution of the word and concept *contributor* for the word and concept *author*. [. . .] Abandoning the concept of author in favor of contributor frees us from the historical and emotional connotations of authorship, and leads us to a concept that is far more in line with the actuality of modern scientific cooperative work. (my italics)[36]

Rennie and his contributors struck a sympathetic chord among other editors and, within two years, leading medical journals like *JAMA*, *Lancet*, *Annals of Internal Medicine*, *British Medical Journal*, and *American Journal of Public Health* implemented versions of their proposal.[37]

According to Rennie and his collaborators, each person who "has added usefully to the work" should be listed as a "contributor."[38] Journals should not limit the number of contributors.[39] Each name should be attached to a verbal description of that person's contribution, and the contributors list should be published on the article's first page. These blurbs are reminiscent of film credits, but are much more descriptive and do not need to make use of standardized job titles. The contributors are asked to write down what they did, without packaging their work into preexisting categories. The team is then asked to ratify these self-descriptions and is also given the opportunity to attach numerical values to each contribution as a percent value.[40] These percentages would not represent absolute measurements of those contributions' value, but only the group's local assessment of them. Collectively, the contributors should also choose the names to be published in the byline if space constraints make that necessary (though both those listed and not listed in the byline are treated as contributors and have their tasks described in the contributors list). The order in which names are listed in the byline should reflect the importance of their contribution, in descending order.[41]

This proposal's goal is explicitly pragmatic: to add transparency to a traditionally opaque process and to reduce its arbitrariness for both authors, editors, and users. The additional information provided by the contributors' job descriptions would give the reader a much better understanding of who did what. Similarly, tenure committees and institutional evaluators would have their work simplified (though not necessarily reduced) by these short narratives.[42] This information would also provide the authors themeselves with some safeguard against arbitrary distribution of credit (because potential credit "usurpers" would have to write down, thereby making explicit, the credit they are taking away from colleagues). For the same reasons, they could also play an important role in assessing responsibilities in the case of fraud allegations by holding the contributors responsible to what they wrote they did. Furthermore, the order of the byline would cease to be tied to local disciplinary customs—a practice that is made increasingly problematic by the confluence of many different subdisciplines and subcultures into large-scale projects.[43]

This proposal introduces important conceptual innovations too. The ICMJE's two-tier distinction between the names of authors and those of people entitled only to acknowledgment credit is virtually erased. The categorical hierarchy between the author as the "creator" of the distinctive traits of the work and the "helpers," who provided only the background conditions for the creator's work, is replaced by different degrees of contributorship. Every person who added something to the project is treated as a contributor (provided he or she is willing to write down what he or she did).

Moreover, while the name of the contributor would continue to work as an entity that constitutes a text as a "work," it would also become simultaneously circumscribed by a description of its own agency. To put it differently, the contributors' names do not work like names of traditional "certifying" authors (like those of IP authors). Rather, they are names of workers whose claims of contributorship should be assessed by the readers (that is, by the "market") based on the description of what they have done. This brings out with some clarity one of the crucial issues we encountered earlier on: Scientific authorship is about rewards, not rights. The author is the producer of the work, but he or she is also "produced" (i.e., recognized and rewarded as such) by his or her peers.

But while this proposal reconceptualizes authorship credit and distances it from the figure of the traditional author, it does a more conservative job when it comes to scientific responsibility. But the innovation, however modest and unarticulated, provides interesting food for thought.

Contributors are to be paired with "guarantors," people whose role seems to resemble that of the traditional and all-responsible scientific author envisaged by the ICMJE:

> All contributors are fully responsible for the portions of the work they performed and have some obligation to hold one another to standards of integrity. At the same time, special contributors must be designated and disclosed as guarantors of the whole work. Guarantors are those people who have contributed substantially, but who also have made added efforts to insure the integrity of the entire project. They organize, oversee, and double-check, and must be prepared to be accountable for all parts of the completed manuscript, before and after the publication. In this way the role of the guarantor is precisely defined and differs from that of the "first author" or "corresponding author" or "senior author."[44]

At first, the proposal seems to put together the two conflicting notions of responsibility put forward by the ICMJE and its critics. Contributors are responsible for their share of the work, but then there is also one or more guarantors who are responsible for all of it. Judging from the reception of the proposal, many readers and editors have had a hard time telling the guarantor and the traditional author apart. Only one journal, in fact, has decided to experiment with the idea of the guarantor.[45]

However, there may be the germ for a new and interesting notion of responsibility somewhere in here, though one that is resisted by Rennie himself.[46] The proposal does a careful job at articulating the role of the contributor, but only offers an example of a "bad" guarantor (Felig) and a "good" one (Collins):

> A Yale advisory committee found that Felig had exercised "poor judgment" in not aggressively investigating charges that his junior had doctored data. In contrast, it seems that Collins, director of the National Center for Human Genome Research at the NIH, responded with dispatch. Accepting responsibility for the aftercare of his work, Collins quickly corrected the published literature by exposing tainted data in 5

> articles thereby preventing other researchers from wasting further efforts in trying to replicate their faulty reports.[47]

While I do agree with Rennie and his collaborators that, under the circumstances, Collins did the right thing, it is not clear how his behavior matches all the features of what they take to be a good guarantor. If the guarantor is supposed to insure the integrity of the entire project and to organize, oversee, and double-check the publication, then Collins failed. And yet he is presented as an exemplar of what a good guarantor should be and do.

There is a subtle but important conceptual difference taking shape here. According to the ICMJE guidelines (but also according to half of the definition of the guarantor), Collins was a "bad" author, or guarantor, because his name appeared on a fraudulent paper. If one sticks to an absolute notion of responsibility, Collins could be said to have been responsible for fraud. If instead one reinterprets the role of the guarantor as that of an auditor, we get a very different picture. Collins may have failed as an auditor (he did not catch the fraud before publication) but that does not make him responsible for that fraud. His responsibility would be limited to the auditing process, but would not extend to the production of the product he is auditing. The latter kind of responsibility should be the contributor's.

Another important difference between traditional notions of responsibility and what we find, in potential form, in Rennie's proposal emerges when we focus on the guarantor's role as the person responsible for the aftercare of the publication (not just the process that lead to its publication). Collins is presented as a good guarantor largely because he cleaned up the mess produced by the fraud. In sum, one could redefine the guarantor as the person who is responsible for (1) the audit (not that which is audited) and (2) for the clean-up operations after fraud allegations are raised (but not for the mess he or she has to clean up).

I don't know whether this interpretation is something scientists and their administrators would accept. What interests me here are the slippages between very different views of responsibility that seem to be happening in this proposal as it tries to define the guarantor—slippages that may be pointing to a speciation developing within the category of responsibility. Moreover, like credit, responsibility appears to be turning into a more operational category and less of an essential feature attached to the name of the author. This turn toward operational views of credit and responsibility seems to be coupled with an increasing subdivision and distribution among different people of the functions that used to be kept together under the all-encompassing figure of the author. Scientific authorship as we knew it may be falling apart, or it may be simply unburdening itself of all those functions it could no longer juggle together.

The Corporate Unburdening of Authorship

Another, much different notion of scientific authorship emerged at about the same time, but in a very different discipline and independently from the debates that had occupied biomedical practitioners and editors. Its introduction was not the result of the kind of heated debates found in biomedicine. The proposal had not even been published, but only distributed electronically and posted on a laboratory's internal webpage. While it still makes use of the term *author*, the concept behind the word is not something an IP lawyer would be familiar with.

A few years ago, a team of high-energy particle physicists working at Fermilab appointed a committee to develop bylaws for regulating their multi-institutional (and multi-million-dollar) collaboration. It was felt that the collaboration had greatly expanded in size and level of complexity, but was still operating according to traditional customs known by a few elderly participants who were now approaching retirement age without having consigned their wisdom to paper.[48] As part of these bylaws, the committee articulated the definition of authorship and the modalities of its management.[49] The proposal was approved in 1998. Similar authorship guidelines are now being considered at other large laboratories, like CERN in Europe.

The CDF (Collider Detector at Fermilab) Collaboration is a consortium of institutions and universities that support and staff the laboratory. Potential members are engineers, students, and physicists who are said to be "blessed" (i.e., selected) by their home institution for work at Fermilab. To be approved for actual membership, a Ph.D.

physicist is required to dedicate at least fifty percent of his or her research time to CDF experiments over a three-year period.[50] Graduate students, instead, are required to work full time in the collaboration, and technical personnel gains membership by "making major contributions to CDF experiment."[51]

The CDF Collaboration has stipulated that every publication emerging from the lab should include all names included in the so-called Standard Author List.[52] This list includes hundreds of names. All of them are to be included in the byline in alphabetical order, independently from what their specific contribution to that paper might have been.[53] The Standard Author List is updated biannually by a committee that reviews the authors' fulfillment of membership requirements in the collaboration.[54]

All members are entered in the Standard Author List, but only after they have have done one FTE-year service work in the collaboration.[55] This simple bureaucratic requirement speaks volumes about the different conceptions of authorship held by CDF and ICMJE. What differentiates a member from an author is not his or her professional hierarchy. Students, technicians, and Ph.D. physicists are all eligible for authorship (while the ICMJE guidelines effectively exclude laboratory technicians). The kind of work they do does not matter either (unlike what we find in the ICMJE guidelines, which restrict authorship only to those in charge of the more conceptual tasks). Instead, at CDF, only a member who has paid his or her dues through labor becomes an author.

The "labor mentality" that seems to characterize CDF (as opposed to the "originality mentality" that frames IP and the ICMJE guidelines) is inscribed in its leave policies. Members are allowed up to a year's leave of absence without losing their author status during that period.[56] This means that, for up to a year, their name appears on all publications produced while they are not there, based on research they may or may not have directly contributed to. Similarly, those who cease to be CDF members remain on the Standard Author List for a year after their departure.[57] This kind of authorship in absentia would be anathema to the ICMJE and to Rennie (and would probably puzzle more than a few IP lawyers). But it makes perfect sense if you think of authorship in terms of credit for accumulated labor. Members do not receive authorship credit until they have worked for a year and maintain author status for a year after they stop working. To use an image that seems ubiquitous these days, members earn "stock options" in CDF and sell them back to CDF when they leave.

These policies suggest that physicists do not think of responsibility in the same terms biomedical practitioners do. The very idea of an absentee (that is, de facto irresponsible) author would be inconceivable in biomedicine. But CDF physicists do not have a lax attitude toward responsibility. Simply, as I will discuss in a moment, responsibility is managed and distributed in ways that make it independent from the presence or absence of an individual author. While both the ICMJE and Rennie's proposal stress individual responsibility, CDF treats it as a corporate matter.[58]

The reasons behind the specific notions of authorship, credit, and responsibility developed at CDF have much to do with the internal structure, physical location, and culture of that community. Biomedical practitioners participating in large clinical trials do not tend to work in the same lab. Like the sources of their data, they may be scattered over hundreds of miles and various institutions. Several of them may be only marginally familiar with each other. Physicists, instead, have only a handful of places where they can detect particles. As a result, CDF represents a kind of collaboration that is tied to a specific apparatus (from which it derives its name). Significantly, its stated objective is:

> [T]o provide the basis for the participation of the Members and Collaborating Institutions in the construction and operation of the Collider Detector at Fermilab, and the analysis of data obtained from the Collider Detector at Fermilab (CDF).[59]

Although they are affiliated with different home institutions, the CDF members work at the same site (which they also helped build) for a substantial portion of their research time. Opportunities for getting to know their colleagues are plenty. And operating in a bureaucratized environment structured by bylaws, committees, and procedures reinforces their sense of corporate identity—one that would be hard to find in biomedicine.

The bureaucratization of the author's name at CDF indicates that authorship credit and responsibility is not crucial in that setting, and it is not crucial because those functions have been taken up by other relations. Authorship has become more of a "fact of life" than a struggle for professional life (as it is in biomedicine). Credit does not reside primarily in one's publication list simply because everyone develops similar lists during the period in which they are part of the collaboration. Credit develops through the professional appreciation one gains from colleagues by working with them on a regular (if part-time) basis throughout the length of the project. Credit seems to travel through letters of recommendation or personal communications more than through publications lists. And given the remarkable size of the collaboration (and the presence of scientists from many different institutions), one's colleagues within the collaboration may already constitute a very large portion of one's disciplinary peers and potential employers. Such a relatively close and inclusive community may reduce the role of the vitee as "professional passport"—a role which, instead, is crucial in more dispersed and less interdependent communities like those of biomedicine.

As with credit, CDF's approach to responsibility is also framed by the structure and scale of its community. Nowhere in the CDF bylaws or in its authorship guidelines can one find the biomedical mantra about the inseparability of credit and responsibility and their essential link to the name of the individual scientists. What one finds, instead, are detailed corporate protocols for the internal review of manuscripts to be submitted for publication. It seems that the physicists at CDF do not need to rely on the name of the scientist as a device to keep credit and responsibility together simply because they are comfortable with the procedures they have developed for managing these two issues separately.

When a subgroup of CDF wishes to publish an article or to present a conference paper, the text goes through three rounds of internal review.[60] The first is a preliminary approval from the publication committee, the last two take place on CDF's internal webpage. The text is posted and all members of the collaboration are asked to comment electronically. After comments are sent and answered, a revised version is posted and the process starts again. After two rounds of revisions, those whose name is on the Standard Author List may withdraw their name from that publication if they are dissatisfied with the end product.[61]

Interestingly, an article carrying fewer names would appear to be less (not more) credible than one with more names—a scenario that is exactly opposite to what happens in biomedicine. Given the remarkable size of the collaboration in relation to the size of the field, most of the competent reviewers are inside, not outside, of CDF. So more names on the byline mean more peer endorsements (especially because those are the names of the peers who would have most to lose if the article turned out poor or, worse, fraudulent). The function of peer review—a function that in biomedicine is constitutive of authorship but is farmed out to colleagues external to the project—is performed internally. While this would be unacceptable in biomedicine (it could even be seen as a clear case of conflict of interest), here it is a non-problem because the inside and the outside of the community of peers overlaps quite substantially.

Like peer review, issues of misconduct are handled internally. CDF members can be involuntarily removed from the collaboration if they are found responsible of professional misconduct. Fraud and misconduct do not seem to have assumed the heated moral connotations they still have in biomedicine. Interestingly, the sanctions leveled against those found responsible of misconduct are exactly the same applied to those who do not live up to their labor commitments.[62] They are simply fired. Misconduct is assessed by specific committees operating according to the rules specified in the CDF bylaws without input from other agencies and institutions.[63]

One might think of expulsion (a form of exile from the community) as a fairly mild punishment. But because there isn't much community outside that community (and because the collaboration includes representatives from many institutions and universities), expulsion is likely to have fatal professional consequences. In fact, I believe that it is precisely because of the community's ability to enforce these sanctions (and because of the effectiveness of these sanctions) that responsibility talk is minimal at CDF. If you can enforce responsibility, you don't need to legislate (or obsess) endlessly about it as seems to be the case in biomedicine.

Conclusions

Despite the vast terminological and substantial differences between the CDF guidelines and those put forward by Rennie and his collaborators I believe they share a common denominator. No matter what names are given to it, scientific authorship is losing (or has already lost) its role as the containment vessel for credit and responsibility (and the vast problems posed by their definitions). The development of large-scale multiauthorship is directly responsible for that. While the names of the scientists remain crucial to the economy of science, the logic of that economy (and the role of the name within it) is changing. The various functions of authorship are being redistributed (among different people within a team) or are taken up by corporate bodies and procedures. The shift from essentialism to operationalism seems clear.

What is also clear is that there are no good or bad definitions of credit or responsibility. My brief description of CDF's protocols may cast it as a success story compared to the apparent chaos found in biomedicine. But CDF's ability to reframe authorship in ways that seem satisfactory to its members is predicated on the very specific internal structure, size, and facility-based nature of that community. The vast differences between their authorship practices and those found in biomedicine can be directly related to their different professional ecologies. I am as certain as I can be that biomedicine (as it is today) could not adopt something like CDF's guidelines.

The inherently community-specific nature of scientific authorship is not a problem but a predicament. We cannot come up with a unified notion of scientific authorship in the same way some would like to achieve the globalization of IP and the notion of author behind it. Scientific authorship is a misnomer, a historical vestige. It is not about legal rights, but about rewards. Similarly, scientific responsibility is not a legal category, but a set of relations among colleagues. As such, they cannot be conceptually unified under legal axioms. It makes sense, therefore, that scientific authorship, whatever shapes it might take in the future, will remain tied to specific disciplinary ecologies.

Notes

1. See, for example, the author list of the CDF Collaboration at <www-cdf.fnal.gov/cdfauthors.html>
2. Different universities have substantially different definitions of scientific authorship. Often, these differ from the definitions adopted by scientific journals, which, in turn, are far from homogeneous.
3. Sheldon W. Halpern, Craig A. Nard, and Kenneth L. Port, *Fundamentals of United States Intellectual Property Law* (The Hague: Kluwer Law International, 1999), 40–44.
4. Pierre Bourdieu, "The Specificity of the Scientific Field and the Social Conditions of the Progress of Reason," *Social Science Information*, 14 (1975): 19–47; Robert K. Merton, "Priorities in Scientific Discovery," *The Sociology of Science: Theoretical and Empirical Investigations* (Chicago: University of Chicago Press, 1973), 294–95, 323.
5. Warren O. Hagstrom, "Gift Giving as an Organizing Principle in Science," in *Science in Context*, eds. Barry Barnes and David Edge (Cambridge, Mass.: MIT Press, 1982), 21–34.
6. Drummond Rennie, Veronica Yank, and Linda Emanuel, "When Authorship Fails: A Proposal to Make Contributors Accountable," *JAMA* 278 (1997): 579–80 (my emphasis).
7. Until recently, the actual workings of peer review in science had received scant attention, a surprising pattern given the fundamental role everyone attributes to it. The most notable exceptions are Daryl E. Chubin and Edward J. Hachett, *Peerless Science* (Albany: SUNY Press, 1990); *Peer Review in Scientific Publishing* (Chicago: Council of Biology Editors, 1991); and *JAMA* 280, no. 3 (15 July 1998) (special issue on peer review).
8. David P. Hamilton, "Publishing by—and for?—the Numbers," *Science* 250 (1990): 1332; Drummond Rennie and Annette Flanagin, "Authorship! Authorship!" *JAMA* 271 (1994): 469; "Are Academic Institutions Corrupt?" *Lancet* 342 (1993): 315 (editorial); Marcia Angell, "Publish or Perish: A Proposal," *Annals of Internal Medicine* 104 (1986): 261–62; Barbara J. Culliton, "Harvard Tackles the Rush to Publication," *Science* 241, 29 (1988): 525; and John Maddox, "Why the Pressure to Publish?" *Nature* 333 (1988): 493.
9. For a very recent assessment of the state of the authorship debate in biomedicine, see Frank Davidoff, "Who's the Author?: Problems with Biomedical Authorship, and Some Possible Solutions," *Science Editor* 23 (2000): 111–19.
10. Richard Smith, "Authorship: Time for a Paradigm Shift? The Authorship System Is Broken and May Need a Radical Solution," *British Medical Journal* 314 (1997): 992; Richard Horton, "The Signature of Responsibility," *Lancet* 350 (1997): 5–6; Richard Horton and Richard Smith, "Time to Redefine Authorship," *British Medical Journal* 312 (1996): 723; Fiona Godlee, "Definition of 'Authorship' May Be

Changed," *British Medical Journal* 312 (1996): 1501–2; and Evangeline Leash, "Is It Time for a New Approach to Authorship?" *Journal of Dental Research* 76 (1997): 724–27.

11. Jeremy Phillips and Alison Firth, *Introduction to Intellectual Property Law*, 3d. ed. (London: Butterworths, 1995), 39–42.
12. Eliot Marshall, "Companies Rush to Patent DNA," *Science* 275 (1997): 780–81, provides a review of recent trends. See also "Gene Fragments Patentable, Official Says," *Science* 275 (1997): 1055. For an earlier overview on these issues see Dorothy Nelkin, *Science as Intellectual Property* (New York: MacMillan, 1984).
13. This foundational assumption of intellectual property law has been challenged by several legal and literary scholars. Examples of this literature are Rosemary Coombe, *The Cultural Life of Intellectual Properties* (Durham, N.C.: Duke University Press, 1998); Jane Gaines, *Contested Culture: The Image, the Voice, and the Law* (Chapel Hill, N.C.: University of North Carolina Press, 1991); James Boyle, *Shamans, Software, and Spleens* (Cambridge, Mass.: Harvard University Press, 1996); Martha Woodmansee, "The Genius and the Copyright: Economic and Legal Conditions of the Emergence of the 'Author,'" *Eighteenth-Century Studies* 17 (1984): 425–45; Jessica Litman, "The Public Domain," *Emory Law Journal* 39 (1990): 965–99; Peter Jaszi, "Toward a Theory of Copyright: The Metamorphoses of 'Authorship,'" *Duke Law Journal* (1991): 455–502; Martha Woodmansee, *The Author, Art, and the Market* (New York: Columbia University Press, 1994).
14. Greg Myers, "From Discovery to Invention: Writing and Rewriting of Two Patents," *Social Studies of Science* 25 (1995): 57–105.
15. See <www.tuxedo.org/esr/writings/>; <www.opensource.org>; and <www.fsf.org/philosophy/free-sw.html>. For a recent discussion of the state of the debate, see Steve Lohr, "Code Name Mainstream: Can 'Open Source' Bridge the Software Gap?" *The New York Times*, 28 August 2000, C1, C7.
16. Phillips and Firth, *Introduction to Intellectual Property Law*, 16–17, 29.
17. Halpern, Nard, Port, *Fundamentals of United States Intellectual Property Law*, 54.
18. *Burrow-Gilles Lithographic Company v. Sarony*, 111 U.S. 53, 4 S. Ct. 279 (1884), as cited in ibid.
19. For instance, the International Committee of Medical Journals Editors (usually referred to as ICMJE or "Vancouver Group") frames its authorship guidelines not in terms of what an author does, but in terms of the kind of contributions that qualify a researcher as an author (ICMJE, "Uniform Requirements for Manuscripts Submitted to Biomedical Journals," *JAMA* 277 [1997]: 928). This subtle difference may be easily lost in the shuffle, but is conceptually crucial. Authorship is seen as a reward, not a cause (not unlike a Ph.D. degree conferred to a student after he or she has fullfilled the appropriate requirements).
20. Sometime, the essential inseparability of credit and responsibility is represented through the figure of the coin (authorship), with two sides (credit and responsibility). See Rennie, Yank, and Emanuel, "When Authorship Fails," 580; and Davidoff, "Who's the Author?" 115.

21. Adapting the False Claim Act of 1865 (developed to curb the delivery of substandard equipment to the army) to sentence scientists with punitive damages up to three times the amount they received from funding agencies shows that the reward system of science cannot prosecute scientific fraud per se, but is forced to step out of itself and adopt the logic of commercial fraud. Paulette V. Walker, "1865 Law Used to Resolve Scientific Misconduct Cases," *The Chronicle of Higher Education*, 26 January 1996, A29. Subsequently, some uses of the False Claim Act have been challenged in court (Paulette V. Walker, "Appeals Court Overturns a Flase-Claim Ruling Against U. of Alabama at Birmingham," *The Chronicle of Higher Education*, 7 February 1997, A37).
22. As puzzling as it may sound, this peculiar image of science has a history. It can be traced back at least to the idea of the "republic of letters"—an imagined community many early modern scientists claimed to belong to.
23. "All persons designated as authors should qualify for authorship. Each author should have participated sufficiently in the work to take public responsibility for the content," ICMJE, "Uniform requirements for Manuscripts Submitted to Biomedical Journals," *JAMA* 277 (1997): 928.
24. A.S. Relman, "Lessons from the Darsee Affair," *New England Journal of Medicine* 308 (1983): 1415–17; R. L. Engler, J. W. Covell, P. J. Friedman, P. S. Kitcher, R. M. Peters, "Misrepresentation and Responsibility in Medical Research," *New England Journal of Medicine* 317 (1987): 1383–93; n.a., "President of Royal College Resigns," *British Medical Journal* 309 (1994): 1530; "Obstetrician Suspended After Research Inquiry," *British Medical Journal* 309 (3 December 1994): 1459; Jane Smith, "Gift Authorship: A Poisoned Chalice?" *British Medical Journal*, Vol. 309 (3 December 1994): 1456–76.
25. Engler et al., "Misrepresentation and Responsibility in Medical Research," 1383–93.
26. ICMJE, "Uniform requirements for Manuscripts Submitted to Biomedical Journals," *JAMA* 277 (1997): 928.
27. For a more extensive discussion of the ICMJE's distinction between "intellectual" and "non-intellectual" contributions see Mario Biagioli, "Aporias of Scientific Authorship: Credit and Responsibility in Contemporary Biomedicine," in Mario Biagioli, ed., *The Science Studies Reader* (New York: Routledge, 1999), 21–41.
28. Raj Bhopal, Judith Rankin, Elaine McColl, Lois Thomas, Eileen Kaner, Rosie Stacy, Pauline Pearson, Bryan Vernon, Helen Rodgers, "The Vexed Question of Authorship: Views of Researchers in a British Medical Faculty," *British Medical Journal* 314 (5 April 1997), 1009–19.
29. Avram Goldstein, "Collaboration and Responsibility," *Science* (23 December 1988): 1623 (but see also Arnold Friedhoff's letter on the same page). Letters by Jay M. Pasachoff, Craig Loehle, and Tobias I. Baskin in "Responsibility of Co-Authors," *Science* 275 (3 January 1997): 14.
30. Domhnall Macauley, "Cite the Workers," *British Medical Journal* 305 (11 July 1992): 6845; Ian W. B. Grant, "Multiple Authorship," *BMJ* 298 (11 February 1989): 386–87. See also the letters to the editor in response to J. B. Kassirer and Mary Angell, "On Authorship and Acknowledgments," *N Engl J Med* 325 (1991): 1510–12 published in *N Engl J Med* 326, no. 16

(April 16, 1992): 1084–85. A few editors have taken these complaints seriously. An editorial in *Lancet* argued that: "Many researchers think this definition [ICMJE's] is out of touch with their own research practice. It leans toward being a senior authors' charter, falling short of providing explicit credit for those who actually do research. [. . .] On balance, the definition seems to fail important tests of relevance and reliability." (Richard Horton, "The Signature of Responsibility," 5–6).

31. Halpern, Nard, and Port, *Fundamentals of United States Intellectual Property Law*, 55 (my emphasis).
32. Ibid., 55.
33. Things are more complicated in the case of scientific multiauthorship because the value of a scientific work is not expressible in a standardized unit of measurement. So, while the joint author of a copyrighted work can at least use money as a unit of measurement in negotiating the distribution of income generated by that work, scientists and their administrators don't have that option (at least not within current definitions of scientific credit).
34. "So, the expansion in numbers of authors per article has tended to dilute accountability, while scarcely seeming to diminish credit." Rennie, Yank, and Emanuel, "When Authorship Fails," 580. While the scarce diminution of credit is cast as a pathology by Rennie and his collaborators, I believe that what they have correctly observed is a structural (not abnormal) feature of scientific authorship.
35. Other factors may contribute to this. Readers or evaluators experience a scientific publication as a whole, not an assemblage of authorial contributions. That has much to do with the way an article is written and printed. The names of the authors are presented at the beginning, but their specific contributions are not flagged within the technical narrative. The voice of that narrative is a unified one, no matter how many people may be behind it. Therefore, the readers' perception of a work as an entity casts its authors as the producers of a whole. Consequently, more names on a byline does not mean more owners of identifiable and quantifiable shares of the work, but more authors of the same whole.
36. Rennie, Yank, and Emanuel, "When Authorship Fails," 582–84.
37. For an early assessment of the experiment, see Veronica Yank and Drummond Rennie, "Disclosure or Researcher Contributions: A Study of Original Research Articles in the *Lancet*," *Annals of Internal Medicine* 130 (1999): 661–70.
38. Rennie, Yank, and Emanuel, "When Authorship Fails," 583.
39. Ibid.
40. Ibid., 582.
41. Ibid., 583.
42. The qualitative information included in the contributor's job description might force the evaluators to stay away from quantitative analyses of the publication list. Their evaluations, therefore, may become more accurate but perhaps even more time-consuming.
43. Usually, the order of authorship is a matter of disciplinary conventions,

though the first and last author tend to be considered the most important ones, leaving the names in the middle somewhat unranked.

44. Rennie, Yank, and Emanuel, "When Authorship Fails," 582.
45. This is the case with *The British Medical Journal* (BMJ).
46. Council of Biology Editors Retreat on Authorship, personal communication, Montreal, May 1999.
47. Rennie, Yank, and Emanuel, "When Authorship Fails," 583 (my emphasis).
48. John Huth, personal communication, November 1998. John Huth (Department of Physics, Harvard, and Fermilab) was heading the committee in charge of formulating the authorship guidelines for CDF.
49. "As it turns out, CDF had absolutely nothing written down on authorship guidelines until I started writing them. What you have is the closest approximation to what I could term an 'oral tradition.' Nonetheless, it is widely agreed upon." John Huth, via e-mail, 9 April 1998.
50. Bylaws of the CDF Collaboration, sec. 3, "Membership." The version of the bylaws of the CDF Collaboration used in this article was received from John Huth, via e-mail, 9 April 1998.
51. Ibid.
52. Guidelines for Authorship in the CDF Collaboration, sec. 0, part ii. Visitors may be added to the list after approval, while people already on the list may elect to have their name not included in specific publications. The version of the guidelines used in this article was received from John Huth, via e-mail, on 4 November 1998.
53. Guidelines for Authorship in the CDF Collaboration, sec. 5.
54. Ibid., sec. 7.
55. Ibid., sec. 1.
56. Bylaws of the CDF Collaboration, sec. 3, "Membership."
57. Guidelines for Authorship in the CDF Collaboration, sec. 8.
58. I am referring to the responsibility of the contributor, not the guarantor, as outlined in the proposal by Rennie, et al.
59. Bylaws of the CDF Collaboration, sec. 2, "Objective."
60. Guidelines for the CDF Publication Process.
61. Guidelines for Authorship in the CDF Collaboration, sec. 3.
62. Bylaws of the CDF Collaboration, sec. 3, "Membership."
63. Ibid.

11.

The Death of the Authors of Death

Prestige and Creativity among Nuclear Weapons Scientists

HUGH GUSTERSON

> We must locate the space left empty by the author's disappearance, follow the distribution of gaps and breaches, and watch for the openings that this disappearance uncovers.
>
> —Michel Foucault, "What Is an Author?"

Introduction

In a 1996 talk at MIT the chemist and "science-in-fiction" novelist Carl Djerassi[1] pointed out that, whereas novelists often eschew personal fame by writing under pseudonyms, it is usually vitally important to scientists to win recognition for their work under their own names. In the words of the narrator of Djerassi's novel *The Bourbaki Gambit*:

> There is one character trait . . . which is an intrinsic part of a scientist's culture, and which the public image doesn't often include: his extreme egocentricity, expressed chiefly in his overmastering desire for recognition by his peers. No other recognition matters. And that recognition comes in only one way. It doesn't really matter who you are or whom you know. You may not even know those other scientists personally, but *they* know *you*—through your publications. (Djerassi 1994, 18–19)

Djerassi was intrigued by a group of distinguished French mathematicians who, playing the exception to the rule, refused science's cult of individual fame by publishing, starting in 1934, under the collective nom de plume Nicolas Bourbaki. (Their aim was, in part, to demonstrate that the truth status of knowledge was independent of the authority of its authors—though, ironically, as "Bourbaki" acquired his own reputation as a mathematician, the experiment fell victim to its own success.) The identities of the mathematicians who made up Bourbaki were kept secret and, in Djerassi's narrator's words, "now people refer to *him*, not *them*" (Djerassi 1994, 18). In Djerassi's novel, the "Bourbaki gambit"—the melding of individual scientists into a collective disguised as a pseudonymous individual—is repeated by an international group of contemporary scientists at the age of retirement who, in Djerassi's fictional narrative, develop the revolutionary biotechnological technique of PCR. At the moment of success, the group fractures as individuals seek to step forward and claim their success.

I want to suggest here that the conditions of bureaucratic secrecy under which American nuclear weapons research has been conducted have created a phenomenon we might refer to as the "Bourbakification" of science. This phenomenon is by no means unique to the world of nuclear weapons science: indeed corporate secrecy in, for example, the biotech industry and the practice, common to Big Science and engineering projects, of assembling large teams to generate new knowledge and develop new technologies is making the discernment of individual contributions to knowledge progressively more difficult in a wide range of science and engineering contexts. However, the world of nuclear weapons science provides a particularly stark instance of a Bourbakified mode of scientific production that is becoming more widely dispersed. In the process of Bourbakification, the distinctive contributions of individual scientists have been repressed or gathered together under the sign of sacralized individuals standing for groups. Unlike the original Bourbaki experiment, this has not been a ruse entered into voluntarily, nor does it derive from an idealistic impulse to show that knowledge can survive independently of the public reputation of its originators. It has been enforced by the conjoint workings of military secrecy and Big Science,[2] both working together to produce the phenomenological death of the scientific author in a way that lends weight to Michel

Foucault's cryptic observation that creative "work, which once had the duty of providing immortality, now possesses the right to kill, to be its author's murderer" (Foucault 1977, 142).

Early and crude examples of Bourbakification in the first heroic decade of American nuclear weapons science are well known. In 1945, for example, after the revelation of the atomic bomb, it was Oppenheimer, the director of the Los Alamos Laboratory and *Life*'s Man of the Year, who received the credit for the bomb—even though the possibility of building such a bomb was first seen by Leo Szilard and the implosion mechanism, crucial in making the plutonium bomb work, was conceived by Seth Neddermeyer (a scientist whose name is hardly well known today), possibly in response to an earlier variant of the idea articulated in Robert Serber's lectures, and was then refined and reshaped with the input of numerous other Manhattan Project scientists, including von Neumann and Kistakowsky, over a period of two years (Hoddeson 1992; Hoddeson et al. 1993; Rhodes 1988).

Seven years later, after the first hydrogen bomb was tested, the media erroneously gave the credit to Edward Teller's new laboratory at Livermore, and scientists at Los Alamos, furious to find their entire institution stripped of credit for its work, were prevented by national security regulations from correcting the error (York 1975, 13).

Edward Teller himself has been known for years as "the father of the H-bomb," even though the key design breakthrough is now widely credited to Stan Ulam,[3] and Teller largely withdrew from the project as it entered the engineering phase. Disquiet among former colleagues at Teller's popular identification as *the* inventor of the hydrogen bomb eventually impelled him, in 1955, to publish his *Science* article, "The Work of Many People," in which he described the H-bomb as "the work of many excellent people who had to give their best abilities for years and who were all essential for the final outcome." He protested that "the story that is often presented to the public is quite different. One hears of a brilliant idea and only too often the name of a single individual is mentioned" (Teller 1955, 267). That individual was, of course, Teller himself and, although in his article he named the other people who were vital to the project, he was not permitted by security regulations to say what any of them actually did. Thus the article, paradoxically, has the effect of reinforcing the appearance of Teller's singularity since, as lone author, he is arbitrator and custodian of others' unknown contributions, which he authorizes.

We see in these examples how secrecy and a mode of production based on teamwork, both characteristic of nuclear weapons research, make it difficult to certify the distinctive contributions of individuals. This can create a situation where credit tends to gravitate toward those, such as Teller and Oppenheimer, who already have established scientific reputations or bureaucratic positions of authority. Thus, in large hierarchical science institutions like nuclear weapons laboratories, intellectual value, or capital, tends to behave in the same way as material value in large capitalist institutions: it is extracted from those on the bottom, who create it through labor, accruing as wealth to those on the top, so that the labor of a Seth Neddermeyer is transmuted into the reputation of a Robert Oppenheimer.

Nuclear Salvage History

The last ten years have seen accelerating attempts to undo the Bourbakification of the inventors of the atomic and hydrogen bombs and to bestow secure identities and lines of credit on those scientists who, as their generation dies, stand between anonymity and immortality. I call this nuclear salvage history. Nuclear salvage history seeks to reverse the phenomenological death of the scientific authors of the first decade of the nuclear era just at the moment when their physical bodies are expiring. This project has been aided by the progressive declassification of the basic weapons design information and by the increasingly urgent desire of the pioneers of nuclear weapons science, now in their twilight years, to record their labors.

The leading practitioner of nuclear salvage history is the indefatigable Richard Rhodes, whose books *The Making of the Atomic Bomb* and *Dark Sun: The Making of the Hydrogen Bomb* have cataloged, in encyclopedic fashion, the personalities and contributions of the principal scientists in the first decade of nuclear weapons science. Rhodes's history is resolutely middlebrow in the sense that it is the story, vividly told, of great men, each a miniature portrait in his own right, acting on the world to change history.[4]

Rhodes's books about weapons scientists are epics of invention in which he is deeply concerned with the documentation and demarcation of individual originality and creativity. Martha Woodmansee points out that the modern conception of authorship is "a by-product of the Romantic notion that significant writers break altogether with tradition to create something utterly new, unique—in a word, 'original'" (Woodmansee 1994, 16). This essentially Romantic trope of originality as an individual gift that strikes in world-changing flashes of inspiration is common in middlebrow science writing, where it resonates with high school textbook accounts of Archimedes' and Newton's discoveries, and it figures prominently in Rhodes's accounts. Some of the most compelling passages in his books describe the exact moment of creative inspiration, which he hunts down with extraordinary determination. Take, for example, the cinematically vivid opening paragraph of *The Making of the Atomic Bomb*, in which he describes Leo Szilard's sudden realization that it might be possible to construct an atomic bomb powered by a nuclear chain reaction:

> In London, where Southampton Row passes Russell Square, across from the British Museum in Bloomsbury, Leo Szilard waited irritably one gray Depression morning for the stoplight to change. A trace of rain had fallen during the night; Tuesday, September 12, 1933 dawned cool, humid and dull. Drizzling rain would begin again in early afternoon. When Szilard told the story later he never mentioned his destination that morning. He may have had none; he often walked to think. In any case another destination intervened. The stoplight changed to green. Szilard stepped off the curb. As he crossed the street time cracked open before him and he saw a way to the future, death unto the world and all our woe, the shape of things to come. (Rhodes 1988, 13)[5]

The same trope recurs in *Dark Sun: The Making of the Hydrogen Bomb*, where Rhodes records Francoise Ulam's memory of her husband's breakthrough in the design of the hydrogen bomb with the same dramatic emphasis on one man's destiny to change history:

> Engraved on my memory is the day when I found him at noon staring intensely out of a window in our living room with a very strange

> expression on his face. Peering unseeing into the garden, he said, "I found a way to make it work." "What work?" I asked. "The Super,"[6] he replied. "It is a totally different scheme and it will change the course of history." (Rhodes 1995, 463)[7]

Michel Foucault (1977, 147) has observed that the modern individualist idea of the author has a "classificatory function," since the author's "name permits one to group together a certain number of texts, define them, differentiate them from and contrast them with others." We see this classificatory function clearly in Rhodes's books, as well as in other accounts of the Manhattan Project,[8] which seek to demarcate the exact contribution made by each of the leading weapons scientists and to rank them. (Rhodes spends several pages, for example, discussing whether Ulam or Teller should get more credit for the hydrogen bomb.) In the process of this enormous accounting operation, Rhodes salvages the contributions, formerly known to few, of less well-known scientists working on the Manhattan Project, saving them from their own premature authorial deaths, and he redefines the contributions of the manager-scientists, of whom Oppenheimer is the obvious exemplar. Oppenheimer's brilliance is displaced in Rhodes's account from scientific invention to recruitment, synthesis, and leadership. For example, Oppenheimer may not have thought of implosion, but he had, in Bethe's words, "created the greatest school of theoretical physics the United States has ever known" (Rhodes 1988, 447), where many of those who made the bomb work were trained. But, above all, Oppenheimer—described by historian Lillian Hoddeson (1992, 266) as "empowered to function like a general in moving his scientific troops around"—was a man who managed and led. Rhodes summarizes his contribution to the Manhattan Project thus:

> Robert Oppenheimer oversaw all this activity with self-evident competence and an outward composure that almost everyone came to depend upon. "Oppenheimer was probably the best lab director I have ever seen," Teller repeats, "because of the great mobility of his mind, because of his successful effort to know about practically everything important invented in the laboratory, and also because of his unusual psychological insight into other people which, in the company of physicists, was very much the

exception." "He knew and understood everything that went on in the laboratory," Bethe concurs, "whether it was chemistry or theoretical physics or machine shop. He could keep it all in his head and coordinate it. It was clear also at Los Alamos that he was intellectually superior to us." (Rhodes 1988, 570)

This evocation of the role of the manager in the big physics laboratories that emerged in mid-century is, incidentally, echoed in Zel'dovich's comment about Oppenheimer's Soviet counterpart, Yuli Khariton, who oversaw the construction of his country's first atomic bomb. Zel'dovich told the young Sakharov, "There are secrets everywhere, and the less you know that doesn't concern you, the better off you'll be. Khariton has taken on the burden of knowing it all" (Holloway 1994, 202).

The Soviet bomb project has produced its own nuclear salvage history, the finest example of which is David Holloway's *Stalin and the Bomb*. Holloway's writing is less novelistic in style than Rhodes's, and it is more deeply informed by an academic grasp of the connections between the unfolding of nuclear science and geopolitical history. Still, like Rhodes, taking an approach that emphasizes the "classificatory function" of authorship, Holloway seeks to discern the contributions made by specific individuals, to rank and compare them, and to mark what was original—though this turns out to be a troubling category.

In producing this history, Holloway faced two special problems. The first was the intense secretiveness of the Soviet state, which had rendered its own nuclear scientists even more anonymous and mysterious, more Bourbakified, than their counterparts in America. Thus, if Rhodes's writing derives much of its power from his ability to show us vivid individual characters and richly textured narratives of scientific work behind Los Alamos's veil of secrecy—to salvage the details of authorship from the well of anonymity—Holloway's accomplishment in salvaging the details of the Russian nuclear story in a much more closed society must be judged still more extraordinary.[9]

Holloway's second difficulty was, in writing his own version of the nuclear epic, to establish the authority of scientists condemned to a repetition. The Soviet scientists were, after all, not only doing something that had already been done; they were, in the case of the atomic bomb at least, doing it with the aid of design information purloined from Los Alamos by the spies Klaus Fuchs and Ted Hall, among others.[10] As Martha Woodmansee (1994) argues, while copying and embellishing the work of others used to be seen as a form of authorship in its own right in medieval Europe, in the context of contemporary copyright law and current ideologies of authorial individualism, copying is now seen as a highly degraded form of creativity. This is especially so in the world of science. Thus, the enterprise of establishing scientific authority in Holloway's nuclear salvage history is enacted in circumstances that call for different, at times more defensive, narrative strategies than Rhodes's. In Holloway's account it is also clear that, given the fusion of technoscientific achievement and nation-building in Soviet nationalist ideology, from nuclear weapons to sputnik, what is at stake in establishing the authorship of these weapons is not only the reputation of individual scientists but also the reputation of the nation these scientists represent.

As far as the atomic bomb is concerned, Holloway's strategy is to remind us that Khariton could not be sure the purloined information was accurate, so that "Soviet scientists and engineers had to do all the same calculations and experiments" as their American counterparts (Holloway 1994, 199). Holloway then details who did what here. In Holloway's narrative, in terms of creativity, the difference between going first (as the Americans did) and going second (as the Soviets did) is minimized and, given the acutely scarce resources of the postwar Soviet state, the obstacles surmounted by the Soviet nuclear weapons scientists were in many respects more formidable than those faced by their American counterparts. As regards the hydrogen bomb, Holloway shows that the information Fuchs gave the Soviets about design efforts in the United States would have misled them since Los Alamos at this time was, under Teller's guidance, pursuing a design strategy that turned out to be a blind alley. Holloway demonstrates that Sakharov and Zel'dovich followed their own design path, in many ways making quicker progress than their American counterparts and that, although the Americans were slightly ahead of the Soviets in creating a full-blown thermonuclear explosion, the Soviets were ahead in learning to use lithium deuteride—the key in making a deliverable bomb rather than an enormously unwieldy thermonuclear firecracker (Holloway 1994, chap. 14).

The stakes attached to originality (even if only the originality of a repetition) here are high, for both individuals and nations. When Hirsh and Mathews published an article in 1990 in a fairly obscure American journal alleging that the Soviets had used fallout from the first American H-bomb test in 1952 to deduce the design breakthrough made by Teller and Ulam,

> it caused some consternation among scientists who had taken part in the Soviet project. Khariton asked that a search be done of the files of those scientists who had been engaged in the detection and analysis of foreign nuclear tests. Nothing was found in those files to indicate that useful information had been obtained from analysis of the Mike test. This was not because of self-denial. Sakharov and Viktor Davidenko collected cardboard boxes of new snow several days after the Mike test in the hope of analyzing the radioactive isotopes it contained for clues about the nature of the Mike test. One of the chemists at Arzamas-16 unfortunately poured the concentrate down the drain by mistake, before it could be analyzed. (Holloway 1994, 312)

Thus did the carelessness of a chemist save the honor of a nation.

The nuclear salvage history of Holloway and others has given names to the scientists behind the Soviet bomb, bestowed epic status on their labors, and enabled them to take their place as individuals in the pantheon of science. In other words, it has saved them from Bourbakification in a way that is nicely evoked by the English physicist Stephen Hawking's quip when he finally met Zel'dovich: "I'm surprised to see that you are one man, and not like Bourbaki" (Holloway 1994, 198).[11]

It is worth noting here that the fate of these American and Russian nuclear weapons scientists has, in their eventual emergence into the pantheon of history, been different from that of, for example, the engineers responsible for ICBMs, the Apollo Program, or the Boeing 747. Anonymity has been the norm for those working on large-scale military-industrial engineering projects, even those in leadership positions, in a way that has not been the case in large team-based physics projects.

In receiving credit for their work as scientific authors, nuclear weapons physicists have finally been treated in accordance with the conventions of the academic science community from which the rules of secrecy had partly severed them.[12]

Interlude

Recent developments in literary theory have destabilized traditional notions of the author. Almost thirty years ago Roland Barthes declared "the death of the author," saying that "the author is never more than the instance writing, just as *I* is nothing more than the instance saying *I*" (155) (emphasis in original). Retheorizing the author not as a centered, willful point of origination for the text but as a medium in some ways created by the text itself, Barthes exploded the Romantic individualist trope of authorship ("the modern scriptor is born simultaneously with the text itself" [156]); turned the author's work into a plural text ("we know that a text is not a line of words releasing a single 'theological' meaning [the 'message' of the author-God] but a multidimensional space in which a variety of writing, none of them original, blend and clash" [156]); and, as a corollary, promoted reading to a form of authorship in its own right ("the birth of the reader must be at the cost of the death of the author" [157]).

In the same year, Michel Foucault's article "What Is an Author?" deconstructed the author in a more historical mode. While echoing Barthes's claim that the unity and coherence of texts is illusory, Foucault was also interested in the historical origins of the author entity itself. He argued that "the coming into being of the notion of 'author' constitutes the privileged moment of *individualization* in the history of ideas, knowledge, literature, philosophy and the sciences" (141) (emphasis in original). More recently Martha Woodmansee (1994b) and Martha Woodmansee and Peter Jaszi (1994), building on Foucault's archaeology of the author, have argued that, in reality, creativity is as often collaborative as individualized and that modern notions of authorship tend to misrecognize "a collaborative process as a solitary, originary one" (Woodmansee and Jaszi 1994, 3). Pointing to collaborative forms of writing and to avowedly derivative forms of artistic creativity of the kind that Henry Jenkins (1992) refers to as "textual poaching," they protest that "most writing today—in business, government, industry, the law, the sciences—is collaborative, yet it is still being taught as if it were a solitary, originary activity" (Woodmansee and Jaszi 1994, 9).

The Death of the Authors of Death

The Livermore Laboratory, which I have been studying as an anthropologist since 1987, was founded in 1952 in order to intensify work on atomic and hydrogen bombs as the cold war escalated. Most parts of the laboratory are off-limits to the public, and access to spaces and to information for its eight thousand employees (almost three thousand of them scientists and engineers with Ph.D.'s) is regulated by an elaborate system of rules and taboos. The laboratory is divided into zones of greater or lesser exclusion related to the system for classifying information and people. A few areas on the perimeter of the laboratory are "white areas" accessible to the public. (These areas include two cafeterias, the Public Affairs Office, the Visitors' Center, etc.) Large parts of the laboratory are "red areas," which are off-limits to the public, although only open research is done there. These red areas serve as a buffer zone around the "green areas," where secret research is done. The green areas constituted roughly half of the laboratory during the 1980s, but have shrunk a little since the end of the cold war. Only those with green badges (bestowed at the end of a lengthy investigation by the federal government) can enter these areas unescorted. They are protected not only by armed guards but also by mechanical barriers such as automated doors that will only open for those with appropriate badges. (As an extra precaution, the badges are magnetically encoded with the weight of their owner, and the access doors to green areas are set within booths that weigh the person seeking entry.) Within the green areas, there are also special exclusion areas, set apart by barbed wire fences and guard booths, accessible only to a few. The plutonium facility, for example, is in an exclusion area, as is the facility where intelligence reports are handled within vaultlike rooms that have built-in counter-surveillance features, such as copper mesh in the walls, to disrupt attempts to intercept electronic activities inside. The laboratory, then, is a grid of tabooed spaces and knowledges segregated not only from the outside world but, to some degree, from each other as well. Red areas, for example, although they are located inside the laboratory's perimeter fence are, in terms of informational flow, functionally a part of the outside world that is separated by informational shielding from the laboratory's green areas—some of which are, in turn, shielded from others (Gusterson 1996, chap. 4).

Unlike academic scientists, Livermore scientists in the green areas are not under pressure to publish in order to keep their jobs. The system of a multiyear probationary period followed by either ejection or permanent tenure that organizes scientific careers in the academy does not apply at the Livermore Laboratory. Here scientists have had near-guaranteed job security as long as they worked conscientiously and kept their security clearances in order, and the laboratory's work ethic, especially in comparison with that of research universities, emphasized teamwork over individual distinction.[13]

Up to the end of the cold war at least, nuclear weapons science was principally organized around the design and production of prototype devices for nuclear tests at the Nevada Nuclear Test Site and around the measurement of these tests. (Measurement was a challenge, since the devices, buried underground with the measuring instruments, destroyed the measuring equipment a few nanoseconds after the commencement of the experiment.) This design and production work was undertaken by enormous multidisciplinary teams of physicists, engineers, chemists, and technicians, with small teams of physicists playing the lead design role and overseeing the tests. The laboratory was divided into various divisions, each of which was responsible for a different part of the nuclear weapons design and testing program. The physicists of B Division, for example, designed the atomic bombs (known as "primaries") that use processes of nuclear fission to produce an atomic explosion. These components serve as triggers for a thermonuclear explosion in a hydrogen bomb. The physicists of A Division designed that part of a nuclear device (known as the "secondary") that, harnessing energy from the primary, uses processes of nuclear fusion to generate a thermonuclear explosion. Within each of these divisions, some physicists primarily focused on the generation of the enormous supercomputer codes that simulated the behavior of different weapons designs, while others took the lead role in designing and troubleshooting devices for testing. Meanwhile, engineers in the laboratory's W Division were responsible for developing prototype devices, in consultation with the physicists of A and B Divisions, while L Division took charge of preparing the enormously complex and subtle diagnostic equipment that measured weapons performance in nuclear tests. Scientists from all of these divisions were assembled through the laboratory's matrix system into large multidisciplinary teams that prepared particular nuclear tests.[14]

Within these teams, and indeed within the laboratory as a whole, the physicists tended to be the elite.[15] The work of these physicists involved calculating the expected performance of the device, often by refining the enormous supercomputer codes used to model nuclear explosions; checking predictions against data from previous tests and, in the process, flagging anomalies that might be resolved by further research; making serial presentations to design review committees; consulting with colleagues whose expertise might improve the experiment; consulting with representatives of the Department of Energy and the armed forces about military requirements; and overseeing the machining of parts and the final assembly of the device and the diagnostic equipment. One weapons scientist, Peter,[16] mentioned in an e-mail message to me that, "while the design activity is genuinely a group effort, neither the contribution to the effort nor the acknowledged credit for the result is evenly distributed. One person may be thought of as the principal architect, while others are given credit for significant components." In particular, the lead designer would get special credit. In the localized face-to-face community of weapons designers, this credit would be established and circulated as much by word of mouth—in gossip and in formal presentations—as through the written documentation of individual contributions and achievements, though there were formal shot reports and supervisors did write evaluations of their subordinates' job performance. The final product of the weapons scientists' labor was as much the test itself as any written distillation of it. It was the test that ultimately clarified the validity of the designers' theories and design approaches, and if we ask what it is that nuclear weapons designers were authoring all those years, we might have to say that it was not ultimately written texts so much as devices and "events"—the weapons scientists' term for nuclear tests.

The world of nuclear weapons science behind the fence is, though not completely informationally impervious to the outside world, fundamentally autarchic. (One weapons designer told me that her first few years at the laboratory felt like the equivalent of a second physics Ph.D. in fields not taught at the university.) Thus, although it is sometimes possible to transform information produced in the laboratory's weapons programs into knowledge that can be traded on the open market outside the laboratory, often this is not the case. Peter described one end of the spectrum in his e-mail message:

> As you know, the people involved in weapons work range from someone like Forest Rogers[17] (who calculates wonderful opacities, but would have little practical understanding of a W or B anything [finished nuclear weapons], to Dan Patterson (who lives and breathes weapons). People at Forest's end of the spectrum can publish the bulk of their work in regular scientific journals. As an example, the first publications of OPAL opacities (OPAL is the code that calculates the opacity) resulted in a paper that for some years was the most cited in astrophysics (fortunately uranium is not important in calculating astrophysical mixtures).

At the other extreme are scientists, the very titles of whose publications are secret, so that their resumes are, to the outside world, surrealistically blank after years of labor. One of these joked during a layoff scare, "If I made a resume there'd be nothing on it." Another physicist, reflecting on current fears of downsizing with some bitterness, characterized the government's attitude to its scientists as: "Thanks for defending the country. It's too bad you don't have a resume, but we don't need you now." And, indeed, when scientists retire, they are not allowed even to keep copies of their own work if it is classified—a "death of the author" of a particularly poignant kind, as his (or her) lifetime's creative work is confiscated and swallowed up by the state at the exact moment it releases his aged body. This reminds us that weapons designers do not own the knowledge they produce—do not even have a guaranteed right of access to it after they have produced it—since it belongs to the state and the bureaucratic organizations that have commissioned it. In other words, weapons scientists, despite their Ph.D.'s, are wage laborers for the state—albeit well-paid ones—

and, in the final analysis, they have little control over the knowledge they build.[18]

This knowledge is often well shielded from the knowledge markets of the outside world. "There was this complete disconnect with the outside world," one scientist told me. Peter's e-mail message says:

> Many [weapons designers] have given up outside publication entirely. Any good academic paper begins by offering a context to show why the particular detail being investigated is of interest. For example, the detailed processes of lithium production in a particular class of stars is pretty boring to most astronomers who are not nucleosynthesis afficionados. It becomes of interest when framed in the context of determining the original baryon density of the universe. The context for much weapons work cannot be provided, and thank the gods that there is no suitable academic journal for the material that they investigate.

Another scientist recalled a colleague who told him he had not been to the library in years because the outside world knew nothing of him and therefore probably had nothing of interest to say to him in its publications. This can induce a twofold sense of erasure: first, one's achievements and hence one's professional person may be completely invisible to the larger scientific community (or even to one's colleagues within the laboratory: one scientist told me that one of his colleagues won the prestigious Lawrence Award for his work, but he was never able to find out what his colleague had done). Second, one's work may be literally written over by the scientific community outside the fence which, in an inversion of the Soviet nuclear scientists' repetition that established itself as original, publishes original work that is unknowingly a repetition. Peter's e-mail message describes the predicament of Livermore researchers in Inertial Confinement Fusion—until recently a highly classified technology because of its applications to thermonuclear design:

> I went to a conference in 1983 at which an academic researcher was discussing hohlraums[19] as a means of smoothing the laser pulse and converting it to X-rays. The lab people had to sit in silence as a colleague re-discovered territory that they had crossed years before.[20]

Until much of the laboratory's work on inertial confinement fusion was declassified and published after the end of the cold war, it did not publicly exist.

But the predicament of nuclear weapons scientists as authors extends beyond their inability to trade their knowledge, and thus to establish their reputations, outside the laboratory. Even within the laboratory, establishing their reputations via written authorship can be complicated. As John Sutton (1984, 208) writes in his own study of Livermore's organizational culture, "Communication within the laboratory is highly compartmentalized—that is, major projects are divided into a number of smaller research tasks, and communication outside the immediate group is only occasional." The laboratory's internal knowledge economy mixes the characteristics of a common market with those of a premodern economy with many separate zones of barter, currency, and taxation. In some ways the national security state has created an intellectual economy analagous to the traditional nonmonetarized African economies described by the anthropologist Paul Bohannon (1988), in which there were separate spheres of exchange that could not be integrated so that, for example, the beads of one family could be exchanged for the cloth but not the food of another family, since beads and food, circulating in different spheres, were untradeable and nonconvertible. Thus, nuclear weapons knowledge was recorded not so much in standardized and refereed articles, as it would be in conventional academic settings, but in reports detailing the results of nuclear tests, new ways of calculating opacities, and so on. These reports, instead of being codified into a uniformly accessible grid of knowledge, were often stored eccentrically. Although a classified library was eventually established at Livermore, the internal compartmentalization of the laboratory's knowledge economy on the one hand and its self-contained informality on the other led to a situation where, as much by accident as by design, knowledge circulated and was stored in less formal, centralized ways. As one scientist described it:

> There was a mill for publishing the results of test shots, the latest methods for calculating opacities and so on. But there was no serious library for these reports in the early days. The reports would get thrown in a room,

> then someone would take one and hold on to it and that article would now be officially "misplaced." (That's why the GAO found that 10,000 secret documents were missing at Livermore. They're not exactly lost. They're not floating around outside the lab. They're in people's offices somewhere.) Old-timers would have safes full of documents inherited from someone else who retired ten years earlier. So, when they retired, you'd get those documents transferred to you, and that was a sort of library.

In other words, even within the laboratory, knowledge could be stored and exchanged in highly localized ways. The circulation of knowledge might be restricted by the semiforgotten nature of a written report, languishing in a colleague's safe, by networks of friendship, or by the assumption that weapons scientists, for national security reasons, should not have access to too much secret information unless it was directly relevant to their work.

This system has its own potential for abuse and manipulation. For example, it was widely believed in the 1980s by weapons designers in A and B Divisions, the two main weapons design divisions, that O Group, a breakaway group of designers ultimately protected by Edward Teller's patronage, manipulated secrecy regulations to protect its work from peer review. O Group was working on, among other things, a nuclear bomb-pumped X-ray laser that was highly controversial both technically and politically and was ultimately canceled.[21] Many weapons scientists complained that they suspected O Group's science was not rigorous, but could not evaluate it because of special levels of classification placed on its reports and briefings.

At its most extreme, the laboratory environment can unmake the very form of writing itself as a means of storing information, creating within one of the most high-tech environments in the world a partial return to the orality that preceded literacy and hence the very possibility of authorship in the modern writing-based conception of the term. Many scientists' reputations rest not on written reports[22] but rather on oral presentations they have given; on insightful questions in design review meetings, on an inventive idea they are locally remembered to have suggested and worked through, on a beautiful component they designed, which was instantly vaporized by the very test whose success it enabled, on huge craters their devices have inscribed upon the surface of the Nevada Desert, and on a socially recognized knack for judgment—a feeling for the devices and how they will behave. Because so much weapons design knowledge is practical knowledge that is unwritten or is thought to be hermeneutic rather than purely factual in nature, it is seen as residing in the designers themselves. (For this reason the laboratory prohibits groups of designers from traveling together on the same plane, in case it crashes.) One of the older designers, Seymour Sack, was described to me as "a walking repository of 500 experiments [nuclear tests]." This unusual emphasis on the oral circulation of knowledge and credit has endured for a number of reasons. First, there is comparatively little need to share knowledge with outsiders—even those at the rival weapons laboratory at Los Alamos. Second, the funding and promotion of individual scientists is not tied to their literary production since, at Livermore, "in contrast to an academic setting, money is awarded to a programme rather than an individual" (Sutton 1984, 206). Third, the small face-to-face settings within which weapons work is largely done at the laboratory have diminished the need to formalize knowledge, creating a system where knowledge tended to be transmitted as much through apprenticeship and oral instruction as through solitary reading. And, fourth, there are advantages to orality in a situation where every classified document that is created requires special measures to store and protect it and cannot be freely copied.

Still, such heavy reliance on oral knowledge entails liabilities, especially as the older scientists with the most extensive knowledge retire. As Peter put it:

> There are so few people genuinely involved in design, you efficiently communicated by other means [than formal writing]. . . . And the formal record suffers from this deficiency. While we have vaults containing the measured results of tests [as well as cutaways of nuclear devices showing their internal "anatomy"], the reason that certain choices were made are not obvious from the materials stored there. This information still exists as oral histories, but the content of this reservoir diminishes as the experience base drops.

The end of the cold war, and the end of nuclear testing in particular, are bringing about changes in the knowledge economy at the laboratory. Managers at the laboratory and at the Department of Energy are worried that, as the most experienced designers retire en masse, they will take with them much of the knowledge, so inadequately recorded, that they have accumulated over the years and that, if the United States needs to again design advanced nuclear weapons at some date in the future, it may find that it has forgotten how to do so. This danger is particularly acute in the absence, now that the testing ranges of the world have fallen silent, of the nuclear tests which, more than written documentation, have enabled the reproduction and transmission of their science. This science has been passed on by means that, in some ways, have more in common with medieval craft apprenticeships than the computerized bibliocentric mazeways of most scientific disciplines at the end of the twentieth century.[23]

Thus the years since the end of the cold war have seen increasing attempts to codify and document what the weapons scientists know and to bring the means by which their information is recorded into greater conformity with the practices of the outside world. This is a form of nuclear salvage work, though it differs from the efforts of Rhodes and Holloway in that it is more interested in the formal codification of knowledge than in the individualization of its authors. Thus, in recent years, Livermore scientists have invested time in cataloguing reports and installing them in a central library, and in making written or videotaped records of the reasons for specific design decisions. Meanwhile the Los Alamos Laboratory has initiated a formal program of instruction in nuclear weapons science for new designers at the laboratory.

In a further attempt to formalize their knowledge, in 1989 the weapons laboratories also started a peer-reviewed classified journal, modeled on those published by university scientists. This journal has not, however, done very well, partly because it runs counter to the comfortable orality of knowledge circulation long established among the weapons scientists. One scientist said the journal was "of little consequence." Another described it as "a strung-out, thin sort of a thing, not conveniently available." He said, "I never tried to publish in the journal because I thought it was pointless. Three people would read it, and then it would disappear forever." He added (echoing the sentiments in the Djerassi quote with which this chapter began) that the point of publishing is to have people who have not met you read about your work but, since his research can only be discussed within a small face-to-face community that already knows about his work, publication would be a futile waste of time.

Conclusion

Michel Foucault and Roland Barthes have both argued that what we recognize as authorship is a social institution that emerged at a particular historical moment defined by social individualism, scientific rationalism, and, we might add, commodification. Over the last two centuries the ideology of authorship has tended to privilege written texts. These have been construed, through the lens of Romantic assumptions about individual creativity, as the products of unique individuals. Especially in the sciences, which Robert Merton (1942) long ago defined precisely in terms of their commitment to the universal circulation and accessibility of texts, these texts have circulated freely and have been collected in libraries that facilitated widespread access to them.

The Livermore Laboratory has developed a mode of scientific production partly at odds with these conventional notions of authorship. Although some knowledge circulates in formally authored texts, much of it circulates orally or via informal publications such as memos and reports. This knowledge is often produced in collaborative teams, so that individual intellectual production is not so highly fetishized as it is in academic circles, where lead authorship and quantity of authorship is so vital a metric in tenure and promotion decisions. And, far from circulating freely, the written knowledge produced within the laboratory often cannot leave the laboratory (unless it is going to Los Alamos) and, even within the laboratory, may lie dormant in safes or travel eccentric routes of exchange marked by chains of friendship rather than being universally available.

What are we to make of this? Martha Woodmansee has argued that the conventional ideology of authorship, which fetishizes the individual and commodifies texts through copyright laws, is a prison-house that inhibits collaborative creativity and forces us to misrecognize the degree to which all intellectual production is, no matter what the copyright lawyers say, inherently social and collaborative. In some ways scientists at Livermore might be said to have escaped this prison-house, liberated by the barbed wire fence around them. The knowledge they have produced largely circulates outside the commodified sphere of exchange regulated and constrained by copyright laws and the academic promotions treadmill. And many Livermore scientists, in a critique of academic culture that is increasingly resonant for this author, criticize the cult of individual assessment in the university and the emphasis in academia on stockpiling refereed articles as commodities, even if hardly anyone reads many of them. Many scientists told me they were attracted to work at Livermore precisely because it emphasized collaborative teamwork and did not force its scientists to publish or perish. As one weapons designer put it:

> I find writing hard, and I don't like the publish or perish business. It's not that I don't like pressure or hard work; I just like to impose my own deadlines rather than jump through other people's hoops. The university is like the military the way it confines you and arranges everyone in hierarchies. . . . I have more freedom at the lab. (quoted in Gusterson 1996, 47–48)

On the other hand, this freedom from the grants and publications treadmill comes at a price, since weapons scientists may lose individual control over the products of their intellectual labor. These scientists may not be allowed to own copies of their own writings once they retire, may not be allowed to circulate their papers—even to name them—to friends, family, and colleagues beyond the barbed wire fence. Indeed, they could be prosecuted for dicussing their own ideas with the wrong people, since their ideas belong to the state. Hence they cannot use their writings to build a public persona as authors conventionally do. Nor, until recently, could they earn royalties if they designed something patentable, since the patent was awarded to the Department of Energy.

There are now signs, however, that the end of the cold war is forcing a revision of authorship practices at the Livermore Laboratory. Just at the moment when it has lost nuclear testing, traditionally a means of consolidating and transmitting weapons design knowledge, the laboratory is increasingly moving to formalize and codify its knowledge, cataloguing and centralizing reports, trying to transcribe oral knowledge, and establishing a peer-reviewed journal for weapons designers. In some ways the laboratory seems to be trying to bring about the (re)birth of the author. It is ironic that weapons scientists should be moving toward the norms of formal, commodified authorship that have prevailed in the wider society just at the moment when, according to many commentators, those norms are increasingly being eroded by corporate practices of secrecy in the increasingly powerful centers of commercial science.[24]

But what are the limits of the (re)birth of the author at Livermore? Can it rupture the isolation of the laboratory and restore its weapons scientists to history, as Rhodes and Holloway have done for Ulam, Neddermeyer, Zel'dovich, and Altschuler? It may be that, unlike the contributions of Neddermeyer and Ulam, the work of today's American weapons scientists lies beyond the retrieval techniques of nuclear salvage history. Working in teams on design tasks seen as routine rather than charismatic, their work shrouded in secrecy and only partly documented, these scientists, known as unique individuals by one another, may be condemned in the knowledge of the outside world to live outside middlebrow history, to always work in what Foucault calls "the anonymity of a murmur."

Foucault finished his interrogation of the author by saying:

> I think that, as our society changes . . . the author-function will disappear. . . . All discourses, whatever their status, form, value and whatever the treatment to which they will be subjected, would then develop in the anonymity of a murmur. We would no longer hear the questions that have been rehashed for so long: "Who really spoke? Is it really he and not someone else? With what authenticity and originality? And what part of his deepest self did he express in his discourse?" Instead . . . we would hear hardly anything but the stirring of an indifference: "What difference does it make who is speaking?" (Foucault 1977, 160)

Notes

This paper was first presented at the "What Is a Scientific Author?" conference at Harvard University in March 1997. My thanks to Mario Biagioli and Peter Galison for organizing the conference, for guiding me to unknown sources in the literature, and for giving me perceptive comments on the first draft of this chapter. I am also indebted to Babak Ashrafi, Roberta Brawer, James Howe, Allison Macfarlane, Abigail O'Sullivan, and Charles Thorpe for clarifying in discussion some of the ideas in this chapter, and to the four weapons scientists who answered my e-mail appeals for information on secrecy and authorship at the Livermore Laboratory.

1. Djerassi, the inventor of the birth control pill, has now completed a trilogy of what he calls "science-in-fiction" novels: novels that take scientists as their principal characters and explain the workings of science to the reader. Apart from the *Bourbaki Gambit* (1994), the other novels are *Cantor's Dilemma* (1989) and *Menachem's Seed* (1997).
2. On Big Science, see Galison (1997), and the essays in Galison and Hevly (1992). Panofsky's essay in the latter volume is particularly apposite to some of the issues discussed here. Sutton (1984) discusses the conjunction of military secrecy and Big Science at the Livermore Laboratory.
3. Ulam thought of making the hydrogen bomb a two-stage device in which the first stage (a fission bomb) would be used to compress, not just ignite, fuel in the secondary. Teller later thought of using radiation rather than neutrons from the atomic bomb to achieve compression (Rhodes 1995, chap. 23). Some weapons scientists have joked that Ulam "inseminated" Teller with the idea and that Teller is in fact the "mother of the H-bomb" (Easlea 1983).
4. This approach also characterizes the biographies of two of the great Manhattan Project scientists: Lanouette's (1992) biography of Leo Szilard and Gleick's (1992) biography of Richard Feynman which, even in their titles (*Genius* and *Genius in the Shadows*) focus on the creativity and uniqueness of their subjects. As the literary theorist David Lodge has observed, commenting on the imperviousness of biography to new literary theories that decenter the subject, "literary biography thus constitutes the most conservative branch of academic literary scholarship today. By the same token, it is the one that remains most accessible to the 'general reader'" (Lodge 1996, 99).
5. Rhodes subsequently revealed the extraordinary labor that went into the research and writing of this paragraph. He had to visit London to see the intersection for himself, and he researched London weather records so that he could evoke the physical setting for Szilard's inspiration as precisely as possible.
6. The "Super" was the hydrogen bomb.
7. If Rhodes's books use, wherever possible, the trope of sudden inspiration to narrate the origins of America's first- and second-generation nuclear weapons, it is interesting that William Broad's (1985) account of the stillborn genesis of third-generation nuclear weapons at the Livermore Laboratory in the 1980s contains exactly the same literary device in its

description of Peter Hagelstein's sudden envisioning of a design for the X-ray laser at a review meeting where he was in a mystical state induced by sleep-deprivation. For a playwright's use of exactly the same literary device, this time to evoke Alan Turing's breakthrough in cracking the Nazi Enigma code during World War II, see Whitemore (1996). The Hollywood film *Fat Man and Little Boy*, in an appalling example of overwrought dramatization, uses the same device in portraying the inception of implosion—attributed in the film to Seth Neddermeyer—during the Manhattan Project.

8. See, for example, Hoddeson (1992), and Hoddeson et al. (1993).
9. This is to speak as if Holloway wrote only about the Soviet scientists and Rhodes only about the Americans. In fact, portions of Rhodes's *Dark Sun* narrate the Soviet bomb project as well, though this part of his work has received less attention, and less acclaim, than his narration of the American hydrogen bomb.
10. In the early 1990s this became a matter of some controversy in Russia as the intelligence services and veteran scientists of the original Soviet atomic bomb project feuded over who should get most credit for the first Soviet nuclear test: the spies who obtained the design for America's first plutonium bomb or the scientists who figured out how to build it (Holloway 1993; Khariton and Smirnov 1993; Leskov 1993; Sagdeev 1993).
11. Hawking meant by this that Zel'dovich seemed to have accomplished too much for one man. The admiration for Zel'dovich, and the sense of him as a great scientist, is also conveyed in a story told to me by a scientist at the Livermore Laboratory: when the Princeton physicist John Wheeler, who had worked on the American hydrogen bomb, finally met Zel'dovich, he presented him with a salt and pepper shaker, one male and one female in shape. Alluding to the greater elegance of the first Soviet H-bomb design compared to its American counterpart, he said that the male represented Zel'dovich and the female, Teller.
12. My thanks to Peter Galison for this point.
13. At the end of the cold war there were fears that military budget cuts would finally destroy the job security of scientists at the weapons laboratories. Although roughly one thousand employees (mostly support staff rather than scientists) were laid off by Los Alamos in 1995, Livermore has had no forced layoffs (as opposed to voluntary early retirement programs) since 1973.
14. The organization of the laboratory and the social production of nuclear testing is described in greater detail in Gusterson (1996).
15. To date only one director of the laboratory, Roger Batzel, has not been a physicist. Batzel was a chemist. Similarly, at Los Alamos only one of the laboratory's directors (Sig Hecker, a metallurgist) has not been a physicist.
16. "Peter" is a pseudonym. Ironically, anthropology's conventional practice of shielding interviewees by giving them pseudonyms in this case becomes another way of killing the authors behind the barbed wire fence.

17. See Iglesias and Rogers (1996); and Rogers, Swenson, and Iglesias (1996).
18. The picture is, in fact, more complicated than this thumbnail sketch allows. Some weapons scientists lead a double life, finding ways to publish in the open literature at the same time as they do their weapons work. This enables them to build intellectual capital and authorial profiles outside the laboratory perimeter in a way that makes them potentially mobile in the scientific job and knowledge markets.
19. A *hohlraum* (German for "hollow room") is a gold chamber inside which sits a pellet of deuterium and tritium in an inertial confinement experiment. When the laser beams strike and enter the hohlraum, it gives off an intense burst of X-rays, which crush and heat the fuel in the pellet, initiating fusion.
20. The Soviets did not classify Inertial Confinement Fusion research to the same degree as the Americans. This could lead to curious situations, such as one at a conference in the 1980s where Livermore fusion researchers were embarrassed that Russian scientists were openly presenting the results of their fusion experiments to an audience that included many Americans without security clearances—even though the rationale for hiding such knowledge from the uncleared was that they might share it with the Russians!
21. For the story of the X-ray laser and allegations of misconduct in its promotion, see Blum (1988), Broad (1992), and Scheer (1988).
22. One interesting example here is Bruce Tartar, the current director of the laboratory. One scientist told me that, curious to know more about his director's scientific career before he became director of the laboratory, he had tried to find what he had written about, but was unable to find a single report or article by him listed anywhere.
23. This has led MacKenzie and Spinardi (1995) to argue that, in the absence of nuclear testing, advanced nuclear weapons design knowledge might more or less fade away.
24. See Benowitz (1996), Blumenstyk (1998), Blumenthal (1997), Cohen (1995), and Marshall (1997).

Bibliography

Barthes, Roland. "The Death of the Author." In *Twentieth Century Literary Theory: A Reader*, edited by K. M. Newton. New York: St. Martin's Press, 1988. 154–58.

Benowitz, Steven. "Is Corporate Research Funding Leading to Secrecy in Science?" *The Scientist* 10, no. 7 (1996): 1, 6.

Blum, Deborah. "Weird Science: Livermore's X-Ray Laser Flap." *Bulletin of the Atomic Scientists* 44, no. 6 (1998): 7–13.

Blumenstyk, Goldie. "Berkeley Pact with a Swiss Company Takes Technology Transfer to a New Level." *Chronicle of Higher Education*, 11 December 1998.

Blumenthal, David, et al. "Witholding Research Results in Academic Life

Science: Evidence from a National Survey of Faculty." *Journal of American Medical Association* 227, no. 15 (1997): 1224–28.

Bohannon, Paul. "The Impact of Money on an African Subsistence Economy." In *Anthropology for the 90s*, edited by Johnetta Cole. New York: The Free Press, 1988.

Broad, William. *Star Warriors: A Penetrating Look into the Lives of the Young Scientists behind Our Space Age Weapons*. New York: Simon and Schuster, 1985.

———. *Teller's War: The Top Secret Story behind the Star Wars Deception*. New York: Simon and Schuster, 1992.

Cohen, Jon. "Share and Share Alike Isn't Always the Rule in Science." *Science* 268 (1995): 1715–18.

Djerassi, Carl. *The Bourbaki Gambit*. New York: Penguin, 1994.

———. *Cantor's Dilemma*. New York: Penguin, 1991.

———. *Menachem's Seed: A Novel*. Athens: University of Georgia Press, 1997.

Easlea, Brian. *Fathering the Unthinkable: Masculinity, Scientists, and the Arms Race*. London: Pluto Press, 1983.

Foucault, Michel. "What Is an Author?" In *Textual Strategies: Perspectives in Post-Structuralist Criticism*, edited by Josue V. Harari, 141–60. Ithaca, N.Y.: Cornell University Press, 1979.

Galison, Peter, and Bruce Hevly, eds. *Big Science: The Growth of Large Scale Research*. Stanford, Calif.: Stanford University Press, 1992.

Galison, Peter. *Image and Logic: A Material Culture of Microphysics*. Chicago: University of Chicago Press, 1997.

Gleick, James. *Genius: The Life and Times of Richard Feynman*. New York: Pantheon, 1992.

Gusterson, Hugh. *Nuclear Rites: A Weapons Laboratory at the End of the Cold War*. Berkeley and Los Angeles: University of California Press, 1996.

Hirsh, Daniel, and William Mathews. "The H-Bomb: Who Really Gave Away the Secret?" *Bulletin of Atomic Scientists*, January/February 1990, 24–26.

Hoddeson, Lillian. "Mission Change in the Large Laboratory: The Los Alamos Implosion Program, 1943–1945." In *Big Science: The Growth of Large-Scale Research*, edited by Peter Galison and Bruce Hevly, 265–89 (Stanford, Calif.: Stanford University Press).

Hoddeson, Lillian, Paul W. Henrikson, Roger A. Meade, and Catherine Westfall. *Critical Assembly: A Technical History of Los Alamos during the Oppenheimer Years, 1943–1945*. New York: Cambridge University Press, 1993.

Holloway, David. "Soviet Scientists Speak Out." *Bulletin of the Atomic Scientists* 49, no. 4 (1993): 18–19.

———. *Stalin and the Bomb*. New Haven, Conn.: Yale University Press, 1994.

Iglesias, C. A., and F. J. Rogers. "Updated OPAL Opacities." *Astrophysical Journal* 464, no. 2 (1996): 943–53.

Jenkins, Henry. *Textual Poachers*. New York: Routledge, 1992.

Joffe, Roland. *Fat Man and Little Boy*. Paramount Pictures, 1989.

Khariton, Yuli, and Yuri Smirnov. "The Khariton Version." *Bulletin of the Atomic Scientists* 49, no. 4 (1993): 20–31.

Lanouette, William. *Genius in the Shadows: A Biography of Leo Szilard, the Man behind the Bomb*. Chicago: University of Chicago Press, 1992.

Leskov, Sergei. "Dividing the Glory of the Fathers." *Bulletin of the Atomic Scientists* 49, no. 4 (1993): 37–39.

Lodge, David. *The Practice of Writing*. New York: Penguin, 1996.

MacKenzie, Donald, and Graham Spinardi. "Tacit Knowledge, Weapons Design, and the Uninvention of Nuclear Weapons." *American Journal of Sociology* 101 (1995): 44–99.

Marshall, Eliot. "Secretiveness Found Widespread in Life Sciences." *Science* 276 (1997): 525.

Merton, Robert. "The Normative Structure of Science." In *The Sociology of Science: Theoretical and Empirical Investigations*, by Robert Merton. Chicago: University of Chicago Press, 1942.

Panofsky, W. K. H. "SLAC and Big Science: Stanford University." In *Big Science: The Growth of Large-Scale Research*, edited by Peter Galison and Bruce Hevly, 129–46. Stanford, Calif.: Stanford University Press, 1992.

Rabinow, Paul. *Making PCR: A Story of Biotechnology*. Chicago: University of Chicago Press, 1996.

Rhodes, Richard. *Dark Sun: The Making of the Hydrogen Bomb*. New York: Simon and Schuster, 1995.

———. *The Making of the Atomic Bomb*. New York: Simon and Schuster, 1988.

Rogers, F. J., F. J. Swenson, and C. A. Iglesias. "OPAL Equation-of-State Tables for Astrophysical Applications." *Astrophysical Journal* 456, no. 2 (1996): 902–8.

Sagdeev, Roald. "Russian Scientists Save American Secrets." *Bulletin of the Atomic Scientists* 49, no. 4 (1993): 32–36.

Scheer, Robert. "The Man Who Blew the Whistle on Star Wars." *Los Angeles Times Magazine*, 17 June 1988, 7–32.

Sutton, John. "Organizational Autonomy and Professional Norms in Science: A Case Study of LLNL." *Social Studies of Science* 14 (1984): 197–224.

Teller, Edward. "The Work of Many People." *Science* 121 (1955): 267–75.

Whitemore, Hugh. *Breaking the Code*. BBC Productions, 1996.

Woodmansee, Martha. *The Author and the Market: Rereading the History of Aesthetics*. New York: Columbia University Press, 1994.

———. "On the Author Effect: Recovering Collectivity." In *The Construction of Authorship*, edited by Martha Woodmansee and Peter Jaszi, 15–28. Durham, N.C.: Duke University Press, 1994.

Woodmansee, Martha, and Peter Jaszi. *The Construction of Authorship*. Durham, N.C.: Duke University Press, 1994.

York, Herbert. "The Origins of the Lawrence Livermore Laboratory." *Bulletin of the Atomic Scientists* 31, no. 7 (1995): 8–14.

12.

"Discourses of Circumstance"

A Note on the Author in Science

HANS-JÖRG RHEINBERGER

This chapter looks into the problem of authorization and the forms of writing in the sciences and asks questions such as: What is a scientific text? Who writes? What is being written? What can be written, and what not? To ask about the author in science thus entices to locate the question, according to a long tradition, on the level of texts. Authorhood and textuality have gone through a long and convoluted common history. This is not the place to recapitulate this history in its entirety.[1] Instead, I will take for granted the junction between text and author for a moment and first sketch out a few reflections about what sorts of texts we actually encounter in contemporary science before I step down the ladder to the materialities of experimentation. The following remarks cover roughly the time period from the end of the nineteenth through the twentieth centuries, and they are written from the perspective of biochemistry and molecular biology. This caveat is necessary since the conventions of writing and authorship differ among different disciplines and research fields.

Texts

Upon closer inspection, the universe of printed texts in contemporary biology and biomedicine reveals a considerable variety of genres. We find a graded transition from conference abstracts to research papers, reviews of different scope and generality, and, finally, textbooks. This notion looks like a pleonasm at first glance. But it reminds us of the fact that the forms of encyclopedic representation in the sciences are not restricted to texts. Besides texts, they also encompass the archive of representations—let us call it the atlas.

The different forms of written fixation of scientific work prevailing today took shape in the course of the nineteenth century, and they evolved in parallel with the development of modern disciplines. There are clear distinctions between them, and there exist unwritten, but subtle and strictly followed rules of authorship and authority that go along with these different genres. Abstracts and poster texts are usually written and signed by the persons who do the work at the bench. These texts report about a piece of research done in one particular laboratory, and they are not subjected to strong peer reviewing.

Research papers, in contrast, are a strictly peer-reviewed genre of scientific text. As a rule, they are the product of a coproduction by at least one junior investigator and a senior scientist. The former occupies the first and the latter the last place in the list of authors for that paper. Research articles report about primary laboratory work and situate it with respect to both the work of the recent past and that of actual competitors. Whereas the single-authored research paper predominated in the nineteenth century, the large projects of today involving several laboratories in cooperative endeavors, such as genome sequencing projects or clinical investigations, result in papers with dozens if not hundreds of authors. In these cases, the subtle game of authorization between a senior and a junior researcher, where the junior investigator testifies to the productivity of the laboratory and the senior scientist grants visibility to the junior's work, is broken, and with it, the whole implicit system of quality control associated with this game. This destabilization has been a major issue in recent cases of scientific fraud.[2]

Writing a review article about a wider research area and topic over a shorter or longer period of time is a privilege of those considered to be the main figures, that is, the "authorities" in that field. To write a review requires invitation from a journal; the initiative is not with the individual researcher. A review grants its author the privilege to promote his or her own specialty, but at the same time expects the author to give a reasonably balanced synopsis of a research field—and thus of the work of a whole group of laboratories—with basic recent findings correctly addressed and allocated.

Finally, textbooks are written by those scientists considered to be major representatives of a subdiscipline or even a discipline. This was different during the nineteenth century, when writing a textbook or handbook used to be the *billet d'entrée* for an academic career. Textbooks expose the actual knowledge of a field for the purposes of introducing newcomers to that field. Textbooks usually drop the names of those associated with the reported findings in favor of formulas such as "it is known that." We see thus that the more or less clear-cut division of genres of scientific literature is paralleled by a marked differentiation in the authorization of the voice of the author. A strange gradient is to be observed between the authority granted to the writer of a text and his or her proximity to the scientific objects. Parallel to the decrease of intimacy and connectedness with the experimental work—from the abstract to the textbook—there is an increase in the authority publicly attributed to those who write about this work.

Regardless of the differences between the texts so far discussed, they have one thing in common. All of them avoid the "I" as nominative case, and often even the pluralistic "we." This was generally not yet the case before 1900. Today, we find no "I" anymore in these texts. Their grammatical structure suggests that the facts or the objects speak to the initiated laboratory workers or to a wider circle of readers. The subject is grammatically silenced. All along the above-mentioned authority gradient is a strict commitment to the passive voice, from which there is no escape. The supposed commitment to objectivity is built right into the language in which the scientist is allowed to speak to his or her fellows and to a wider audience. Therefore, and in a certain sense, authorship as a warranty to speak appears to be, in scientific writing, always already crossed out.

The active voice is only permitted at the outer fringes of the spectrum of writing science. The scientist as a person enters the stage when it comes to historical reflections and anecdotes on the occasion of congressional openings and commemorations, and, of course, in scientific autobiographies. Here, the scientist may take the freedom to expose his or her personal view, something that has no place in the regular canon of scientific writing. A sharp distinction between the function of an author of a scientific text and the function of the scientist as a subject is in operation here. The author is the one who has stripped off the subject function. Paradoxically, whereas the author-function accumulates credit on a particular person's name, the subject-function, along with the permission to say "I," is a strict result and a derived mode of that credit. In contemporary scientific discourse, a subject cannot stand up without the accumulated credit of making objects speak for themselves and, as a consequence of that, having already vanished as a subject. Whereas today the author-function has a tendency to become collective, the subject-function—"I, the scientist with this particular name"—remains singular, but it also is and remains derived. Even before the formation and specialization of genres, the anonymization of the producers of science was already inherent in the modern scientific text.

These different sorts of texts are located in different strata and occupy different nodes in the network of the discursive practices of writing science. Such a differential view brings up the problem of the fragmentation of the one who writes science, his or her dispersed and distributed nature. When scientists write research articles, they act as primary mediators; that is, they pretend to let their objects speak to the small community of those working on similar matters. When they write reviews, they act as secondary mediators; besides promoting their own work, they report about work done by others, both collaborators and competitors. These scientists are sorted, their work is weighted, and a place is attributed to them in the landscape of knowledge. When they write textbooks, they act as tertiary mediators; reviews of reviews are produced and hereby the voice of those who did the primary work usually becomes completely silenced. The anonymization of the knowledge producers inherent in the production of modern scientific texts is epitomized by those textbooks that are rewritten by successive editors, sometimes over many generations. Paradoxically, in an act of inverted eponymicity, these textbooks tend to retain the name of the first writer, although virtually nothing of the original text may be left over after a few re-editions. These textbooks no longer carry any vestiges of the research work with which their first authors were occupied when they were first written.

Michel Foucault is right when he claims that modern scientists represent a kind of author "captivated in the network of all those who talk about the 'same thing,' contemporaries and successors: a network that wraps them all up in dispatching those huge patterns without civil state which one calls 'mathematics,' 'history,' 'biology.'"[3] And Foucault certainly thought of the sciences when he made the general claim that "today we have to find out how an individual, a name can act as the support of an element or a group of elements which, integrating itself into the coherence of discourses or the indefinite network of forms, effaces this name or at least renders empty and useless this individuality of which nevertheless it carries the mark up to a certain point, for a certain time, and in certain respects. We have to conquer the anonymity, to justify the enormous presumption to become anonymous one day." And he concludes: "It is of our days to efface one's proper name and to embed one's voice in that grand anonymous murmur of ongoing discourses."[4] What a year later, in 1968, Roland Barthes had claimed for contemporary literature, namely the "death of the author" and the takeover by writing,[5] Foucault here seems to generalize into an epochal event of discursive takeover shaped in accordance with the characteristics of modern scientific activity. The few names that escape anonymity usually do so on the grounds of priority claims and for reasons connected to activities other than their scientific work itself. Even Nobel Prize winners tend to be known today only by insiders. What a distance from and discrepancy with those early modern "authors" that Thomas Sprat, in his 1667 history of the Royal Society, praised and hailed as "discoverers" and even "inventors of nature."[6] What a reversal of emphasis! As Bernhard Fabian has shown in his paper on the natural scientist as "original genius," the scientists of the seventeenth century were indeed the models for the "original geniuses" of eighteenth-century art and literature![7]

Scribbles

Thus far, I have been looking at varieties of scientific texts that, in their neat distinction, address different sorts of readers. Scientific authorship has revealed itself as a problematic issue already on that level. Now, I will go a step further and see how it looks below the level of these texts, in the realm of the literary practices of the research process itself, in the realm of the production of marks and traces that do not immediately address an audience. Exemplars of this kind of scientific writing are laboratory notebooks, or research notebooks in general. As far as these literary practices are not on the order of public communication, the question as to what kind of author-function they are subjected to is posed in a different way. To whom is it that the researcher speaks when taking notes? Here, there is writing, but since there is apparently no reading interpreter other than the writer himself, is there an author of these texts? Who authorizes the entries of the researchers at the laboratory bench? Do they authorize themselves through the very act of inscription?

Research scribbling, as a special kind of scientific recording, is a field that has so far not received much attention, neither from a historical nor from a systematic point of view. And yet, there is an immense variety of primary written research traces. These traces reach from jotting down ideas to drawing sketches of experiments, recording data, arranging data, processing data, interpreting experimental results, trying out calculations, and designing instrumentation. All these and many more comparable activities circumscribe a space that lies *between* the materialities of the experimental systems and the various written communications that are eventually released to the scientific community. This intermediate space is carved by individual idiosyncrasies on the one hand and by local, national, and even epochal regimes of scientific recording on the other. It belongs thus to a particular discourse formation, but it also escapes it and has paradoxical features. In one respect, these scribblings are much nearer to the materialities of scientific work than are research communications, insofar as the scribbles are quasi parts of the research objects, and therefore have a share in what will become the passive voice of those objects. In another respect, the scribbles carry an element of subjectivity, unruliness, and privacy that they are supposed to leave behind if they are to become elements of a scientific text. Without such a tension the objects in question would not be able to function as epistemic things. It is here, in this intermediate space where the objects of research have not yet become paper and where the paper—the protocol, the note—is still part and parcel of a materially mediated engagement, that the subjectivity of the scientist develops itself and plays out its potentials. It is here that the individual style of scientific discovery is shaped. Here we have to look for the equivalent to that inventive uniqueness, which today we exclusively associate with the oeuvre of a modern literary author or a creator of works of art. Research notes display the full range of possible subject-functions of the scientist that become suppressed when assuming the author-function. For the author-function consists, not in developing this subjectivity to its extreme and purest expression, but in denying it by receding behind the objects that are supposed to speak for themselves in the full scope of their proper order. On this level, therefore, we gain a completely different idea of science in the making than on the level of texts and the possibilities of analysis they offer. Here we find ourselves in the space of the prenormative, of the assay in the deeper sense of this notion which is constitutive for the making of science.

As François Jacob remarks, scientists, when going public, "describe their own activity as a well-ordered series of ideas and experiments linked in strict logical sequence. In scientific articles, reason proceeds along a high road that leads from darkness to light with not the slightest error, not a hint of a bad decision, no confusion, nothing but perfect reasoning. Flawless."[8] Research notes are the documentary traces, the immediate products of what Jacob, in contrast to the well-ordered "day science," calls "night science." "By contrast, night science wanders blind. It hesitates, stumbles, recoils, sweats, wakes with a start. Doubting everything, it is forever trying to find itself, question itself, pull itself back together. Night science is a sort of workshop of the possible where what will become the building material of science is worked out. Where hypotheses remain in the form of vague presentiments and woolly impressions. Where phenomena are still no more than solitary events with no link between them. Where the design of experiments has barely taken shape. Where thought makes its way along meandering paths and twisting lanes, most often leading nowhere."[9]

What is a scientist? When does a scientist assume the function of an author? Is the author the one who moves around errantly and lives his life in that insecure, frightening, and startling world of incipient traces in all his unsheltered subjectivity and hesitation? Or is he or she the one who finally writes a paper, erases himself or herself grammatically, and in all that oblivion convinces the competitors that there is only one way to see the logic of "the facts" after these facts themselves have spoken? Do we have to submit to an order of discourse in which the distinction has become insurmountable between, on the one hand, the epistemic subject in its intimate engagement with its epistemic objects, and, on the other, the logical subject usually called the author only when disentangled from its former epistemic intimacy?

On the one hand, we have the scientist as the author-originator, as the subject of an epistemic activity, as a novelty-producer, as somebody who has, by definition, no authoritative voice in his struggle with the murmur of events at the point of their emergence. That there is no authority at this point is at least the implicit assumption of those who believe that the core of science is the emergence of novelty—in contrast to naive realists, for whom authority here always already sides with nature, and in contrast to naive constructivists, for whom the will of knowing always already supersedes the resilience of matter. It is here that the core of what Gaston Bachelard calls the "epistemological obstacle" resides. The epistemological obstacle is not an "external obstacle such as the complexity or the fugacity of phenomena, or [the] weakness of the senses and of human mind. Within the act of gaining knowledge itself, in its innermost agitation, inertia and entanglement make their appearance according to a kind of functional necessity."[10] And it is here that we have to locate what on another occasion Bachelard calls his project of a "non-Cartesian epistemology," an epistemology of messiness and unprecedence; not an epistemology of the pure, but one of purification, where "the method remains incorporated in its application."[11] "By essence, and not by accident," therefore, a non-Cartesian epistemology is and remains "in a condition of crisis."[12] A non-Cartesian epistemologist must be interested in the texture of this crisis and in the ways and means of translating this permanent state of crisis into the order of a scientific text.

On the other hand, we have the scientist as the author-master of a double game of representation: The game is, first, that of playing out the logic of the object in the absence of all circumstantiality, and second, that of presenting its logic in the translucent style of a scientific prose, which, by definition, reaches its perfection at the very point where it rids itself of the appearance of being simply written, of being a literary text. At the very peak of textual construal it is bound to appear as its opposite, as the most "natural" of all possible accounts. With Bachelard, again, we may conclude: "[Empirical] reasoning is clear only *après coup*, when the apparatus of explanation has been set going."[13] That goes for all novel forms of scientific reasoning: "*Après coup*, they project a recurrent light on the obscurities of knowledge incomplete."[14]

Media

Let us now ask whether there are "collective" equivalents to these "private" forms of scientific scribbling. The question amounts to an exploration of these graphisms that can be seen as no-longer-simple materials but not-yet-definite arguments, these intermediate forms of scientific representation located in the twilight between the laboratory bench and the organized public discourse of the scientific community—to speak with Foucault, an exploration of the "discourse-objects" of a laboratory archaeology.[15] In this intermediate realm we find again different categories of writing, of preserving traces and marks.

One of these categories comprises lists, tables, and other forms of scientific bookkeeping. These technologies of *numeracy* serve, in the research process, as registers from which to retrieve the items, data, or figures that are necessary for assembling an experimental setup, or that have to be chosen in a particular experimental situation. In addition, they constitute databases into which research results can be entered and thus made available beyond a local laboratory space. They serve as media and mediators for the exchange of primary data. Today, they have largely taken on electronic forms of storage, retrieval, display, and communication. Prominent examples are the DNA sequence databases on which molecular geneticists and gene technologists rely in constructing their probes and comparing their results and into which they feed their sequencing products—at least in the ideal case of federally funded research. The information entered in these registers is not recognizably individualized, although care is taken that errors be retraceable to the contributors. The items that compose these pools of information constitute quasi-anonymous or at least anonymized products of the research process. In consulting them nobody asks for names, nobody wants to know who did the work unless by default, and contributing to them is seen as a service at best. And yet, they are one of the primary sources from which new questions spring, driven by the inadvertent power of synopsis.

Another category in this realm contains standardized protocols and laboratory manuals, that is, technologies of *literacy*. They consist of written-up procedures that have proven to be robust and reliable enough to be applied in a more or less routinized form. They usually are marked by the collective idiosyncrasy of a local laboratory community. They preserve for generations of experimenters what has proven to be successful. They are the forms of life into which newcomers are socialized, and they constitute a particular laboratory identity. It is very tempting to see in these conserved, written, mimeographed, chronically overwritten forms and formats the laboratory itself emerging as a collective author. Here we encounter an author created by and existing in the form of a laboratory tradition, of a particular collective, yet identifiable, way and style of doing experiments in which many people can participate precisely because of these protocollary reifications. This constellation could be called the "laboratory-function." It is different from the ordinary author-function, which may be, but does not need to be, an immediate product of the former. The laboratory-function represents more than the mere fact that a group of people have collaborated in order to arrive at a particular result. It is rather the choreography of how to get at results, the collective form of an epistemic subject-function, the way in which personality and style in science take on the form of interpersonal work, in the competitive as well as in the collaborative mode. Such a laboratory-function is at the base of what for quite some time in science studies and the history of science has been discussed as research traditions or research schools.[16] Instead of concentrating on the sociological features of these schools or traditions, such as the strong leader, the special opportunities of the local setting, or the disciplinary junctures in a particular laboratory or institute, I would like to claim that it will be worth investigating in more detail the material circumstances and the embodied gestural repertoire of the epistemic foundation of these phenomena. Research traditions are shaped in a process of material reproduction, in which the reified idiosyncrasies of the laboratory, such as recipes, procedural advice, log sheets, standardized experimental designs, and adapted software, play a major role.

What I am looking for, in a very tentative manner, in this medial realm between semi-matter and semi-print, are forms of scientific numeracy and literacy that are not of the order of publicly issued texts and that do not display a clearly individualized author-function. They take their shape from a sort of collectively accumulating memory and communalized experience, from one laboratory generation to the next. The question is what precisely these forms can tell us about that strange but epistemically crucial form of half-authorized subjectivity and of half-private objectivity, something that is different from "signature" writing, something I would like to call, in contrast to authorization, "authorifaction." What does it take epistemically to make a researcher into a part of a knowledge-gaining collective? As what kind of figure, in precisely what kind of function, does the researcher act at the bench? Who speaks to whom and through which media in the process of research, particularly in that space and time where things are no longer private dreams but not yet sanctioned facts, that semipublic realm where communion supersedes communication? The mechanisms of reinforcement that hold a knowledge-producing community such as a laboratory together, in both the synchronic and the diachronic axes, are materialized in a special kind of laboratory discourse with a unique laboratory-function somewhere between the dense and impenetrable subject on the one hand and the articulated author on the other. Just as languages of art, in their capacity as systems of symbols, oscillate between density and articulation, between picture and text,[17] authorifaction can assume all kinds of hybrids, mixtures, and blends between the seamless plenitude of an acting subject and the punctuated detachment of a signature, between the jargon of a recipe with its performative, almost private, language barely intelligible for the noninitiated with its object-signs, and the codified and punctuated argumentation of a research paper. Bachelard's project of a psychoanalysis of knowledge must be located somewhere in this space—a workplace "where one can love what one destroys, continue the past by negating it, admire his master by contradicting him."[18]

Materials

Let me finally move to a few questions concerning the structure of the scientific activity itself in its connection to systems of experimentation. Here, *authorship* has to be regarded on the level of the experimental practices themselves, that is, the material sites of data production and the carving out of epistemic things. Is there such a thing as the "material" author, an author of and out of research practice? This question touches the perennial philosophical *problematique* of the relation between subject and object. Who does the work? Every practicing scientist is familiar with the experience that the more he or she learns to handle his or her experimental system, the better it brings its own possibilities into play. To a certain degree it becomes independent from the wishes of the researcher, not because he or she does not interfere, but, on the contrary, just because he or she has shaped it with all possible skill. Here, at the deepest level of experimentation, we encounter a kind of dynamics that entertains a strange resonance with the "I" prohibition at the highest level of scientific texts.

Such "intimate exteriority," or "extimacy,"[19] such self-empowered entanglement between the epistemic thing and the epistemic tinkerer results in a relationship where the question of on which side do authority and agency reside no longer makes good sense, because what effectively takes place is a mutual coproduction. There are two complementary modes of extimacy involved in this coproduction. The tacit knowledge of the epistemic subject is complementary to the technical apparatus of the experimental system, whereas the epistemic object around which the experimental system revolves is complementary to what Michael Polanyi calls the "subsidiary" awareness, or attention of the knower.[20] From the perspective of a non-Cartesian epistemology, a profound historicity is built into the very process of knowledge production, with no authority of a perennial method to characterize the scientific author: "The concepts and the methods, everything is a function of the domain of experience/experimentation; every scientific thinking must change before a new experience; a discourse about the scientific method shall always be a discourse of circumstance—*discours de circonstance*—can never describe a definitive constitution of the scientific spirit."[21]

On several occasions,[22] I have tried to show that epistemic objects such as, for instance, transfer RNA or messenger RNA, are not the product of a deliberate search for these entities, but that such entities are the experimental effects of an ongoing "discourse of circumstance." The early test tube history of protein synthesis is one of many telling examples in which the history of molecular biology abounds. The experimental system of in vitro protein synthesis was established out of cancer research and eventually became part of molecular genetics. None of the inaugurators of this trajectory from oncology to biochemistry to molecular biology would have been able to anticipate its turns, not even those moments that revealed themselves as the most decisive in retrospect. The trajectory was the result of a recurrent experimental signification process, and the entities that emerged from the depths of the cellular space to be rendered manipulable in the test tube underwent a permanent re-signification over decades. This re-signification process became possible not in spite of but precisely because it relied on the iterations of one particular experimental system continually developed over time.

Just as scientific objects are constituted by the recurrent action of experimentation, that is, by the iteration of experimental systems and not by the deliberate will of psychological subjects, so too are scientific authors constituted by recurrence, that is, by the more or less circumstantial ascription and fixation of merits of novelty and/or priority—or by the accusation of fraud, which is only a consequence of the quest for priority. I hope that I have been able to make clear that before answering the question of what it means to be the author of a scientific writing, we will have to find an answer to the question of what it means to master—and who is the master of—the game of experimental investigation, without yielding to the hypostasis of either the metaphor of a self-determined and self-consciously acting subject or that of "nature" itself, with its uncontested "facts" as the ultimate author and arbiter.

In his *Margins of Philosophy*, Jacques Derrida proposes a generalized view of writing as a process of iteration whose typology has barely begun to be sketched.[23] According to Derrida, writing is characterized by the structural possibility of becoming weaned either from its putative originary referent, from that which the writing means, or from its putative origin, from the one who writes. The author who puts his or her signature under his or her writing is the product of the impossible wish to prevent risking this double loss—loss of immediacy and loss of presence. But it is exactly in the possibility of such a loss that the historical productivity of writing resides. The same holds for the historical process of gaining knowledge, for productive experimental systems with their multiple grafts, displacements, demarcations, disseminations, distributions, submersions, and exclusions, as well as their multiplicity of inscriptions. The forces they enact and the sutures by which they are structured are no longer those of Cartesian egos; they are those of knowledge producers inextricably melded with material, medial, scribbled, and written textures. Accordingly, we have to revise our views of what it means for scientists, as authors, to be engaged in the process of knowledge production.[24]

Notes

The title of this chapter is drawn from Gaston Bachelard, *Le nouvel esprit scientifique* (Paris: Presses niversitaires de France, 1968), 135.

I thank Peter Geimer and Michael Hagner for valuable suggestions, and I am grateful to Colin Nazhone Milburn for editing the English text. A German version appeared in Norbert Hass, Rainer Nägele, and Hans-Jörg Rheinberger, eds., *Liechensteiner Exkurse IV: Kontamination* (Egginger: Edition Isele, 2001), 63–79.

1. For a comprehensive overview, see Henri-Jean Martin, *Histoire et pouvoirs de l'écrit* (Paris: Librairie académique Perrin 1988).
2. See Mario Biagioli, "The Instability of Authorship: Credit and Responsibility in Contemporary Biomedicine," *FASEB Journal* 12 (1998): 3–16; and Biagioli, this volume.
3. Michel Foucault, "Sur les faisons d'ecrire l'histoire," *Les Lettres françoises*, no. 1187, 15–21 June 1967, 6–9. Reprinted in *Dits et ecrits 1954–1988*, tome I, 1954–1969 (Paris: Gallimard, 1994), 594.
4. Ibid., 596.
5. Roland Barthes, *La mort de l'auteur*, in *Le bruissement de la langue. Essais critiques IV* (Paris: Seuil, 1984), 63–69.
6. Thomas Sprat, *The History of the Royal-Society of London: For the Improving of Natural Knowledge*, London, 1667.
7. William Duff, "An Essay on Original Genius; and Its Various Modes of Exertion in Philosophy and the Fine Arts, Particularly in Poetry," London, 1767. For the historical development of the notion of the scientist as original genius, see Bernhard Fabian, "Der Naturwissenschaftler als Originalgenie," in *Europäische Aufklärung. Herbert Dieckmann zum 60. Geburtstag*, eds. Hugo Friedrich and Fritz Schalk (Munich: Wilhelm Fink Verlag, 1967), 47–68.

8. François Jacob, *Of Flies, Mice, and Men* (Cambridge, Mass.: Harvard University Press, 1998), 125.
9. Ibid., 126.
10. Gaston Bachelard, *La formation de l'esprit scientifique* (Paris: Vrin, 1969), 13.
11. Bachelard, *Le nouvel esprit scientifique*, (Paris: Presses Universitaires de France, 1968) especially chap. 6, 136.
12. Ibid., 162.
13. Bachelard, *La formation de l'esprit scientifique*, 13.
14. Bachelard, *Le nouvel esprit scientifique*, 8.
15. Michel Foucault, *The Archaeology of Knowledge* (New York: Pantheon Books, 1972), 140.
16. For a comprehensive overview, see Gerald L. Geison and Frederic L. Holmes, eds., *Research Schools: Historial Reappraisals. Osiris 8* (Chicago: University of Chicago Press, 1993).
17. Nelson Goodman, *Languages of Art* (Indianapolis: Bobbs-Merrill, 1968).
18. Bachelard, *La formation de l'esprit scientifique*, 252.
19. Jacques Lacan, *Séminaire VII, L'éthique de la psychanalyse* (Paris: Seuil, 1986), 122.
20. Michael Polanyi, *Knowing and Being,* ed. Marjorie Grene (Chicago: University of Chicago Press, 1969), especially Part 3.
21. Bachelard, *Le nouvel esprit scientifique*, 135.
22. See, e.g., Hans-Jörg Rheinberger, *Toward a History of Epistemic Things. Synthesizing Proteins in the Test Tube* (Stanford, Calif.: Stanford University Press, 1997).
23. Jacques Derrida, "Signature événement contexte," in *Marges de la philosophie* (Paris: Editions de Minuit, 1972), 365–93.
24. Rheinberger, *Toward a History of Epistemic Things*, epilogue.

13.

The Collective Author

PETER GALISON

I. The Transcendental Author

In the formation of collaborations, there are practical questions that press upon us. How is an individual scholar to be evaluated for hiring and promotion? How can error be detected when every member of a team is not in a position to judge the final publication? But these and related questions are not mine here, at least in the first instance. I am after something different, I want to explore what it means, quite literally, for a collaboration to know something about the world, and I want to ask this question of the largest, most intricately technical scientific collectives ever established—the detector teams surrounding colliding beam accelerators at the end of the twentieth century. With collaborations mounting to over a thousand participants, it hardly takes algebraic topology to reckon that quite soon only a handful of these teams will embrace the careers of nearly all of the seven thousand experimental particle physicists expected to be employed at the end of the century. But to ask in what sense a collaboration can know, argue, or show something, it may be useful to consider a Kantian analogy.

At the very center of Kant's project of a *Critique of Pure Reason* is an argument directed in equal measures against the empiricists and against Descartes. While Descartes begins his attempt to secure knowledge by the cogito, I think therefore I am, Kant wants to interrogate the "I" itself. What, Kant demands, are we doing when we assume there is a unified self out of which the *I* speaks?

> There can be in us no modes of knowledge, no connection or unity of one mode of knowledge with another, without that unity of consciousness which precedes all data of intuitions, and by relation to which representation of objects is alone possible. This pure original unchangeable consciousness I shall name *transcendental apperception.* (A 107)

Kant here contends that all our representations of the world have to refer back to some common consciousness; without that funneling back to a single point of awareness the bits and pieces of our perceptions would remain disconnected and the objects around us would be nothing to us. Here is a metaphor (not Kant's): without communication back to some*one*, the myriad of individual weather observers, each privately recording hourly temperatures, would never come to recognize the existence of a weather front. Only when there is one or more observers who can view the spatial combination of these isolated data into isotherms or isobars does the cold front, as a concept, enter. Without the unity of apperception, each one of us would be like such an unintegrated amalgam of private, uncorrelated observers. But in the absence of the unity of apperception our world would lack far more than cold, warm, or occluded fronts; it would lack the very concept of an object.

Kant's insight was this: the unity of our individual consciousness is a necessary precondition for the unity of any appearance of an object, indeed that unity of consciousness is necessary for there to be *for us* any object at all. As the weather front metaphor already suggests, my concern here is not with the traditional Kantian question so much as the correlate of this unity of apperception in the functioning of a manifestly collaborative inquiry. I want to ask here: What does it take for a phenomenon to be something to a collaboration? That is, what are the specific mechanisms used to vouchsafe the existence of the "we" invoked when the collaboration speaks to the existence of a new entity or effect in science? Who—or rather what—is speaking?

In late twentieth-century physics we are faced with collaborations such as the four detector teams at the CERN Large Electron Positron (LEP) accelerator, where each is staffed by a team of some five hundred physicists from fifty institutions along with hundreds of technicians and engineers. In the trash bins of recent history lie the two collaborative detector teams that were to have constructed their machines at the Texas Superconducting Supercollider with a thousand physicists each. Even larger is the CERN-based Large Hadron Collider (LHC) with its twin collaborations of somewhere between fifteen hundred and two thousand physicists apiece. What, we can ask, does it mean for such concatenations of institutions *collectively* to have found a particle or confirmed a theoretical contention? I want to know what the *we* already presupposes in the collaboratively-produced document. *Where* is the information, *who* has it, *what kind of unity* of the collective is already assumed when the collaboration rules something to be the case about the physical world?

In moving from the conditions under which "I" can be uttered to the conditions of possibility for a "we," violence has, of course, been done to the Kantian position in several ways. First, it is clear that my concern is at a much higher level in the hierarchy of concepts—not in the conditions necessary for us humans to say "I see a pen," or "what is needed for us to have the notion of an object in general," but rather the conditions under which it is possible for it to be said: "The OPAL collaboration has measured the Z width," or "The UA1 collaboration saw the first W decay." Second, for Kant, the "transcendental" analysis of the unity of apperception signals two features. First—and I do follow Kant in this—the transcendental argument asks what is already taken for granted: in his case, "What is already built into the thinking individual self?" here, "What is already built into the collective self?" At the same time, however, Kant employs another meaning of *transcendental* when he takes these conditions of possibility to be a priori (that is, before any experience at all). For against the empiricists, throughout these sections of *Critique,* Kant argues that we could never extract the unified "I" from experience, just as we could never get to notions of necessity from an encounter with experience. On the contrary (so he continues), the unity of apperception is needed for there to be appearances in our experience in the first place, just as the very possibility of perception (intuition) presupposes that we already have some sense of space and time. In sum, Kant took his unity of apperception to be a priori true, and since it was therefore not learnable from individual experience, it was a fortiori not changeable within history. By contrast, it is part of my argument here that (a) collaborations, even at a given time, structure their sense of "self" variously, and (b) there is a broad and clear shift in the nature of the *collaborative self* from the bubble chamber work of the postwar period to the huge colliding beam collaborations of the 1980s and 90s, and then an even stranger shift now visible on the horizon with the advent of so-called mobile agents.

Despite these synchronic and diachronic *dis*analogies between the collective "we" and the individual "I," the spirit of Kant's question remains. In the extraordinary richness of high-energy physics, what is presupposed about the unity of the "we" that lies behind the pronouncements of a collaboration? What is the process, so painstakingly worked out by these collaborations, that lies behind what one might call the constitution of the collective self? What is the "we" and how does it relate to the knowledge claims that people outside the collaboration are invited to accept?

II. The Pseudo-I

That the collective experimenter differed from previous scientific authorship was already apparent in the 1960s, as bubble chamber physics began driving the size of collaborations from single digits to fifteen or twenty. Brookhaven National Laboratory's Alan Thorndike, then the leader of one of the most prominent hydrogen bubble chambers in the world, put it this way in 1967:

> Who is "the experimenter" whose activities we have been discussing? Rarely, if ever, is he a single individual. . . . The experimenter may be the leader of a group of younger scientists working under his supervision and direction. He may be the organizer of a group of colleagues, taking the main responsibility for pushing the work through to successful completion. He may be a group banded together to carry out the work with no clear internal hierarchy. He may be a collaboration of individuals or subgroups brought together by a common interest, perhaps even an amalgamation of previous competitors whose similar proposals have been merged by higher authority. . . .

> The experimenter, then, is not one person, but a composite. He might be three, more likely five or eight, possibly as many as ten, twenty, or more. He may be spread around geographically, though more often than not all of him will be at one or two institutions. . . . He may be ephemeral, with a shifting and open-ended membership whose limits are hard to determine. He is a social phenomenon, varied in form and impossible to define precisely. One thing, however, he certainly is not. He is not the traditional image of a cloistered scientist working in isolation at his laboratory bench.[1]

In this extraordinary text, Thorndike sketches the collaboration-as-author and it is just this feature that strikes me as central. One could ask other questions, questions about how individuals made their decisions to join the group or how each climbed the career ladder, but it is the much more radical import of Thorndike that intrigues me—his situation of the collaboration not as a collection of experimenters, but rather his identification of the collaboration-as-experimenter. Just in virtue of this fact, the experimenter becomes "a social phenomenon," an entity with indeterminate limits, geographical dispersion, varying form, and aleatory internal structure. As grammatically awkward as this may seem, Thorndike has captured something crucial about postwar physics when he says the experimenter has become "composite."

Despite this compositeness, the experimenter remained, in the 1960s, under the authorial name of an individual. Everyone knew the largest hydrogen bubble chamber collaboration at Lawrence Berkeley Laboratory as the Alvarez Group. Similarly, other bubble chamber groups at LBL were known by their leadership as the Trilling-Goldhaber Group, or the Powell Group. And at Brookhaven—where Thorndike was—no one would have had any difficulty locating the central figure of the Thorndike Group. For though the complex operation of a bubble chamber required expertise of various sorts, all of these expert subgroups reported back to a single center. Alvarez was as much in charge of the data-processing dominion as he was of the cryogenic engineers or the physicists. In the end, all decision-making about physics results to be published came back to him, and all funding *into* the group passed by him as well. For all these reasons, I take seriously the fact that the group carried the name of its single leader: the collaboration of the 1960s was modeled on a quasi individual, a single person who may have taken his actions in consultation and ultimately through others, but (at least to the outside) when the Alvarez Group found a new particle, it was, in a sense, as an extension of Alvarez himself. Alvarez therefore stood as the name-giving center of the group. Even while the actions of the team had already mutliplied into separate cryogenic, scanning, analysis, and mechanical subgroups, "Alvarez" referred two ways, both to the individual and to the composite, "pseudo-I," of the group as a whole.

III. Hierarchies and the Absent Center

Along with the growth in size (from experiments costing on the order of $10 million to collaborations with equipment alone in the range of $100 million to $1 billion), came many other alterations in the structure of collaboration. Bubble chamber groups, like those of Alvarez or Thorndike, had a single, leading institution: Lawrence Berkeley Laboratory in the case of the former, and Brookhaven National Laboratory in the latter. No one working at LBL in a group originating from Johns Hopkins would have had any doubt in 1967 that collaborating did not mean sharing equally in authority; LBL was the first among the many institutions with which it shared work. As the detectors shifted in scale, this local dominance could not be maintained: no longer could individuals, groups, or even nations join a collaboration in a fixed subordinate position. This fact was evidently not simply a feature of the political economy of particle physics, it reflected changes in the relations among universities in the United States, among the United States and the countries with which it collaborated (e.g., Japan), and among the various countries collaborating at CERN outside Geneva.

There is another dimension to this multiplication of centers that is at once technical and symbolic. In the bubble chamber, the apparatus itself had a certain unity—essentially a vat of liquid hydrogen, the structural and thermal integrity of the whole constituted the principle engineering difficulty. Could the chamber withstand the millions of compressions and decompressions that would allow bubbles to begin forming and then squish them back out of existence to prepare for the next round of interactions milliseconds later? Could the hydrogen be kept at a uniform temperature throughout to avoid convection currents that would make spurious curves out of "truly" straight particle trajectories? Technology shifted away from bubble chambers towards the hybrid electronic detectors that were designed to capture the detritus of colliding beams of particles and antiparticles. With that change, the technological unity of the whole began to disintegrate as well.

Two forces met. On the one side, the technical hybridization of the chamber made it easier in many respects to partition the work of planning, construction, maintenance, and analysis to different laboratories. One LEP collaboration, OPAL, divided their detector such that Bologna and Maryland took on the forward detector, the hadron calorimeter, and the beam pipe; Chicago took the electromagnetic barrel presampling detector; and Tokyo adopted the electromagnetic barrel calorimeter, the central detector, and the trigger system. Other groups—and there were tens of them—divided up the myriad of remaining detector components. On the other side, this division of "property" had a symbolic dimension as well: each group needed to have something to show, an identifiable piece of real estate that could be exhibited in slides and reports to funding agencies and, in some cases, national scientific ministries. While analytically it is helpful to separate these two dimensions (the technical and the symbolic), in the real world conduct of physics, such division is not sharp: the partition of the technological effort into identifiable component parts is at one and the same time a political-symbolic act and a practical-technical one. The bottom line was this: no one group could command the whole of a multiinstitutional, increasingly multinational collaboration. Scientific politics and work in the late 1990s would not allow a major detector to be "American" or "German," much less "Alvarez's" or "Thorndike's." And with that splintering, the group as quasi individual began to give way to a more federal association of parts; the collective self as unity disaggregated.

This shift was reflected in the leadership name, from the author name of a prominent physicist to the explicitly corporate "Executive" or "Collaboration" Council. In the detailed structure of scientific and technical protocols we can see, in detail, the de-centered authorship that mirrored these new conditions of production. Let us focus attention on some specific sites and authorship practices. It is, I believe, *only* in following the specific mechanisms by which the collaboration comes to assign its name to a result that we can see both what the collaboration is as an internal structure and how it can carry conviction to the outside world.

IV. The Protocols of Authorship

At the end of the two-mile-long Stanford Linear Accelerator Center (SLAC) lies the Stanford Linear Detector (SLD), which opened in the early 1990s. Not wanting to be caught off guard when they opened for business, SLD spelled out their author policy in a publication that predated by several years any actual measurements. Their first and simplest specification fixed who should be included. When a publication allowed only limited space for author names, authors were to be listed in alphabetical order; when printing in circumstances of unlimited space, institutions would be put alphabetically, and authors given alphabetically within those institutions.

> There should be no exceptions to the above, such as placing the student's name first if the paper originated as a thesis, as our first priority should be the coherence of the group and the de facto recognition that contributions to a piece of physics are made by all collaborators in different ways.[2]

If the thesis was written by a single person, then that might well be indicated in the first footnote. But as this stricture indicated, group coherence—the stability of the collaboration as such beyond the contribution of any individual—became a factor in the order of the author list from the get-go.

"Who Is an Author?" asked the SLD Collaboration Council innocently enough in July of 1988. But the answer had important implications not only for the individuals in the collaboration but for the very process of writing and certifying. The council: "For physics papers, all physicist members of the collaboration are authors. In addition, the first published paper should also include the engineers." Now, as was evident from the specification of "physics papers," it was clear that there were to be other forms of writing with different authorship protocols. For example, there were reports to be made on the hardware, and here the protocol divided the cases into three. If a system-wide innovation was at stake—for example, to the WIC—then the relevant physicists and engineers should both sign the paper. The same was true for a contribution to the construction or functioning of subsystems. The pad electronics, by way of illustration, was a component part of the wire chamber, and the subgroup of physicists and engineers responsible for the work could sign it. Indeed, the protocol indicated no objections to either system or subsystem hardware reports being authored by engineers alone, should that prove appropriate; anyone *not* contributing to one of these efforts was encouraged to delete his or her name. Finally, there were "individual" contributions that were "envisaged as rare exceptions where one person has produced an individual 'invention.'" The system manager must first agree that the proposed paper merits such a classification. Then the author would circulate a memo to all system physicists (those participating in that particular system) declaring the intention of producing a paper and "inviting those interested to contact the author to help in writing. A draft of the paper should be circulated to provide other system physicists an opportunity to ask that their name(s) be added."[3]

As the various types of hardware papers indicate, the collaboration "speaks" in different registers or modes depending on both content and intended audience. When it comes to the physics itself, this differentiation of registers became even more refined, running from internal memos to the most crucial physics paper staking a claim to new results. We can paraphrase the genres of SLD literary production as follows:

1. *Internal memos* were not for public or even full SLD distribution. The writer(s) could freely decide on who would count as authors and what would be contained as content.
2. *SLD notes* circulated to the full SLD collaboration, and could, when appropriate, be allowed to reach the public. Author(s) were invited to circulate such writing to all people involved in pertinent data or apparatus. SLD notes required approval of the relevant system manager.

3. *Conference Proceedings on SLD Physics* were to be signed in the form "The SLD Collaboration, presented by Isaac Newton," with a footnote naming all of the SLD authors.
4. *Conference Proceedings on Instrumentation Research and Development* would list group, if well defined, and acknowledge SLD collaboration. Here too system manager approval would be required.
5. *Reviews of SLD Physics or Detector Design* would be treated as type 3.
6. *Review papers*, including but not restricted to SLD results, assuming they were rapporteur talks and not talks of type 3, would be allowed to cite unpublished SLD results "as individual efforts" and therefore identified as publications from the author's home institution.

For our purposes, the significance of these various modes is that the regime of authorship is a function of both the scope of the audience *and* the knowledge claim. When the scope of the knowledge claim is highly restricted—to the functioning of a piece of hardware, for example—the author list can simply be the writer(s) and may omit physicists altogether if the work was done by engineers. Or when the audience is sufficiently restricted, as in an "internal memo," an individual physicist may stand as a single author. By contrast, when a principal physics claim is made, such as the discovery of an anomalous decay, and is to be disseminated to the world at large, the collaboration *had* to be the author, with the individual writer relegated to the role of "presenter." Even subinstitutional "ownership" of an analysis was to be avoided—constituent groups were forbidden from giving a cover number to memos because they might be read as "coming from institution X" (rather than the collaboration as whole).

Indeed, in just the central case of a physics publication, considerable care is taken to define what is to count as a constituent author within the collaboration. It is a definition that alternates between the practical demands of the career structure of the participants and notions of what kind of work counts as author-making.

> [F]or [physics] papers, an author is defined as a physicist who has contributed (by running shifts, doing analysis, building hardware, etc.) to the results which are the subject of the paper. In normal circumstances, this means that anyone joining SLD full-time should become an author

> essentially immediately. To accommodate dry periods when no data are taken, a person who has been a member of the collaboration for at least one year is automatically included on all papers, even those based on data taken before he joined the group. Since joining the collaboration is not necessarily synonymous with joining an institution, the precise starting date is left to the integrity of the individual and to the Collaboration Council member for his institution.

Note a certain tension here. On the one hand, there are certain kinds of work that serve as a necessary precondition to authorship (shift work, analysis hardware construction); at the same time, the criteria recognize the vicissitudes of life around an accelerator (the hazards of "dry periods") by allowing someone to be an author of results obtained before he or she joined the group. On similar grounds, "A person leaving the collaboration remains an author for a time equal to that for which he was active on the experiment," unless that person requests otherwise.

When it actually came time to write a paper, the process would begin by some set of writers producing a detailed memorandum. This would then be presented at SLAC, followed by the formation of a committee consisting of five to seven people, which, upon approval, would bring the paper to the collaboration as a whole, which would then have two weeks to comment. After taking any criticisms into account, a "group reading will be scheduled," normally for three hours. "At this public reading, it is in order for all present to comment, argue about conclusions, etc. (The creative ferment stirred up in this way generally leads to improved papers.)" Out of the public reading would come a new draft with two more weeks in which criticisms could be registered, a final draft circulated for one week, at which point only corrections of "errors of fact or of blunders in English or typos are permitted."[4] Finally, when a major discovery was thought to be in hand, a press conference or press release was to be made, authorization to do so had to come from the SLAC director, and the SLD cospokesmen, with advice from both the Collaboration Council and the Advisory Group. For cause, the bona fides of the experiment were guarded by more than one set of doors.

The complexity of these rules stems in part from two desires that pull in opposite directions. Pulling towards *inclusiveness* is the desire to make the collaboration as complete and unified as possible; anyone left out might undermine the authority of the claim. Readers might ask *why* someone's name failed to appear. This is explicit in the protocol of the DØ collaboration at Fermilab, where the rules state near the very top of the document: "Withdrawing individual names because of a lack of close involvement in some particular aspect of the analysis will tend to undercut the impact of any publication and is therefore to be strongly discouraged."[5] As we saw, even a student's thesis work was to be instantly and seamlessly absorbed into the collaboration as a whole. Pulling toward *exclusiveness* is the desire to make each name stand for command of and agreement to the work. Both tendencies (inclusiveness and exclusiveness) are tied to the issue of credibility: the collaboration must function with sufficient unity for its name to stand for something. The credibility of a fragmented "we" would have the effect on the outside world not unlike an individual experimenter a hundred years ago who said that a certain substance both was and was not magnetic–the contradiction would essentially erase him or her as a contributing member of the research community.

Unity is important. Senior particle physicists remember all too clearly the flack the E1A experiment received when different members went public with preliminary and contradictory claims about the existence or nonexistence of weak neutral currents. Rapidly—and tragically for their credibility—the superb experiment with some of the most stunning results of the last half century were derided as having discovered "alternating neutral currents."[6] The fate of the collaboration and its results ride together.

In the 475-physicist ALEPH Collaboration at LEP, one of the spokesmen, Gigi Rolandi, recently wrote, "The general principle is that no ALEPH result can be presented in public without the approval of the Collaboration. Aleph can have only one official result for a given analysis."[7] They achieve this one result by a process not unlike that of SLD. A physicist or physicists present an analysis at a regularly scheduled Thursday meeting of the collaboration, with suitable prior advertisement. At that meeting, the collaboration—or more precisely, its representatives—would vote to allow either a public presentation or a paper preparation. If the latter is chosen, the writer(s) present a draft to the chair of the editorial board (a person chosen by a collaboration spokesman), who then designates some referees. After one or two drafts, the paper may come before the collaboration for a second time; but whether or not it does, in all cases the editorial board votes on it for final approval. One could describe this process in two different but very closely related ways. On the one hand, it is a matter of the collaboration finding a way for the collective to know something—getting the paper right. On the other hand, it is a matter of structuring the collaboration's output in such a way that they have a single, persuasive message for the outside world.

In DØ, one of the two massive colliding beam detectors at Fermilab's Tevatron (a collaboration of 424 physicists and growing), the authorship document insists that all "serious" participants ought be on *all* publications. The snag, not surprisingly, is defining *seriousness*. With certain exceptions, their March 14, 1991, policy on authorship demanded the following to be an author: one had to be a senior graduate student or above, and work for a year prior to submission of a paper for publication. As in most of these collaborations, there was a provision allowing authorship to continue for a year after departure. But a putative author must have (1) done one half the average number of shifts on data runs relevant to the paper in question. And beyond that, he or she must have (2) fulfilled one of the following two requirements: " (*a*) spent at least (the equivalent of) one person-year, at one's home institution or at Fermilab, working on the implementation of part of the detector or of the software used in acquiring the data from the run(s) on which the paper was based; or (*b*) made a major contribution to the analysis of the data from the run(s) on which the paper is based. This includes writing software, performing analysis, producing DST's, writing the paper, internal review of the paper, etc."[8]

These criteria in some ways embody what might be thought of as an updated version of earlier experimentation, as criterion 1 required physical presence on site. This restrictive clause soon died as it became clear that the work structure of an experiment in the 1990s could not be tied to that specific form of physical labor. On June 2, 1994, the collaboration revised its "Rules of Authorship" toward a far broader concept of what would make one an experimenter. Now the demand became the more broadly construed demand that, to any combination of the following potpourri of activities, an author must contribute a total of twelve hours per week:[9]

1. Design, construct, or debug the detector or the DØ test facility.
2. Write software, for example, utility packages, Monte Carlo simulation of detectors; process or analyze DØ data.
3. Process Monte Carlo events or data obtained at DAB or any DØ test facility.
4. Run shifts at DAB or at the DØ test facility. (Four per month were thought to be needed to "prevent memory loss and the need for extensive retraining.")
5. Manage personnel issues for DØ, administer contracts, grants; serve as a "physics convener," or convener of a technical topic such as electron identification.
6. Write or review DØ papers; advise graduate students on DØ matters.
7. Take part in DØ meetings, workshops, or discussions; analyze physics simulations or physics analysis for paper or Ph.D. thesis.

(The authorship rules specify that items 6 and 7 require special authorization from the group superego in the form of the spokesman-appointed Committee on Authorship.)

As it came time to actually produce a paper, the DØ 1994 procedure went like this: When someone had an "imminent" physics draft note, the cospokesmen of the collaboration appointed a custom-built editorial board to review the results. That is, for each case, the editorial board was composed of the author(s), plus four other physicists: an advisor (also known as the "Godparent"), someone from the same group (that is, the same physics or algorithm group), and two other collaboration members. While one draft physics note went to the editorial board, another copy was released to the full collaboration—any member could comment via the electronically distributed DØNews. By putting the author(s) on the editorial board, the collaboration deliberately broke with the referee model employed in other groups in order to "facilitate productive exchange," to "avoid misunderstandings," and to eschew "confrontations." While harmony might be fostered by this more inclusive editorial board structure, it was at the same time the basis for a more probing inquiry. Unlike the refereeing process in other groups, the rules of authorship in DØ specifically mandated that the editorial board was to have access to all backup materials including theses, backup analyses, DØ notes, and other items as required.

Assuming that the editorial board approved, a waiting period followed: four days posting for a DØ physics note; ten days posting for a possible publication. Normally, about ten percent of the collaboration comments on the posted note. In addition, the authorship rules demanded a "public reading" at a DØ Physics Analysis Meeting, a General Collaboration Meeting, or a specially scheduled session. Finally, at the end of the waiting period, assuming the editorial board could assure itself that all objections had been addressed, the vetted physics note gets a number (unlike the unreviewed DØ notes) and is entered into the publicly available disk space of the collaboration. Those destined not to be published would be marked "Preliminary Results from the DØ Collaboration," while those headed for the world of print were launched forward toward their target journals.

My final example is the OPAL Collaboration at CERN's Large Electron Positron collider. OPAL (Omni Purpose Apparatus for LEP) consisted of some twenty-four groups distributed in their responsibilities over some fourteen "subdetectors." As with all of these very large colliding beam detectors, maintaining integration of the whole became in every respect the central and most difficult problem. When it came time to author papers, OPAL, like SLD, ALEPH, and DØ, aimed to pull the candidate publication into line with some notion of the collaboration as a whole:

> OPAL operates a rigorous internal review procedure for all physics results which are to be published or shown outside OPAL, whether final or preliminary. The aim is to ensure that OPAL results are reliable, of high quality and well presented. Results should never be discussed outside the collaboration before members of OPAL have had an adequate chance to examine, criticize and approve them.

In a procedure altogether similar to the other groups we have examined, someone with a physics idea would ask one of the physics coordinators to appoint an editorial board, a body consisting of the authors along with four other OPAL physicists, among whom should be: a native speaker of English, an expert in the specific area of the paper, a nonexpert to ensure that it is comprehensible more broadly, and someone located outside CERN. When the editorial board is satisfied—and writer(s) must leave the board at least a week to look it over—the board electronically launches the draft to all of the OPAL laboratories scattered around the world using a program called DISPATCH. For at least two weeks the paper must remain posted in this way so that comments or criticisms can flow back from the various sites. At the end of this criticism period, the editorial board, satisfied that due consideration has been given to any objections raised by the broader collaboration, schedules a public reading. This event—attended by the editorial board members as well as other interested members of the collaboration—is the occasion for last, substantive corrections before the draft goes to the collaboration for final approval.

Also, like the other collaborations, OPAL writes in different registers. *Physics Notes* are defined as "internal OPAL documents which allow the reader to understand and judge the reliability of the analysis."[10]

> Since physics notes describe the analysis of results intended for the public domain, they are made accessible to physicists outside OPAL on a restricted basis. Single copies are given on request to interested parties who have a valid reason for wanting one but they are not sent to any non-OPAL mailing lists. They can also be used, for example, by OPAL collaborators in discussions with students and colleagues outside OPAL.[11]

Note that here, as in the other "constitutions," draft journal papers require public reading—an open OPAL meeting at CERN, following which the authors consult with the editorial board and prepare a draft for final approval. And again, the wider the audience, the deeper the article must penetrate into the collaboration itself. Presentation to the outside and the creation of a "we" inside enter together.

Toward the end of the twentieth century, when the Department of Energy commissioned a report on future modes of high-energy physics research, the problem looked different than it did even a few years later. Thousand-strong collaborations struck the committee as unreasonable, as they skeptically looked ahead to the SSC. By the time the LHC had amassed an army of two thousand physicists, discontent had set in even more deeply: when the European Committee for Future Accelerators polled members of the big teams in 1995, they found that some seventy-five percent of their respondents disliked present publication habits, largely because it damaged the possibility of career advancement and of receiving credit for their work, and some sixty-seven percent wanted change. Yet few wanted to limit the author lists—though they did want more weight placed on internal publications.[12] Indicative of things to come was the publication of the first Fermilab claim for the existence of the top quark: some eight hundred authors signed off on the initial papers.

Unease stemmed as well from the perception of outsiders. In 1988, the Department of Energy committee uncomfortably contemplated the consequences of being inclusive in the author lists, as such endless rosters gave physicists from other fields "the impression that all individuality is submerged in high energy physics," a not irrelevant image when questions of hiring, tenure, and even field support arose. Worse, the committee feared that it was becoming impossible to know who was responsible and who understood the experiment in detail. Indeed, at the end of the day, the very length of these lists of names radically devalued the worth of the publication on an individual's cv. As a direct consequence of this publication inflation, the committee noted that evaluation increasingly was being based on recommendations, with all the problems that letters brought, and not the work itself. One idea—an idea we saw considered in many of the author regulations—was to reduce authors after first publication. But this met with strong objections, for example by those people who maintained a calibration needed for subsequent results. One young physicist put it bluntly when he said, "these experiments truly are the result of the work of 100 people and it would be fundamentally dishonest to pretend otherwise."[13]

In February 1996, Roy Schwitters—former director of the now defunct SSC—suggested in the *Chronicle of Higher Education* that new guidelines be adopted in the big teams. These included listing all members of the team when "planned" discoveries were made—discoveries anticipated at the time of construction. He also called for changes that would encourage people to publish in the area of work to which they had contributed. Machine builders would publish technical reports about component parts, experimenters would publish their analysis of data; and software engineers would produce reports directed at their peers. Finally, Schwitters advocated wide intra-experiment circulation of reports before publication, and the publication of more than one published interpretation if opinion divided within the experiment.[14]

A real-life example of explicit, published dissent from within a collaboration occurred in the spring of 1995, around research prosecuted at the Los Alamos Meson Facility. The majority of the collaboration published a paper strongly bolstering the idea that neutrinos had mass (more specifically that there were good candidate events showing oscillation between anti-muon-neutrinos and anti-electron-neutrinos). University of Pennsylvania graduate student James E. Hill disagreed—and *Physics Review Letters* published back-to-back his dissenting paper: "An Alternative Analysis of the LSND [Liquid Scintillator Neutrino Detector] Neutrino Oscillation Search Data on [anti-muon-neutrino → anti-electron neutrino]." As far as Hill was concerned only two of the nine purported oscillation events were truly good candidates. Immediately the collaboration plunged into some real soul-searching, asking questions like these: Could someone dissent from the collaboration this publicly? If there was dissent, who ought to be signatories on the paper? What were the obligations of journals like *Physical Review Letters* toward the collaborations that produced the data? Though lauded by some, including Schwitters, and condemned by others, such open clashes did little to resolve the fundamental issue of authorial splits within the pseudoindividual of a collaboration.

In fact, one can see written into the authorship protocols of large collaborations a fundamental tension between condensing authorship around an individual or small group and the equally powerful drive to diffuse authorship around the entirety of the collaboration. These opposing forces are apparent in the new literary category of "Scientific Note" that ATLAS (one of the CERN Large Hadron Collaboration detectors) promulgated in February 2000. Titled "A New Class of Publications to Recognize Individual Contributions to Future Large Experiments," the ATLASIANS posed the problem this way:

> The career advancement of Experimental High Energy Physicists at Universities and Research Institutes has become harder in the last ten years due to the large number of authors appearing on each publication within the field. This large number of authors makes it harder to evaluate the individual contribution when comparing with other fields in science. Collaborations associated with forthcoming LHC experiments are typically several times larger than existing experiments. Thus, if no action is taken, the problem of recognizing individual contributions to experiments will become even more acute.[15]

Caught between the twin exigencies of representing the experiment as a whole and individuals in particular, the ATLAS team proposed a delicate authorial dance. First, they insisted that it was "understood" that the LHC experiments would present their scientific results "under the name of the full collaboration." But immediately they went on to introduce a new literary form, the scientific note that would lie halfway between a full-bore scientific article (authored by the ATLAS writ large) and the commonly and more roughly produced "ATLAS Notes," which were directed within the bounds of the collaboration.

According to the ATLAS Collaboration Board, scientific notes would emphasize a "clarity, completeness and style" appropriate to ordinary scientific papers, would appear in refereed journals (not just in posted web and internal publications) and yet would not impinge on the territory of full–collaboration scientific results. When the full collaboration did publish their scientific results, the Board urged its ATLAS colleagues to cite, explicitly, the achievements of individuals that had been established in scientific notes. Individuals with specific names and responsibilities ought to author these notes—"subsystem communities" should not invoke this person-highlighting form. Of course for the new form to work, editors would need to restructure their pages to accommodate the new literary object: "The editors of scientific journals will be contacted to establish a new class of publications under the name of technical or scientific notes. These notes will contain results of analyses, detector development and improvements, detector and physics simulations, software, algorithms and data handling." Above all for scientific notes to boost the careers of particular persons, the collaborative habits of inclusion would need to be curtailed, as the section "Authors" made clear: "Scientific Notes should represent the work of a single individual or a small group and be signed only by the direct authors. Naturally, such work will often benefit from contributions, past or present, of persons other than the direct authors. When so, such help should be duly acknowledged, but not necessarily lead to inclusion into the author list."[16]

If the refereeing process is part of the mechanism that constructs the author, so to speak, then the statutory fate of scientific notes is quite revealing. One final piece of the mechanism had to do with content control, and, in principle, the following would govern its evaluation. First, at least optionally, the scientific note would pass through the normal procedures for the (internally aimed) ATLAS Note. Second, the scientific note would follow the procedures established for full-collaboration papers, included internal refereeing as guided by the ATLAS editorial board and subject to final approval of the spokesperson. Finally, the scientific note would be refereed like any other journal submission in physics–by evaluators outside the collaboration.

Reading the authorship protocols of the many collaborations discussed here, including ALEPH, SLD, DØ, OPAL, and ATLAS, complicates Schwitters's considerations. Some collaborations (as we have seen) actively *discouraged* the withdrawal of names from the author list because such actions could be seen as dissent undermining the force of the argument. Others specifically forbade an individual or group of individuals from signing a publication under their own names. The coherence of the group counts for much: for credit, yes, but also for the continued legitimacy of both the group and its productions. Yet other collaborations banned a university group from publicly assigning a number to the document. To print on a particular preprint the line "University of Michigan 97–23" on a collaboration document would (on the view expressed in some of the authorship protocols) be to arrogate both credit and responsibility. Even when, as with the newly coined Scientific Notes of ATLAS, the collaboration sought to reinstate the individual with partial-group authorship, the collaboration not only demanded final say on the paper, it exempted main-claim scientific results of the collaboration from being treated by a handful of physicists authoring a scientific note. Disunity of authorship appeared to many if not most participants in these large collaborations as tantamount to epistemic subversion. All these gestures of control served to create both an internal and external social-epistemic unity: they aimed at making the knowledge embodied in physics claims come from the group as such, not from its component parts. In short, they aimed to secure the integral structure of "we," in ways that often cut against the grain of Schwitters's proposals. The whole of these massive authorship protocols aim to form the "self" of a monster collaboration so that the "we" of the collaboration can produce defensible, authored science.

V. Mobile Agents Confront Kant

So far we have followed an economic system of credit that pitted the individual I of a single authorial name reward against the pseudo-I of the group. Some of the procedures (like the Schwitters proposals or the renegade publication at the Los Alamos Meson Facility or the compromises implicit in the ATLAS scientific note procedure) aimed to reinforce the individual's claim to accomplishment. At the same time, groups inaugurated other mechanisms to *prevent* individuals from claiming identifiable contributions—among these were protocols to discourage people from removing their names, procedures to enforce a group endorsement of individual publications, regulations to control who can speak and where.

But even to pose the credit economy in this way is to assume that individual contributions could, in principle, be isolated. Formally, and this is stamped into the software, we are dealing here with a more or less fixed network of contributors, dispersed institutionally, of course, with complementary specializations. In a large colliding beam experiment, one group, say from U.C. Riverside, might control one component, say a muon calorimeter, and be responsible for writing and maintaining the software that collects and formats data collected by that component of the larger machine. That hardware and its software image constitute the group's claim to authorship in the collective.

In the factory-style structure of the postwar laboratory, the center (such as Alvarez's LBL Bubble Chamber), was in every sense the centerpiece of the collaboration. LBL was where the center control on everything from software to detector control to scientific judgment took place, even if occasionally Alvarez distributed bubble chamber film to be analyzed and even published elsewhere. But even as that center-directed structure dispersed from 1975 forward, there remained a certain legacy of the centered collaboration in the handling of data. In particular, groups continued to interact through client-server relationships. By the late 1990s, if the center computer needed to distribute computational workload, many large systems were outfitted in such a way that it could download certain programs; in other, applet programs, the user could deliberately download a program. Yet there was no question, at the end of the twentieth century, that the computers at CERN (or Fermilab) constituted the centers of their respective collaborations. Hierarchies of computational capacity, access, and control filtered the process: from CERN through national computers, down to laboratory, group, and to the individual's workstation.

Each element of this rigid regulation of capacity, access, and control presented problems for early twenty-first century collaborations. One response within the largest collaborations has been to introduce *mobile agents*—self-propelled programs capable of leaping from one computer to the next, with the ability to dissolve the hierarchical relations of access and control.[17] Acting like solvents on the posts and beams of classical collaborations, these agents complicated, in striking new ways, the constructed subject position of the experimenter. Let me explore this line of development further, as it bears directly on the future of massive-collaboration authorship.

In the 1990s, four of the largest, most data-intensive experiments then under construction allied themselves in a metaexperiment known as GriPhyN, the Grid Physics Network.[18] This project did not only seek to alter the four projects (LIGO, the Laser Interferometer Gravitational Wave Observatory; SDSS, the Sloan Digital Sky Survey; along with the two huge CERN Large Hadron Collider collaborations discussed earlier, CMS and ATLAS). Each of these four collaborations expected in the early years of the century to be shuffling hundreds of petabytes of data, where a petabyte is a thousand terabytes and a terabyte embraces a thousand gigabytes. Such massive data sets, along with vast associated computational needs, outstripped the memory and computational power of any computational network expected to be workable in a reasonable time. In response, the GriPhyN (meta)collaborators intended to introduce mobile agents, where these mobile agents could hop from computer to computer anywhere within a network at times it, rather than following a path that a central computer, deemed appropriate. In hopping, the mobile agent could reproduce itself or could simply leap, taking stock of where it is in its computation so it could resume computation in the new host.

There is more. Mobile agents work around the usual hierarchies that ordinarily segregated the highest level centers (like CERN) from national centers (like France) from laboratories and, in turn, from the individual uses. Just because they must establish, in each locale, means of coordinating different priorities, security arrangements, performance, reliability, and so on, the mobile agents do not resemble the procedures dictated from above. GriPhyN's goal was to use the myriad of self-activating wandering programs within the four collaborations to create a model of a coherently managed distributed system, "where national and regional facilities are able to interwork effectively over a global ensemble of networks, to meet the needs of a geographically and culturally diverse scientific community."[19] The effect of these agents is to render different kinds of equipment and protocols transparent, from the massive processors at laboratory or regional centers all the way down to workstations. Take the Sloan Sky Survey, a project to map a large fraction of the northern sky to faint magnitudes (including about ten million galaxies with highly processed images of each one), along with a comprehensive survey of each object's spectroscopic signature. Scientifically, SDSS could hardly have a broader ambition, and as such was more a platform for collaborations than a well-defined collaboration per se. The Sky Survey would provide the basis for a statistical survey of the galaxies, stars, and quasars in such a way that it would shed light on issues ranging from cosmogenesis to galactic structure. Imagine that an astronomer user (or group of users) of the Sloan Digital Sky Survey wanted to examine correlations in galaxy orientation induced by the gravitational lensing effect of intergalactic dark matter. Data needed for this task could be stored in a network cache, in a remotely located disk system, or in a deep and compressed archive. For her project, the astronomer would need a computer system that could find all these data and images, produce any images not previously constructed from pixels, and then actually compute the correlations—shuffling terabytes of data and computational programs back and forth, and possibly manipulating an even larger simulation file to compare against the actual data.

Suppose GriPhyN succeeded in providing a truly centerless, scalable, heterogeneous computer resource for these four multipetabyte collaborations. Suppose that these mobile agents could so effectively wander through the system that no one need care about the dimensions of the collaboration. That is, assume that as groups and individuals join, withdraw, or move to other tasks, their computers continue to provide partial time storing, and computing and to recreate data. Who or what is the experimenter emerging here? Something is in construction that no longer quite fits either the "I" or even the well-defined, bounded "pseudo-I" that we expect to find as the presupposed subject of the statistical sky object. This new subject is coordinated but not commanded from a point, functioning more like a hive than a hierarchy.[20] Asked *where* the data are or *where* the data are being reduced, we would have to answer: in the hive-I of the Grid. If this is right, then the knowing subject presupposed by the establishment through GriPhyN's version of SDSS of a gravitationally induced lensing of galaxies is truly without fixed boundaries. Whatever fictions are demanded by the apparatus of prizes, promotions, and publication, there would be neither a unified individual nor even a bounded team at the (metaphorically) small end of the telescope. In the place of Kant's transcendental unity, we would have an ever-fluctuating mobility of apperception. The "amorophous-I" is a new species of author whose claims will require new forms of evaluation.

VI. Conclusion: Authorship and the Collective Self

Intriguingly, during the last several decades, two strands of inquiry into the nature of authorship have existed side by side without interacting. In addition to the scientists' own efforts to grapple with the problem, there is the literary-philosophical literature of authorship to which the French have contributed so extensively, especially the work by Roland Barthes that was so strikingly reconceptualized by Michel Foucault. In part, Foucault, though in an utterly different idiom, was also grappling with the problem of individuation of the authorial self, and it is worth considering the relation of these two sets of consideration to one another.

For Foucault, one set of problems involves the establishment of what counts as the "work" of an author. He asks: Are we to attribute the status of work to everything he or she wrote, and if so what will count as "everything"? Observing that only certain bits of speech are seen as singular, that is differentiated from everyday remarks that could have been uttered by anyone at all: "What time is it?" except in special cases, is not part of the language we call authored.

> The author's name serves to characterize a certain mode of being of discourse: the fact that the discourse has an author's name, that one can say "this was written by so and so" or "so-and-so is its author," shows that this discourse is not ordinary speech. . . . On the contrary, it is a speech that must be received in a certain mode and that, in a given culture, must receive a certain status.[21]

Assigning authorship to a certain body of texts or utterances has consequences; authored speech is characterized by a certain "mode of existence, circulation and functioning of certain discourses within a society."[22] But Foucault cut science out of his analysis. In his view, after the seventeenth century, the author's name no longer conferred authority on scientific texts, since the truth of the sciences was in principle always "redemonstrable." Foucault argued that author names served only to label theorems and otherwise decorate the *results* of science. Given the exceptional lengths to which both individuals and collaborations go to protect their "good name," however, Foucault's diametrically opposed categories of science and the rest seems, on the face of it, to fly in the face of the lived world of scientists long after 1700.

But suppose we omit Foucault's demarcation of scientific authorship from authorship more generally. Two questions then emerge from his analysis. First, one could ask how works are associated back to a given author, how, in a historically specific way, authentification actually functions. Quoting St. Jerome, Foucault showed how, in Jerome's time, there were rules that included or excluded particular writings. Were certain works of notably lesser quality, for example? Were there certain places where doctrine flew in the face of well-attributed assertions made elsewhere? Were there sections or works with references to times subsequent to the putative author's death? These each were signs of imperfection, of the failure of a particular text to have been by the author in question, and as a result, Jerome struck them from the canonical list of authentic productions. No doubt the problem of authentification could be extended into more recent periods, and perhaps even in some domains of science—the whole minifield of scientific misconduct would be informative here. But for various reasons (including the economic viability of the objects studied, the very different structure of team research, the availability of individuals' work for inspection, the modes of honorary authorship, and perhaps even the scale of collective authorship), particle physics, astrophysics, and observational cosmology—unlike immunology, clinical epidemiology, molecular genetics—has not been caught up in scandals of fraud and the powerful institutional framework for its detection and prosecution. Indeed, I do not know of a single instance in high-energy physics where fraud, fabrication, or authentification of authorship became a pressing issue.

But there is another, and much more interesting direction in which Foucault proceeded: not from author to work, but from work to author. Beginning with the work itself, he asked: What kind of author does this work presuppose? In posing such a question, Foucault took up a variant of the fundamental Kantian question with which I began this essay, though now about the individual author rather than the "I" per se, and now in a historicized frame, not in the form of an a priori structure. Foucault asked of that which was written:

> What are the modes of existence of this discourse? Where has it been used, how can it circulate, and who can appropriate it for himself? What are the places in it where there is room for possible subjects? Who can assume these various subject functions? And behind all these questions we would hear hardly anything but the stirring of an indifference: What difference does it make who is speaking?"[23]

Here the author protocols of contemporary physics *do* intersect Foucault's questions. The modes of discourse are variable, and importantly so: draft physics notes, physics notes, scientific notes, technical publications, rapporteur talks, conference presentations, physics publications—each had its characteristic content, form of review, specified author names, and intended range of circulation. Who could speak was also strictly regulated; each authorship committee (or its equivalent) precisely determined what could and could not be said "publicly" (where public was itself variable). And finally, who counted as someone who could participate in the author list was itself a finely tuned affair—from temporary participants to masthead members; from engineers allowed to sign on initial papers and technical reports and forbidden from signing physics; from authors left off papers to others actively dissuaded from removing their imprimatur.

What distinguished turn-of-the-century authorship in high-energy physics and astrophysics from other domains of collaborative scientific work is the confluence of three factors: First, there was the raw scale of the collective author, in particle physics moving upwards from five hundred to two thousand people. The metacollaboration of GriPhyN must, in some sense, be understood as a collaboration of more than five thousand scientists and an equal number of engineers and technicians. Second, there is the highly structured system of control over what can be said, when, and to whom. Finally, there is in many colliding beam experiments a fundamental *heterogeneity* that makes the collaboration as supraindividual author more than the additive sum of many individuals all executing similar tasks. The team supplanted the individual not because the individual was just articulating a widespread murmur of the group. No, the team replaced the individual because the individual did not (could not) *know* the length and breadth of the experimental problem. When the spokesperson spoke, she did not necessarily articulate the general consensus, she spoke of things that no one in particular could ever possibly fully know, but that the group could, in the end, assemble. That assembly was two-fold: a construction of the group and a construction of the argument it presented.

Perhaps the distinction might be put this way. After Foucault's lecture "What Is an Author?" Lucien Goldmann, the great philosopher-literary historian, stood up to say that he understood how the author as such had died. After all (so claimed Goldmann), his account of Pascal in *Le Dieu Caché* showed Pascal to be uttering something belonging in a sense to a group, not to the products of an isolated and genial mind. In light of this, Goldmann concluded, he—and Foucault—were both saying something similar: A focus on the group, not the individual, was necessary to understand the broad and deep manifestation of a single rule-governed collective voice.[24] Foucault, of course, did not here or elsewhere claim that the author was dead—his own interest was in using the fact that people were claiming that the author had died to understand the way that "author-hood" had altered its functioning in recent culture. And it is in that spirit of philosophically-motivated empirical research into the function of being an author that I am intrigued by the physicists' concerted efforts to define and shape the idea for a kind of writing that has no precedent.

At this point, the obvious needs to be restated: all groups are not alike. The high-energy physics collaboration functioned not at all like a collection of homogeneous agents of which one could be the spokesman just because of his typicality. Indeed, it was precisely because of the heterogeneity of the collaboration that the fundamental practical paradox of authorship arises. Each subgroup *is* necessary precisely because its special function is needed. If authorship means having contributed work that is a sine qua non for the result as a whole, then indeed each subgroup can and must be counted among the indispensable. But at the same time, when the question is asked: "Who did this work, that is, who is fully in command of this particular analysis and all on which it depends?"—the answer must always be deferred. It is entirely possible, even likely, that no one individual (much less a group of individuals) is entirely in control over the full spectrum of justificatory arguments that feed all the way down into the guts of the forward hadron calorimeter, the analysis code, and the calibration methods. Even that degree of instability is incomplete: with mobile agents and perimeters, transparent hierarchical levels of collaborations, even the collaborative pseudo-I, yield to the hive-I of open-ended coordination. The answer to "Who Is We?" in the context of a two-thousand-strong fluid collaboration must always remain unstable, oscillating between the desire to make scientific knowledge the issue of a single conscious mind and the desire to recognize justly the distributed character of the knowledge essential to any demonstration.

In a sense, every detail of these complex authorship protocols is part of a never-ending struggle to stabilize this instability and to reconcile these irreconcilable goals of centralization, distribution, and open-endedness. It may well be that experimental knowledge in the age of massive collaborations never comes back to a single center, but rather only to partially overlapping, complicated, inchoately bounded assemblages. Yet given the circumstance that the fates of individuals, groups, departments, and even national scientific efforts ride on the apportionment of credit, the attempt to localize authorship is not likely to end soon. On one level, then, the authorship struggle might be relegated to the special configuration of this sector of the physics community. My own suspicion, however, is that the conundrum of the massive collaborations now forming around particle physicists, astrophysicists, or theoretical biologists is not so atypical, after all. Rightly conceived, the tension between the felt need to condense scientific work to the single point of a pseudo-I and the recognition that knowledge is piecewise interconnected into a broad, blurred reservoir of expertise is not a parochial difficulty. It characterizes an unremovable instability in the securing of knowledge itself.

Notes

1. Allan M. Thorndike, Brookhaven National Laboratory. From *Bubble and Spark Chambers*, vol. 2, ed. Shutt (New York: Academic Press, 1967): 299–300. See Galison, *Image and Logic* (Chicago: University of Chicago Press, 1997), esp. chapters 1, 5, 7, 9.
2. "SLD Policy on Publications and Conference Presentations," 1 July 1988, 1.
3. Ibid.
4. Ibid., 3–4. Note that here, in contrast to the "invisible technicians" of the seventeenth century, the first papers were statutorily to include the names of engineers who had contributed substantially to the development of the apparatus.
5. DØ Executive Committee, "Criteria for Authorship of DØ Physics and Technical Papers," 14 March 1991. Hereafter, DØ 14 March 1991.
6. Cf. Peter Galison, *How Experiments End* (Chicago: University of Chicago Press, 1987), chap. 4.

7. Rolandi to Peter L. Galison, e-mail, 30 January 1995.
8. DØ 14 March 1991.
9. DØ, "Rules on Auhtorship of DZero Publications," 2 June 1994: "Retirement and Severance Benefits" allows authorship to be extended beyond departure for a period prorated at six months to a year service up to a maximum of three years. The "masthead" is defined as those members of the collaboration who are active along with those who are "quiescent" (coming aboard); the masthead and the authorship list are not coextensive. For example, someone who leaves may during his extended presence be an author but not on the masthead; someone coming on board may be on the masthead but not yet an author.
10. Dave Charlton and Peter Maettig, "Analysis: Basic Guidelines for OPAL Physics Notes and Papers. General Editorial Policy and Procedure," 10 January 1997, 3.
11. Ibid., 4.
12. ECFA 95/171, "ECFA Report on Sociology of Large Experiments."
13. "Report of the HEPAP Subpanel on Future Modes of Experiemental Research in High Energy Physics," July 1988 (DOE/ER: 0380), 50–51.
14. Roy Schwitters ["The Role of Big Science," *Chronicle of Higher Education*, 16 February 1996, B1–B3.]
15. "ATLAS Guidelines for the Publication of Scientific Notes," approved by the ATLAS Collaboration Board, 11 February 2000. My thanks to Bertrand Nicquevert for this information.
16. Ibid.
17. On mobile agents, see, e.g., David Kotz and Robert S. Gray, "Mobile Agents and the Future of the Internet," <http://www.cs.dartmouth.edu/~dfk/papers/kotz:future2/>, and the less optimistic view of David Reilly, "Mobile Agents–Process Migration and Its Implications," <wysiwyg://zoffsitebottom.12/http: //www.da. . .com/topics/software_agents/mobile_agents/>. The discussion of mobile agents in the context of large physical science collaborations is principally based on the GriPhyN project cited below along with the (massive) documentation independently fielded by the constituent collaborations (LIGO, SDSS, CMS, ATLAS).
18. For a discussion of GriPhyN, see <http://www.phys.ufl.edu/~avery/mre/proposal_final.pdf>.
19. Ibid., 4.
20. I have adapted the term *hive-I* from William Gibson's partially related notion of hive mind in his trilogy that began with *Neuromancer* (New York: Ace Books, 1994) and also from *multiplicity* in the work of Gilles Deleuze: "Multiplicity must not designate a combination of the many and the one, but rather an organisation belonging to the many as such, which has no need whatsoever of unity in order to form a system." From Deleuze, *Difference and Repetition*, trans. Paul Patton (New York: Columbia University Press, 1994), 182.
21. Michel Foucault, "What Is an Author," In *The Foucault Reader*, ed. Paul Rabinow. New York: Pantheon Books, 1984), 107.

22. Ibid., 108.
23. Ibid., 109. "A reversal occurred in the seventeenth or eighteenth century. Scientific discourses began to be received for themselves, in the anonymity of an established or always redemonstrable truth; their membership in a systematic ensemble, and not the reference to the individual who produced them, stood as their guarantee. The author function faded away, and the inventor's name served only to christen a theorem, proposition, particular effect, property, body, group of elements, or pathological syndrome."
24. Goldmann's intervention is in Foucault, *Dits et écrits: 1954–1988*, vol. 1 (Paris: Gallimard, 1994), 812–16.

PART IV
Commentaries

End Credits

TOM CONLEY

Sunken in the soft seats that furnish the rooms of most multiplex movie theaters we inhabit on Saturdays and Sundays, our limbs extended and skulls cushioned by the spongy headrests of lounge-like chairs, we survive the two-and-a-half-hour feature movie—any of thousands made over the past twenty years—by gazing at the end credits. Fatigued and mildly bored, we listlessly stare at an endless scroll of names spelled out in a narrow column of bold white characters, which slowly descend from the top of the black screen to the bottom. Attracted to the information for reason of our dedication to the seventh art or merely for the sake of discipline and duty, seeking to learn where the film was made, the songs it played, or perhaps who might have been the key grips, caterers, and drivers, we scan the image for names. A glimpse of someone strange, we reason, might break the spell of monotony. Trying to read at pace consonant with the speed of the scroll (many names disappear in the empyrean above the screen before we can decipher them), we watch what seems to be a moving page of a cinematic telephone book. Below the list of the leading players on the top, like dead souls, hundreds of nameless names pass before our eyes. The legend of the film becomes a testament to anonymity. (If only, we murmur, we could fast-forward the list as we do at home with the videocassettes we rent from the Blockbuster archive.)

The disgruntling effect of end credits might serve as an emblem for the mosaic reflections that comprise the sum of *Scientific Authorship*.

The scientific author is like the mostly anonymous (but, on rare occasions, fortuitously recognizable) names in movie credits, figures in projects that owe their form to the work of hundreds of active contributors, hired hands, and caretakers. Credit is given to the star, to the director, but also to those who clasp the sound booms, tinker with the backlights, scaffold the stage, and even those who butter the sandwiches for hungry teams of editors, the people who, late in the dark nights of production, in front of their computer terminals, review the rushes of the day. In the early modern age of cinema (roughly, with the coming of sound), a modest number of credits were fused with the establishing shots of the narrative or else were painted on panels crafted by nameless titlers and graphic designers. Perhaps then, in what we would like to believe were kinder and gentler times, economy of enumeration made recognition and fame easier to gain. For American cinephiles, the short list of credits caused a variety of otherwise unknown names to become so familiar and famous that in our minds they stood strong next to Isaac Newton and Alfred Einstein. Alfred E. Neuman figured in many productions, as did Nunally Johnson. Gowns were done by Irene, photography by Rudolph Maté, and music by Max Steiner or Ezy Morales; second-line actors whose names we loved to read included Alan Hale, Robert Newton, Joyce Compton, Barton MacLaine, C. Aubrey Smith, George Tobias, and a host of others.

The authors of this volume who work from the bias of modern history show how the scientific auteur, like that of cinema, came into view. Mary Terrall shows that French savants of the *Lumières* became visible when they were able to mix science and wit in the arena of the salon. Rob Illife studies how, in order to rival with God, Isaac Newton stippled his *Principia* with equations aspiring to the language of hieroglyphs. Andrew Warwick, author of "A Very Hard Nut to Crack," adduces how the fame and legibility of James Clerk Maxwell's *Treatise on Electricity and Magnetism* depended on specific and local relations between the author, his friends, and his students at Cambridge. Myles Jackson affirms that a pragmatic and gifted artisan, Josef von Fraunhofer, never obtained the fame he deserved because his technical work was felt inferior to that of contemporary intellectuals in the German Republic of Letters. What a pity, the reader infers, that in his life and time Fraunhofer never obtained the glory of his namesake, "Frenhofer," the mad painter of Balzac's miniature chef d'oeuvre "The Unknown Masterpiece," who, in 1832, anticipated the style and vision of astract expressionists of the order of Willem de Kooning!

In our day, as many of the scholars of the second half of *Scientific Authorship* demonstrate, the name of the author often dissolves into the ferment of think tanks, where corporate investors generate new technologies prior to being adapted for the ends of enterprise. The scientific genius can be shrouded in oblivion when military contractors assign secrecy to research led in nuclear and advanced military science. And, as teams of scientists grow with the cost and size of laboratories, the names of gifted and genial thinkers become as anonymous as those in the multiplex that pass before our eyes. The scientific author of our time belongs to systems of citation that run from groups of names beneath the titles of articles in specialized journals to résumés, onto webpages, and through indexes of quotations in libraries and electronic media.

The anonymization of authors, which we see in the evolution of movie credits, follows a course of development parallel to what Hugh Gusterson and others discern in the administration of laboratories. Although distant in space, many share experiments and exchange the fruits of their researchers and technicians, but without there being any connection made with the diurnal world at large. The effect is especially visible in genetic engineering, shows Marilyn Strathern, where the ostensively chemical bases of the "author" are so tested and recombined that the consequences of what is yielded in research on human eggs arches back to the extensive and complicated genealogies of medieval and early modern times. The sheer cost and mass of a standard linear detector machine and an operation requires a cast of thousands of scientists that exceeds, notes Peter Galison, the limits of our imagination. The laboratory has no epic counterpart in the cinema, the arts, or even public architecture.

If any binding conclusions can be drawn from the impressive variety of articles it would be, in the words of Mario Biagioli, that today scientific authorship has become a "misnomer," indeed a "historical vestige" of romantic times past. They also might be summed up in Peter Galison's remarks to the effect that the idea of the collaborative author ratifies Kant's observations about a common consciousness informing any single perception of experience. Collective authorship of scientific research depends on a grounded consciousness that embraces what both the "we" and the "I" are capable of observing. The way that research is led determines the death of the author insofar as it had been thought to be the motivating cause for the effect of its work.

Readers opening *Scientific Authorship* and scanning the pages for the first time will have probably taken note (surely before happening upon these words) first of the acknowledgments, then the footnotes, and finally the index. From there a geography, if not a community, is given to be established. But also, in a gesture of protective narcissism, the same readers may be looking for the presence of their own names, thus affirming the predicament of the scientific author who must be at once lost in the pages of writing but also hold to the illusion of being sovereignly visible, reiterable, and, as a result, immortalized and encrypted within the book and its history. In different ways, all of the authors betray similar dilemmas by the way they treat their own names and what they might feel is the inalienable property of their remarks. (Under the title of the draft version of "The Collective Author," Peter Galison opens a parenthesis in which he sternly warns his reader, "Not for quotation or citation without the written permission of the author.") Other chapters, like that of Adrian Johns on "The Ambivalence of Authorship in Early Modern Natural Philosophy," eternize themselves by placing a © before their given names and reiterating the same formula at the bottom of every page of their manuscripts. The self-enclosing and self-sustaining effect of the bubble like sign of "copyright" might reveal the doubt that an author versed in the history of science may harbor about the fluid, volatile, or ephemeral substance of their research. Numerous contributors disperse themselves into their field by self-quotation or by the usual flourish, "as I have shown elsewhere," a phatic sign signaling the advent of a footnote to the author's other printed writings. The gesture attests to a security of reference grounded in doubt ("Yes, I am a scientific author . . . because, well, I must remind both you and me, too, that I am in print") and the insecurity of the need to produce the illusion of a self-given presence ("Maybe my reader is unaware that I am, as I hope my gesture of authentification will attest, yes, I mean, a *bona fide* 'author'"). Under the title of acknowledgments other contributors specify the context of the crafting of the article and the names of the colleagues who assisted in its writing or who inflected the discussions that led it to its conclusions. In these instances, as in the paragraph inserted between the conclusion of Marilyn Strathern's "Emergent Relations" and her long scroll of notes, the reader is invited to imagine the context of its performance and writing, indeed, what Biagioli later calls "relations among colleagues," which are now becoming the rule of authorship.

A cursory glance reveals that every chapter in *Scientific Authorship* bears a name and an affiliation; none is folded into a collective murmur, stands behind a pseudonym, or deploys a tactic of the kind of anonymity often being discussed. In a vital way the disposition of the book makes clear the point that Hans-Jörg Rheinberger recalls from Jacques Derrida's "signature événement contexte," the decisive conclusion appended to *Marges de la philosophie* of 1972, in which it was shown how the "author" is a signature owing to the performance of its writing (as the poetic underside of the title indicates in the homonym "signature, événement qu'on texte," or "the signature, an event that is written"). "The author," says Rheinberger by way of Derrida, "who puts his or her signature under his or her writing, is the product of the impossible wish" that would present a double loss of *immediacy* and of *presence*. To be an author entails losing the presence and authority of one's speech through the mediation of writing. The act that would "copyright" the words is symptomatic of the anxiety that is felt about loss and distortion of an illusion of meaning that would define a person's creation. The pleasures and fears concerning the erosion and constitution of scientific authority, the very topic of *Scientific Authorship*, are thus seen in the margins of every chapter.

In all events Michel Foucault and Martha Woodmansee are two points of critical reference for almost all of the essays. Roger Chartier's ambivalence, in "Foucault's Chiasmus," about the author of "What Is an Author?" is a symptom, perhaps, of an oedipal structure inhering in the "authority" of one of Chartier's intellectual fathers. Upbraiding him for having failed to locate the emergence of the self-named writer of scientific truth before the age of print culture, Chartier later bends his knee in reverence to Foucault for having allowed him to redress and correct the master. Speaking in the name of the father ("Foucault though . . . ," he utters on two occasions), Chartier arrogates an omniscience that betrays ambivalence about the circumstantical nature of historical and scientific truth. The play of subject-positions affirms thus that the chiasmus in the title is the authorial "X" marking the spot where the author-son exhumes a site occupied by the corpse of the father. In his final analysis of collective authorship at the other end of *Scientific Authorship*, Peter Galison varies on Foucault's figure of the anonymous "murmur" of modern voices. In the labor of collective enterprises he notes "the tension between the felt need to condense scientific work to a point and the recognition that knowledge is piecewise interconnected into a *broad, blurred reservoir* of expertise" (stress added). The ring of the metaphor is telling. Elsewhere Foucault, in "Le langage de l'espace" (1964), following reflections on the novelists Michel Butor and Claude Ollier, argued that any statement contains in its speech and writing a "lacunary reserve" that cannot quite be put in words. The turn of Galison's telling expression recalls the context in which the question of the identity of the author had been conceived.

For Foucault's favorite authors, indeed, his progenitors and fathers, the creative author of the age of triumphant capitalism and science, the post-revolutionary moment that he would later call "the beginning of modern times" (seen in the subtitle of *Surveiller et punir*) sought to turn away from anonymity and find immortality in the aura of a name inscribed over summas melding poetry and science. For the novel it was Balzac who applied the art of scientific nomenclature to the vital but formless substance of everyday life in *La Comédie humaine*. For poetry it was Victor Hugo who concocted a pseudoscientific *Légende des siècles*, an epic that, with his own name rivaling that of God, accounted for the history of man and his works. Modern authors, authors in the wake of Flaubert, Foucault attested, sought to be dissolved in the languages and forms that had never been theirs in the first place. By becoming an indifferent *murmur* they walled themselves (the resounding echo of *mur* in the French *murmure* is obvious) into a carceral realm that would become the architecture of institutions of our time: hospitals, offices, military barracks, prisons, schools and universities, and even—thanks to what the authors of this volume bring forward—scientific laboratories.

A reader of Foucault's master texts (especially Raymond Roussel, Kafka, Maurice Blanchot, and the human scientist Claude Lévi-Strauss) quickly discovers the origins of the concepts of *discursive* and *visible* formations. If by discursive formation Foucault meant that a subject or speaker is spoken by the discourse into which he or she is born; and if, too, by a visible formation he suggested that we are ordered and classified by the blackboards, maps, period tables, city plans, and movie houses that chart many of the ways we imagine ourselves living in the worlds we inhabit: he implies that our scientific or creative identities are wedded to a history of things becoming forever *other*. In the author, the *auteur*, which most of the authors of this volume affiliate with the etymology of *augere* and with the success of performative signatures (the author as *actor*), also resides the inflection of the *autre*. To create by othering is to be authoring. Othering might be comparable to the art of reiterating the words and ways of inherited discourses and modes of observation by citational means. A collective and authorial, if not authoritative, tenor is obtained by a careful mimicry. A creative difference would reside in muteness, that is, in the selection of tones and tenors of speech, or even in the classical simplicity of prose clearly written and elegantly argued that is betrayed by some kind of almost imperceptible "lacunary reserve" that logic cannot control. The blurred reservoir, murmur, or reserve is indeed, in different and variegated ways, the topic of most of the chapters of *Scientific Authorship*.

On this score Peter Jaszi and Martha Woodmansee mark the political implications of the singular and collective identities of the author by drawing our attention to masses of people in the world to whose murmur many scientific and political communities in dominant economies remain inaudible. Their communities, if they are to survive, need a way of "finding their own voices." The issues concerning what the two authors call "cultural sustainability" may have their most immediate manifestation in what in São Paulo the anthropologist Darrell Posey has been advocating for denizens of the Brazilian rain forest. In order to stave off the destruction of the biosphere and to preserve the diversity of the milieu, he and a team of colleagues have sought to *copyright* the knowledge of the flora and fauna owned by the shamans of the forest. His discovery that pharmaceutical companies have used the mantle of anthropological inquiry to garner information about the medicinal virtues of plants has led to a countertactic of the kind that reach "beyond authorship": the shamans ought to be granted the privilege of owning knowledge, in the form of software as it were, which would be shared in the world at large for the price of the conservation of the precious forest. Yet, insofar as the shamans are declared not to be citizens of their country on the grounds they do not have social security numbers, they are unable to apply a "©" to their knowledge. How to work out of that kind of double bind is a question facing local cultures all over the planet. How can they acquire authority that will sustain them without being consumed by the logic of the individual name and free enterprise becomes an issue paramount not only for themselves but for the future of the world.

In deference to the traditional cultures that Woodmansee and Jaszi champion in their remarks about cultural sustainability, it might be wise to conclude this postface in the absence of the coda of a signature. By recalling the end credits that assail us in the multiplexes carved into the sprawling malls bulldozed and paved over places on four of the five continents of the world, and after speculating about how many proper names are seen within these air-conditioned theaters in the space of a day, we cannot fail to be tempted to seek the anonymous murmur or the soothing solace of the "broad, blurred reservoir" of expertise we cultivate in our local communities, even if they are ridden with dilemmas that run a gamut between self-promotion and collective secrecy. It may aver that the communities that we would wish to be would be affiliated with those invoked in "Beyond Authorship." In our own way, in what Mario Biagioli calls our "disciplinary ecologies," we would wish to produce communities of relations and productive dialogue (certainly not like the sniping scenarios that Corynne McSherry recounts in her study of the passage from archaic to capitalistic economies in the confection of syllabi), in other words, communities with authors whose names are merely points of reference in a labor seeking dialogue and diversity. May the name of the author of this commentary dissolve into the many reflections of this community dedicated to the fortunes *Scientific Authorship*, and be, like the end credits of today's movies, a piece of oblivion.

What Is Not a Scientific Author?

MARK ROSE

Scientific Authorship. The title of this very suggestive collection invokes Michel Foucault's seminal essay, "What Is an Author?"—the piece that Roger Chartier also discusses in his contribution. Chartier focuses in particular on the *chiasmus* that Foucault identified as a phenomenon of the seventeenth or eighteenth centuries. Before this watershed moment, Foucault suggested, scientific texts were guaranteed by the name of the author, whereas literary texts circulated anonymously, valorized by their reputed antiquity. In the early modern period, however, a reversal occurred in which literary discourse came to be guaranteed by the "author-function," whereas the rule of anonymity now came to command the production and accreditation of scientific statements. Of course Foucault was cautious in his description of this reversal, or chiasmus, because he was aware that the distinction between science and literature was neither universal nor stable, and therefore he employed provisional expressions such as "these texts we call literary" or "texts that we now call scientific," and he was also rather vague about the exact moment when the chiasmus occurred.

Chartier acknowledges that he is working within the general field of discourse—the history of authorship—that Foucault opened up, but he challenges Foucault's proposal by making a number of concrete observations. Chartier points out, for example, that literary authorship—the emergence of the author-function with respect to contemporary literary works—is significantly older than Foucault supposed.

As evidence, Chartier cites the *libro unitario*—a manuscript book embodying the works of a single modern author—which began to be produced as early as the fourteenth century. The Ellesmere Chaucer manuscript of about 1410 would be, as I understand it, an example of such a *libro unitario* and an indication that the contemporary figure Chaucer was being treated as an author. Conversely, Chartier notes, the Scientific Revolution of the seventeenth century did not mean the expulsion of proper names from knowledge claims.

Chartier thus challenges many of the particulars that form the foundation of Foucault's chiasmus. Moreover, Chartier proposes that we need to understand the author-function not only as a matter of discourse, but also as a function of the materiality of the text—that is, as a product of such concrete material developments as the production of manuscripts like the Ellesmere Chaucer. It is not entirely clear to me, however, whether Chartier means to challenge the entire idea of the chiasmus or only the historical particulars as Foucault—vaguely and generally—presented them. Did a reversal between scientific and literary discourse in fact occur, even if not in precisely the terms that Foucault suggested? I note that Chartier certainly does not reject the foundational concept of discourse as a historical form, but only insists on the importance of textual materialities. Nor does Chartier directly challenge the more fundamental assumption that seems to me implicit in Foucault's sketch of a dancelike exchange of positions between scientific and literary discourse—namely, the idea that, whatever the particulars, scientific and nonscientific discourses are in some way systematically related.

But what is nonscientific—or literary—discourse? I note that writers such as Charles Darwin and Sigmund Freud, not to mention such early figures as Francis Bacon and David Hume, are regularly taught in literature courses and so, of course, are a host of other materials not generally considered literary. Until some thirty or so years ago, *literature* was an honorific term that evoked such concepts as the "best that has been thought and said." The literary canon—or canons, since the concept of literature was also rooted in nationalist practices naturalized through the ideas of language and untranslatability—was certainly not an unchanging collection, but it was a recognizable entity. That is no longer the case. The feminist movement, the various ethnic movements, the disintegration of the cohesive national ideal and other major social phenomena, together with the explosion of critical theory in the 1970s and the subsequent emergence of such intellectual styles as cultural studies and new historicism, have made any notion of canon difficult to sustain. Literary studies can now legitimately include the study of practically anything from comic books to law cases, and excellent doctoral dissertations can be written on such topics as the representation of the kitchen in advertising and popular culture in the postwar era. Moreover, a number of departments, my own included, have recently begun to develop specializations in the emergent field of digital culture.

My point is not to lament the death of literature—though I do confess to some nostalgia for the bad old days when more was taken for granted—so much as to express my honest confusion as to what the term "literary discourse" can now be supposed to encompass. So far as my personal practices go, I am *not* confused. I currently teach mostly Shakespeare—often in conjunction with films based on Shakespeare such as Akira Kurosawa's *Ran* or Peter Greenaway's *Prospero's Books*—and I currently write and research mostly on the history of copyright. But as to how my practices and those of my colleagues in literature departments add up to something that might be called "literary discourse" I am unclear. What is clear to me, however, is that the field in which I work, if it is a field, has undergone truly enormous changes in my own professional lifetime.

Foucault's chiasmus implies a systematic relationship between the orders of literary and scientific discourse. In principle, then, major changes in one field of discourse ought to be detectable in the other, perhaps like the stellar perturbations through which extra-solar planets have recently been detected. But is there any relationship between these changes in literary discourse and changes in either the substance or the form of scientific discourse? In the present period of withdrawal from grand theories of all kinds, I am inclined to think not. In his contribution to the present volume, Mario Biagioli argues that even the idea of scientific authorship is something of a misnomer because scientific authorship practices are tied to specific disciplinary ecologies and have little to do with authorship in other areas. But then what happens to Foucault's underlying assumption about a general system of discourse? And if that notion of a general system in which discourses interact like gravitationally bound bodies turns out to be chimerical, then what, if anything, can be said about the relationship between scientific and literary discourse? Can Roger Chartier's skeptical but respectful critique of Foucault's chiasmus be taken further and the whole notion of a dance between discourses be discredited? Alternatively, can Foucault's brilliant—and, I suspect, aesthetically motivated—assumption about the dance be documented with something like the concreteness, clarity, and persuasiveness that Chartier can document the emergence of the modern author in the fourteenth-century *libro unitario*? Perhaps such a project seems at present the equivalent of the nineteenth-century attempt to document the existence of spirits through photography. But on the other hand we know that gravity, the force that relates one moving physical body to another, also once seemed occult.

Contributors

Mario Biagioli is Professor of the History of Science at Harvard University. He is the author of *Galileo, Courtier* (1993) and the editor of *The Science Studies Reader* (Routledge, 1999).

Roger Chartier is Directeur d'Etudes at the Ecole des hautes Etudes en Sciences Sociales in Paris and Annenberg Visiting Professor at the University of Pennsylvania. He is the author of numerous books, including *The Order of Books: Readers, Authors, and Libraries in Europe between the Fourteenth and Eighteenth Centuries* (1994); *Forms and Meanings: Texts, Performances, and Audiences from Codex to Computer* (1995); *On the Edge of the Cliff: History, Language, and Practices* (1997); and *Publishing Drama in Early Modern Europe* (1999).

Tom Conley is Professor of Romance Languages and Literatures at Harvard University. Among his publications is *The Self-Made Map: Cartographic Writing in Early Modern France* (1997). He has also translated several booklength works.

Peter Galison is Mallinckrodt Professor of the History of Science and of Physics at Harvard University. He is the author of *How Experiments End* (1987) and *Image and Logic: A Material Culture of Microphysics* (1997). He has co-edited *Big Science* (1992); *The Disunity of Science* (1996); *Picturing Science, Producing Art* (Routledge, 1998); *The Architecture of Science* (1999); and *Atmospheric Flight in the 20th Century* (2000).

Hugh Gusterson is Associate Professor of Anthropology and Science and Technology Studies at MIT. He is the author of *Nuclear Rites: A Weapons Laboratory at the End of the Cold War* (1996) and co-editor of *Cultures of Insecurity* (1999).

Rob Iliffe is Reader in History of Science at Imperial College, London; editor of the journal *History of Science*; and editorial director of the Newton Project.

Myles W. Jackson is Associate Professor of History, History of Science, and Humanities at Willamette University. He is the author of *Spectrum of Belief: Joseph von Fraunhofer and the Craft of Precision Optics* (2000).

Peter Jaszi teaches at the Washington College of Law of American University in Washington, D.C., where he also directs the Glushko-Samuleson Intellectual Property Law Clinic. He is the co-author of a standard copyright textbook and has edited, with Martha Woodmansee, *The Construction of Authorship* (1994).

Adrian Johns is Associate Professor in the Department of History and the Committee for Historical and Conceptual Studies of Science at the University of Chicago. He is the author of *The Nature of the Book: Print and Knowledge in the Making* (1998).

Corynne McSherry holds a Ph.D. in Communication from the University of California, San Diego, and is now a student at Stanford Law School. She is the author of *Who Owns Academic Work? Battling for Control of Intellectual Property* (2001).

Hans-Jörg Rheinberger is Director at the Max Planck Institute for the History of Science in Berlin. Among his books are *Toward a History of Epistemic Things: Synthesizing Proteins in the Test Tube* (1997) and, co-edited with P. Beurton and Raphael Falk, *The Concept of the Gene in Development and Evolution: Historical and Epistemological Perspectives* (2000).

Mark Rose is Professor of English at the University of California, Santa Barbara. Among his books are *Heroic Love: Studies in Sidney and Spenser* (1968); *Shakespearean Design* (1972); *Spenser's Art* (1975); and *Alien Encounters: Anatomy of Science Fiction* (1981). *Authors and Owners: The Invention of Copyright* (1993).

Marilyn Strathern is Professor of Anthropology at the University of Cambridge. She is the author of many books, including *Women in Between* (1972); *Kinship at the Core* (1981); *The Gender of the Gift* (1988); *After Nature* (1992); *Reproducing the Future* (Routledge, 1992); as co-author, *Technologies of Procreation* (Routledge, 1993); and *Property, Substance and Effect: Anthropological Essays on Persons and Things* (1999).

Mary Terrall is Assistant Professor of History at UCLA. She is the author of *The Man Who Flattened the Earth: Maupertuis* and *Sciences in the Enlightenment* (2002).

Andrew Warwick is a Senior Lecturer in the History of Science in the Centre for the History of Science, Technology and Medicine at Imperial College, London.

Martha Woodmansee is Professor of English and Comparative Literature at Case Western Reserve University where she also directs the Society for Critical Exchange. Her books include *The Author, Art, and the Market* (1994), the collection *The Construction of Authorship: Textual Appropriation in Law and Literature* (1994), co-edited with Peter Jaszi; and, as editor, *The New Economic Criticism: Studies at the Intersection of Literature and Economics* (Routledge, 1999).

Index